T0250799

Problem Books in Mathematics

Series Editor:
Peter Winkler
Department of Mathematics
Dartmouth College
Hanover, NH 03755
USA

More information about this series at http://www.springer.com/series/714

Ralucca Gera • Stephen Hedetniemi • Craig Larson
Editors

Graph Theory

Favorite Conjectures and Open Problems - 1

 Springer

Editors
Ralucca Gera
Department of Applied Mathematics
Naval Postgraduate School
Monterey, CA, USA

Stephen Hedetniemi
School of Computing
Clemson University
Clemson, SC, USA

Craig Larson
Department of Mathematics
Virginia Commonwealth University
Richmond, VA, USA

ISSN 0941-3502   ISSN 2197-8506 (electronic)
Problem Books in Mathematics
ISBN 978-3-319-81159-8   ISBN 978-3-319-31940-7 (eBook)
DOI 10.1007/978-3-319-31940-7

Mathematics Subject Classification (2010): 05Cxx, 01-02, 01-08, 03Fxx

Printed on acid-free paper

This Springer imprint is published by Springer Nature
The registered company is Springer International Publishing AG
The registered company address is: Gewerbestrasse 11, 6330 Cham, Switzerland

Introduction

R. Gera, S. Hedetniemi and C.E. Larson

1 Conjectures and Open Problems

This book has its roots in the idea that conjectures are central to mathematics and that it is useful to periodically identify and survey conjectures in the various branches of mathematics. Typically, the end results of mathematics research are theorems, the most important and famous of which show up in textbooks, which in turn are taught to students. This often gives students the impression that theorems are the most important things in mathematics. The popular press reinforces this idea; when mathematics is in the newspapers, it is most often to report a *proof* of some well-known, unsolved conjecture or problem.

However, as every research mathematician knows, progress in mathematics involves much more than proving theorems, and its practice is much richer. Mathematics research involves not only proving theorems but raising questions, formulating open problems, and stating the conjectures, the solutions to which become the new theorems. Mathematics research also involves the formation of new concepts and methods, the production of counterexamples to conjectures, the simplification and synthesis of different areas of mathematics, and the development of analogies across different areas of mathematics.

The three editors of this volume happen to be graph theorists, or more generally discrete mathematicians, which explains the major focus of the following chapters. In this collection of papers, the contributing authors present and discuss, often in a storytelling style, some of the most well-known *conjectures* in the field of graph theory and combinatorics.

Related to conjectures are *open problems*. Conjectures are either true or false. But what counts as the resolution of a problem is often less clear-cut. Nevertheless, a conjecture clearly specifies a problem—and many problems can be naturally formulated as conjectures. For example, it is a famous unsolved problem to determine whether or not the class **P** of decision problems is equal to its superclass **NP**. The famous **P=NP** problem is one of the seven Millennium Problems identified by the Clay Mathematics Institute, whose resolution carries a $1 million dollar prize [1]. For any problem like this, having a yes or no outcome, the associated conjectures are either that the problem can be resolved in the positive or that it cannot be. Many or most mathematicians, for instance, conjecture that $P \neq NP$. But a few, including Bela Bollabás, conjecture that $P=NP$.

While most mathematicians are most likely to be known for their theorems, some are known for their conjectures. Fermat and Poincaré, while famous for their theorems, are also known for their conjectures. Graph theorists know the name of Francis Guthrie only for his conjecture that planar maps can be colored with four colors [12].

The late, world-famous mathematician, Paul Erdős, is an exceptional example. While he is known for, among other things, his development of Ramsey theory, the probabilistic method, and contributions to the elementary proof of the prime number theorem, he is perhaps equally famous for his conjectures and problems. He traveled with these, talked about them, worked on them with hundreds of collaborators, and even offered monetary prizes for the solutions of many of them. Some of his graph theory conjectures are collected in [6]. His conjectures and prizes have inspired considerable research and numerous research papers, and still 20 years after his death in 1996, his conjectures continue to have considerable influence.

Not all conjectures are of equal importance or significance, and not all will have the same influence on mathematics research. The resolution of some conjectures will impact textbooks and even the history of mathematics. The resolution of others will soon be forgotten. What makes a conjecture significant or important? A few mathematicians have recorded their thoughts on this question.

The famous British mathematician, G. H. Hardy, the early twentieth century analyst and number theorist, discussed this question in his 1940 essay, *A Mathematician's Apology* [10], which is a biographical defense of mathematics as he saw and practiced it. Hardy is often remembered for discounting the practicality or utility of mathematics.

Laszlo Lovász has discussed the question of what makes a good conjecture [11]. He says that "it is easy to agree that" the resolution of a good conjecture "should advance our knowledge significantly." Nevertheless Lovász wants to make room for some of the conjectures of Erdős that don't obviously satisfy this criteria, but are "conjectures so surprising, so utterly inaccessible by current methods, that their resolution *must* bring something new—we just don't know where."

Lovász also discusses experimental mathematics as a source of conjectures, a specific example of which being Fajtlowicz's *Graffiti* [7], a computer program which makes conjectures, many of which are in graph theory. It is easy to write a program to produce syntactically correct mathematical statements. The difficulty in writing a mathematical conjecture-making program is exactly how to limit the program to making interesting or significant statements. When Fajtlowicz began writing his program, he would ask mathematicians what constituted a good conjecture. John Conway told him that a good conjecture should be "outrageous." Erdős, in effect, refused to answer, telling Siemion Fajtlowicz, "Let's leave it to Rhadamanthus."

We won't here give a definitive answer to the question: what makes a good conjecture? Fame is neither a necessary nor a sufficient condition for a conjecture to be considered good. Sociology plays some role in fame. The nonexistence of odd perfect numbers is probably more famous due to its age, dating back to Euclid and later to Descartes, than its importance [13]. But many conjectures of famous mathematicians are worked on because of their intrinsic importance to mathematics. We certainly expect there to exist little known but significant conjectures. The history of mathematics contains numerous examples of important research which was not recognized in its own time. The work of Galois, for instance, is a well-known example.

Our thought is that there may not be any better way of identifying good conjectures than to ask the experts, people who have tilled the mathematical soil for some time and know best which seeds will sprout.

There are *internal* reasons why a given conjecture can be important, strictly mathematical reasons related to the furtherance of mathematics research, and the question, how would resolution of this conjecture advance mathematics?

Of course the goal of mathematics research, and what research mathematicians are paid to do, is to *advance* mathematics. Mathematics is seen by the public as a tool for the sciences—they would have much less interest in paying mathematicians to be artists than they would as researchers who may play a role in improving their lives.

But how can a conjecture play a role in advancing mathematics? In particular it may seem that we have a new question to address: how does mathematics advance?

A conjecture can be said to advance mathematics if the truth of the conjecture yields new knowledge about a question or object of existing mathematical interest.

Furthermore, the advancement of mathematics requires not just new concepts, conjectures, counterexamples, and proofs (uniquely mathematical products) but also effective communication. Lovasz writes:

"Conjecture-making is one of the central activities in mathematics. The creation and dissemination of open problems is crucial to the growth and development of a field". Lovasz, in his 1998 reflection "One Mathematics" [11], writes: "In a small community, everybody knows what the main problems are. But in a community of 100, 000 people, problems have to be identified and stated in a precise way. Poorly stated problems lead to boring, irrelevant results. This elevates the formulation of conjectures to the rank of research results." Conjecturing became an art in the hands of the late Paul Erdös, who formulated more conjectures than perhaps all mathematicians before him put together. He considered his conjectures as part of his mathematical œuvre as much as his theorems. One of my most prized memories is the following comment from him: "I never envied a theorem from anybody; but I envy you for this conjecture."

2   About This Book

Earlier, we mentioned that this book grew out of the idea that conjectures are central to mathematics and that it is useful to identify and survey conjectures in the various branches of mathematics. This idea is not novel, and this volume has many predecessors, even in our own field of graph theory. Erdös, of course, regularly gave talks on his favorite problems in graph theory and the other fields in which he worked. Bondy and Murty's classic text [5] included papers listing conjectures. A recent version was compiled in 2014 by Bondy [4]. The *Handbook of Graph Theory* lists conjectures and open problems as well. [9].

The conference *Quo Vadis* in Fairbanks, Alaska, in 1990, was also an inspiration, when John Gimbel assembled many leading graph theorists to talk about the future of graph theory [8].

A nice collection of open problems is present online in the "Open Garden Problem"[2].

In this volume, we aim to contribute to the identification and distribution of the outstanding problems in graph theory. We started this by co-organizing three special sessions at AMS meetings on the topic "My Favorite Graph Theory Conjectures." These sessions were held at the winter AMS/MAA Joint Meeting in Boston in January 2012, the SIAM Conference on Discrete Math in Halifax in June 2012, and the winter AMS/MAA Joint Meeting in Baltimore in January 2014. At these three sessions, some of the most well-known graph theorists spoke. All sessions were highly popular and extremely well attended. At the Boston session, there was standing room only for a series of 12 talks, and at the Halifax session, people were sitting on the steps, and there were rows of people at the door listening in. The speakers and the titles of their talks at these sessions can be found at http://faculty.nps.edu/rgera/conjectures.html [3].

In this volume, we asked the contributors to write informally, to share anecdotes, to pull back the curtain a little on the process of conducting mathematics research, in order to give students some insights in mathematical practice. Thus, all chapters are written in a much less formal style than that which is required in archival journal publications.

The Editors
Ralucca Gera, US Naval Postgraduate School, Monterey, CA
Stephen T. Hedetniemi, Clemson University, Clemson, SC
Craig Larson, Virginia Commonwealth University, Richmond, VA

References

1. http://www.claymath.org/millennium-problems/rules-millennium-prizes
2. http://www.openproblemgarden.org/category/graph_theory
3. http://faculty.nps.edu/rgera/conjectures.html
4. Bondy, A.: Beautiful conjectures in graph theory. Eur. J. Comb. **37**, 4–23 (2014)
5. Bondy, J.A., Murty, U.S.R.: Graph Theory with Applications, vol. 290. Macmillan, London (1976)
6. Erdős, P.: Problems and results in graph theory and combinatorial analysis. In: Proceedings of the Fifth British Combinatorial Conference, pp. 169–192 (1975)
7. Fajtlowicz, S.: On conjectures of Graffiti. Discret. Math. **72**(1–3), 113–118 (1988)

8. Gimbel, J. (ed.): Quo Vadis, Graph Theory? Challenges and Directions. North Holland, New York (1993)
9. Gross, J.L., Jay Y. (eds.): Handbook of Graph Theory. CRC Press, Boca Raton (2004)
10. Hardy, G.H.: A Mathematician's Apology, Cambridge University Press, University Printing House, Cambridge CB2 8BS, First printing 1940, March 2012
11. Lovasz, L.: One Mathematics, pp. 10–15. The Berliner Intelligencer, Berlin (1998)
12. Maritz, P., Mouton, S.: Francis Guthrie: a colourful life. Math. Intell. **34**(3), 67–75 (2012)
13. Voight, J.: On the nonexistence of odd perfect numbers. MASS Selecta, pp. 293–300 (2003)

Contents

Chapter 1
Highly Irregular

Gary Chartrand

Abstract Over the years, there has been considerable interest in graph theory concepts that have dealt with items that are all the same – such as the degrees of the vertices in regular graphs, subgraphs in isomorphic decompositions of graphs and subgraphs whose edges are colored the same in an edge coloring of a graph. More recently, there has been interest in opposite concepts, that is, irregular concepts, such as graphs whose vertices have different degrees, subgraphs in decompositions all of which are non-isomorphic and subgraphs in an edge-colored graph, all of whose edges are colored differently. It is concepts such as these that are discussed in this chapter, together with related conjectures and open questions.

Mathematics Subject Classification 2010: 05C07, 05C70, 05C45, 05C15, 05C12, 05C05

The three graphs shown in Figure 1.1 are the unique connected graphs of orders 4, 5, and 6 having only two vertices of the same degree. That is, these three graphs are "nearly irregular." Over the years, much research has been done in graph theory concerning concepts dealing with "all things the same." In fact, what might be considered the first purely theoretical paper in graph theory, Julius Petersen [24] authored a paper in 1891 on regular graphs (all degrees the same). In recent decades, there has been considerable research on concepts of a somewhat opposite nature (all things different). This has led to a number of concepts, results, conjectures, and open questions that have attracted the attention of many graph theorists.

G. Chartrand (✉)
Department of Mathematics, Western Michigan University, 1903 W. Michigan Ave., Kalamazoo, MI 49008-5248, USA
e-mail: gary.chartrand@wmich.edu

© Springer International Publishing Switzerland 2016
R. Gera et al. (eds.), *Graph Theory*, Problem Books in Mathematics,
DOI 10.1007/978-3-319-31940-7_1

Fig. 1.1 The only connected graphs of order n, where $4 \leq n \leq 6$, having exactly two vertices of the same degree

1.1 Introduction

When I began my graduate work at Michigan State University, the Chair of the Department of Mathematics was J. Sutherland Frame. He was not only the Chair, he taught a course and was the director of the local chapter of Pi Mu Epsilon. Because of his promotion of and numerous contributions to Pi Mu Epsilon, a lecture in his honor is presented each year at MathFest, a mathematics conference hosted annually by the Mathematical Association of America. I became a member of Pi Mu Epsilon when I was a junior and never missed a meeting when I was a junior, senior, or graduate student. Professor Frame told my fellow graduate assistants and me that it was not only our responsibility to teach (two courses each quarter except for my last year) but to attend all colloquium talks. He said it was good for us to know what mathematicians work on. We will probably only understand the first few minutes of each talk, but it's not our fault if we don't. It's the speaker's fault. I faithfully attended each colloquium talk.

Early in 1962, I had nearly completed my coursework for a PhD. One day there was a note in my mailbox asking me to see the Chair of the Graduate Committee (Professor Bonnie Stewart). This was a bit nerve-wracking for me because I didn't know what this meant. What Professor Stewart wanted to tell me, however, is that I had never selected an area for my dissertation or an advisor. Of course, I knew this. I told Professor Stewart that although I enjoyed (almost) all the courses I had taken, I didn't see myself doing research in any of these areas. He told me to talk to some professors and think seriously about any suggestions they might have for me. Fortunately, I had become acquainted with several faculty members at Michigan State.

Luckily, during fall quarter 1962, I attended a colloquium talk given by Professor Edward A. Nordhaus. His talk was on graph theory (specifically Ramsey numbers). I had never heard of graph theory before but was immediately fascinated by it. The very next day, I visited Professor Nordhaus and asked him if I could take a reading course in graph theory from him. He told me that the only book on graph theory with which he was familiar was one written in German (by Dénes König [18]), but he was willing to give me a reading course in lattice theory. I told him that I'd like to think about this. It wasn't long afterward that I saw an advertisement in the *Notices of the American Mathematical Society* for a new book (in English) titled

Theory of Graphs [22], written by Oystein Ore of Yale University. I immediately ordered a copy of this book, and when I received it, I took it to Professor Nordhaus for him to see it. In this book, a theorem on graph colorings [20] due to Professors Nordhaus and Gaddum (also a faculty member at Michigan State) was mentioned. I showed this to Professor Nordhaus, and he said, "Let's both read the book." So I took a reading course in graph theory during winter quarter 1963 – although I had already started reading the book during the preceding holiday break. After two or three weeks, I knew that I enjoyed this subject and so I asked Professor Nordhaus if he would be willing to be my advisor in graph theory. He agreed. He then told me to find some problems to work on.

During the latter part of my graduate program, I had adopted a different way of studying the courses I was taking. I had become accustomed to asking questions of my own while I was working on assignments. This helped me a great deal in my studying and would prove to help me when I started doing research. I ran into the concept of line graphs in Ore's book (although Ore called these *interchange graphs*) and chose this for my dissertation topic. Unfortunately, I was working in an area in which I had essentially no background and came up with questions that I didn't know were interesting or interesting enough or even new. I thoroughly enjoyed this area, however. What I worked on in my dissertation varied widely. I kept encountering questions that had little or nothing to do with line graphs.

In October 1963, I learned that Professor Nordhaus knew a mathematician in graph theory from the University of Michigan: Frank Harary. I had run across his name often from articles I had read in the library at Michigan State. I became very familiar with that library. Professor Nordhaus called Frank Harary, who said that he was willing to meet me. Professor Nordhaus and I drove to Ann Arbor so I could meet Harary. Although it's a long story, I earned a PhD from Michigan State in June 1964, became a faculty member at Western Michigan University, and was given the opportunity to attend the graph theory seminar in fall 1964 at the University of Michigan.

One of the many facts I stumbled into while working on my dissertation was that no nontrivial graph contained vertices with distinct degrees.

Theorem 1.1. *Every nontrivial graph contains two vertices having the same degree.*

Proof. Suppose that there exists a graph G of order $n \geq 2$ with distinct degrees. Since $0 \leq \deg v \leq n - 1$ for every vertex v of G, there must be exactly one vertex having each of the degrees $0, 1, \ldots, n - 1$. In particular, there exists a vertex x of G with $\deg x = 0$ and a vertex y of G with $\deg x = n-1$. This is impossible, however. ∎

While completing my dissertation, Professor Nordhaus took on a second doctoral student: Mehdi Behzad. I told Behzad about the observation that no nontrivial graph had vertices with distinct degrees, and he thought that this might make an interesting note. While I thought it was interesting, I felt certain that this observation must be well known even though I had never seen it mentioned anywhere. (Later I saw that Dirac noticed this years before.) I then thought that if we could come up with something about graphs containing only two vertices with the same degree, then this would add some substance to a joint note on this topic.

We showed that for every integer $n \geq 2$, there are exactly two graphs of order n having only two vertices of the same degree – and that these graphs are (obviously) complements of each other, one connected and one disconnected. The unique connected graph of order n having exactly two vertices of the same degree is that graph G with vertex set $V(G) = \{v_1, v_2, \ldots, v_n\}$ such that $v_i v_j \in E(G)$ if and only if $i + j \geq n + 1$. For $n = 4, 5, 6$, these are the graphs shown in Figure 1.1. For the disconnected such graph of order n, $v_i v_j$ is an edge if and only if $i + j \geq n + 2$. If $n = 2r \geq 4$ is even, then for $U = \{u_1, u_2, \ldots, u_r\}$, $W = \{w_1, w_2, \ldots, w_r\}$ and $V(G) = U \cup W$, the graph G can be constructed by letting $G[U] = K_r$, $G[W] = \overline{K}_r$ and $u_i w_j \in E(G)$ if and only if $1 \leq i \leq j$ for $j = 1, 2, \ldots, r$. If $n = 2r + 1 \geq 3$ is odd, then the construction is the same with $|U| = r + 1$ and $|W| = r$.

I thought of calling a graph G "perfect" if no two vertices of G have the same degree. That way the note could be given the humorous title "No Graph Is Perfect" [5]. My idea for this title came from the final lines of the movie "Some Like It Hot" (in which actors Jack Lemmon and Tony Curtis had been masquerading as members of a female band) when Joe E. Brown asks the Lemmon character to marry him. Then Lemmon takes off his wig and says he's a man. Joe E. Brown then says, "Well nobody's perfect." Of course, later I would learn that Claude Berge used the term "perfect" to mean something entirely different in graph theory (see [6]).

Graph theory is filled with concepts and topics dealing with things that are the "same," including regular graphs (where the degrees of all the vertices are equal) and isomorphic decompositions (where every two subgraphs in such a decomposition are isomorphic). Later my use of "perfect" to mean the opposite of "regular" necessarily had to be changed – and it was. Perfect became *irregular*. So for some 50 years, from time to time, this topic would occur to me when I asked myself what would be the opposite of "the same" in whatever concept I was currently considering and whether it might be interesting enough to study. That a concept or topic is "interesting" was always critical.

I was fortunate to be given the opportunity to spend 1965–1966 at the University of Michigan with Frank Harary. He would become a good friend of mine, and later we would often discuss ideas we had for research topics. One day, while discussing whether a topic was interesting enough to study, I recalled asking Harary whether a topic being interesting is simply subjective. I've always remembered his response, "That's why we have to have good taste."

One topic that came up a number of times is that of attempting to look at a concept in a variety of ways. So a graph G is *irregular* if no two vertices of G have the same degree. It's not only rather easy to prove that no nontrivial graph is irregular (as we saw in Theorem 1.1), it didn't seem all that interesting. At one time, two ideas occurred to me: (1) Give a different interpretation to the word "graph" in the definition of "irregular graph." (2) Give a different interpretation to the word "degree" in the definition of "irregular graph."

1.2 The Irregularity Strength of a Graph

It was easy to see that multigraphs exist in which every two vertices have different degrees. The multigraph G in Figure 1.2(a) has this property. For example, if G is a connected graph with $E(G) = \{e_1, e_2, \ldots, e_m\}$, $m \geq 2$, then the multigraph H obtained from G by replacing each edge e_i ($1 \leq i \leq m$) by 2^{i-1} parallel edges has this property. Since multigraphs with many parallel edges were difficult to draw, it would be easier to draw these multigraphs as (edge) weighted graphs, where the weight $w(uv)$ of an edge uv is the number of parallel edges joining u and v. A weighted graph in which every two vertices have distinct degrees (where the degree of a vertex is then the sum of the weights of its incident edges) was initially called an *irregular network* rather than an *irregular weighted graph*. In fact, every connected graph G of order 3 or more can be converted into an irregular weighted graph by an appropriate assignment of weights to the edges of G (see Figure 1.2(b)). The *strength* of a multigraph G is the maximum number of parallel edges joining any two vertices of G. This gave rise to the *irregularity strength* $s(G)$ of a graph G, defined as the smallest positive integer k such that if each edge of G is assigned one of the weights $1, 2, \ldots, k$, then an irregular weighted graph can be produced. I decided that this subject might make an interesting talk at the 250th Anniversary of Graph Theory Conference held at Indiana University-Purdue University Fort Wayne in 1986. A paper [11] on this topic was written and appeared in the proceedings of this conference, which was published in 1988.

Another paper [25] was published in 1988 containing a connection to the concept of irregular graphs. The British mathematician David Wells has written extensively on mathematics education. The fall 1988 issue of the journal *The Mathematical Intelligencer* contained an article [25] written by Wells and titled *Which is the Most Beautiful?* In this article, Wells listed 24 theorems and asked the readers to vote for the theorems they considered the most beautiful. In 1990, he wrote a follow-up article [26] titled *Are These the Most Beautiful?* giving the results of this survey. The theorem that finished on top was Euler's $e^{\pi i} = -1$. Second place was another theorem due to Euler: For every polyhedron with V vertices, E edges, and F faces, $V - E + F = 2$. The theorem stating that there are infinitely many primes came in third place. The theorem coming in at #20 caught me by surprise:

At any party, there is a pair of people who have the same number of friends present.

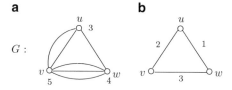

Fig. 1.2 (**a**) An irregular multigraph and (**b**) irregular weighted graph

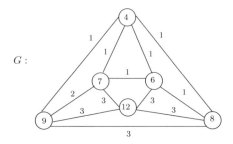

G :

Fig. 1.3 A graph with irregularity strength 3

First, I was surprised that it was even considered a theorem and also that it was among the 24 theorems listed by Wells. Although not stated in terms of graphs, what this theorem clearly says is that no graph is irregular. So perhaps the concept of irregular graphs was better known and of more interest than I thought.

The problem of determining the irregularity strength of a graph would become rather popular, with many research papers written on this topic. In fact, the book *Color-Induced Graph Colorings* [30] by Ping Zhang contains a chapter on irregularity strength. In this book, the structure of those graphs containing exactly two vertices having the same degree is shown to play a useful role in the proofs of certain results.

As an illustration, the 4-regular graph $G = K_{2,2,2}$ shown in Figure 1.3 has irregularity strength 3. In this figure, each edge is assigned one of the weights 1, 2, 3 with the weighted degrees shown within the vertices. Since these degrees are all distinct, $s(G) \leq 3$. Since no graph is irregular, $s(G) \geq 2$. Therefore, to verify that $s(G) = 3$, it only remains to show that there is no way to assign the weights 1, 2 to the edges of G to produce an irregular weighted graph. Although there are several ways to see this, suppose, to the contrary, that an irregular weighted graph can be obtained by assigning the weights 1, 2 to the edges of G. Let H be the spanning subgraph of G all of whose edges are assigned the weight 1. Then H contains two vertices u and v having the same degree, say $\deg_H u = \deg_H v = k$. Since all edges of G incident with u and v not belonging to H are assigned weight 2, the weighted degrees of u and v in G are $k + 2(4 - k) = 8 - k$, a contradiction.

Aigner and Triesch [1] and Gyárfás [15] are among many who have obtained bounds on the irregularity strength of a graph and exact values for certain classes of graphs.

Theorem 2.1 ([11]). *For each integer $n \geq 3$, $s(K_n) = 3$.*

Theorem 2.2 ([11, 15]). *If G is a regular complete multipartite graph of order at least 3, then*

$$s(G) = \begin{cases} 4 & \text{if } G = K_{r,r} \text{ where } r \geq 3 \text{ is odd} \\ 3 & \text{otherwise.} \end{cases}$$

Theorem 2.3 ([1]). *If G is a connected graph of order n ≥ 4, then s(G) ≤ n − 1.*

One problem that interests me is that of finding a short, easily understood proof of Theorem 2.3. While no formula has been obtained for the irregularity strength of every tree, the following conjecture by Ebert, Hemmeter, Lazebnik, and Woldar [13] has been made. The notation $\Delta(T)$ denotes the maximum degree of a tree T.

Conjecture 2.4 ([13]). *For a tree T, let n_i be the number of vertices of degree i in T for $i = 1, 2, \ldots, \Delta(T)$ and let*

$$\lambda(T) = \left\lceil \max_{(i,j)} \left\{ \frac{(n_i + n_{i+1} + \cdots + n_j) + i - 1}{j} : 1 \le i \le j \le \Delta(T) \right\} \right\rceil.$$

Then $s(T) = \lambda(T)$ or $s(T) = \lambda(T) + 1$.

1.3 The 1-2-3 Conjecture

As the title of Zhang's book [30] suggests, years later assigning weights to edges would be looked at in terms of assigning colors to edges, and the resulting degrees would become vertex colors, where then the color of a vertex is the sum of the colors of its incident edges. Zhang then came up with the idea of demanding that every two *adjacent* vertices have different colors (a proper coloring) instead of requiring that *all* vertices have different colors. It soon appeared that for every connected graph of order at least 3, this vertex coloring condition could be accomplished using only the colors 1, 2, 3 for the edges. But neither Zhang nor I could prove this. Zhang suggested writing a paper mentioning this problem but I kept delaying this, hoping that we could prove more than we had. Then we learned of a paper [17] on the exact same problem and conjecture, which eventually acquired a rather catchy name. The following conjecture is due to Karoński, Łuczak and Thomason [17].

The 1-2-3 Conjecture For every connected graph G of order 3 or more, each edge of G can be assigned one of the colors 1, 2, 3 in such a way that the induced colors of every two adjacent vertices are different.

This conjecture thus states that there is a nonproper edge coloring of every connected graph of order 3 or more using only the colors 1, 2, 3 that gives rise to a proper vertex coloring of this graph. This is illustrated for the three graphs in Figure 1.4.

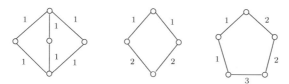

Fig. 1.4 Illustrating the 1-2-3 Conjecture

Karoński, Łuczak and Thomason [17] showed that there is an infinite class of graphs for which the 1-2-3 Conjecture holds.

Theorem 3.1 ([17]). *For every connected graph G of order 3 or more having chromatic number at most 3, the 1-2-3 Conjecture holds.*

While it is not known if the 1-2-3 Conjecture will become the 1-2-3 Theorem, Maciej Kalkowski, Michał Karoński, and Florian Pfender [16] showed that there is, in fact, a 1-2-3-4-5 Theorem. Therefore, even a 1-2-3-4 Theorem appears to be elusive.

Theorem 3.2 ([16]). *For every connected graph G of order 3 or more, each edge of G can be assigned one of the colors 1, 2, 3, 4, 5 in such a way that the induced colors of every two adjacent vertices are different.*

1.4 F-Irregular Graphs

Another way that an "irregular graph" might exist is if the term "degree" were defined in a different way. While the standard degree of a vertex v in a graph G is defined as (1) the number of vertices in G that are adjacent to v or (2) the number of edges of G that are incident with v, it can also be defined as (3) the number of subgraphs of G that contain v and are isomorphic to K_2. But what if K_2 was changed to some other graph?

For a connected graph F of order 2 or more and a graph G, the *F-degree* of a vertex v of G, denoted by $F \deg v$, is the number of subgraphs of G isomorphic to F containing v. Therefore, in the case where $F = K_2$, it follows that $F \deg v = \deg v$. This concept was described in [9] by Chartrand, Holbert, Oellermann, and Swart, and some of the results obtained are given below.

Theorem 4.1 ([9]). *Let F be a graph of order $k \geq 2$ and let G be a graph. If G contains m copies of the graph F, then $\sum_{v \in V(G)} F \deg v = km$.*

Corollary 4.2 ([9]). *Let F be a graph of even order and let G be a graph. Then G has an even number of vertices with odd F-degree.*

Of course, a special case of Corollary 4.2 is that every graph has an even number of odd vertices. It was mentioned that every nontrivial connected graph can be converted into an irregular weighted graph – except for one graph, namely, the graph K_2. In the case of F-degrees, there may be a single exception here as well, and it may be the same exception. An *F-irregular graph* is a graph in which no two vertices have the same F-degree. We know that there is no F-irregular graph when $F = K_2$, but for many other choices of F, there was success in constructing an F-irregular graph. For the path P_3 of order 3, the graph G of Figure 1.5 is P_3-irregular, where the P_3-degree of each vertex is shown in the figure.

Theorem 4.3 ([9]). *For every integer $n \geq 3$, there exists a $K_{1,n}$-irregular graph.*

G :

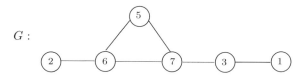

Fig. 1.5 A P_3-irregular graph

Theorem 4.4 ([9]). *For every integer $n \geq 3$, there exists a K_n-irregular graph.*

As the graph G of Figure 1.5 illustrates, there are P_3-irregular graphs. However, it was shown in [9] that there is no regular P_3-irregular graph. Erdős, Székely, and Trotter (Personal communication, 2014) showed that there are, in fact, infinitely many K_3-irregular graphs, that is, no two vertices belong to the same number of triangles. Whether there exist regular K_3-irregular graphs is unknown.

Problem 4.5. *Does there exist a regular K_3-irregular graph?*

The primary conjecture on this topic is the following, however.

Conjecture 4.6 ([9]). *For every connected graph F of order 3 or more, there exists an F-irregular graph.*

1.5   Highly Irregular Graphs

Every four years during the period 1968–2000, Western Michigan University hosted an international conference on graph theory. A faculty member who played a major role in hosting these conferences was Yousef Alavi. Among the many mathematicians who attended these conferences was Ron Graham, who became good friends with Alavi. A common comical phrase used by Alavi was "highly irregular." To Alavi, many things that occurred were "highly irregular." Graham suggested the idea of introducing a concept to be called "highly irregular graphs" and to write a paper with Alavi on this topic. All that was needed was a definition of "highly irregular." Ortrud Oellermann suggested a definition that we (Fan Chung, Ron Graham, Paul Erdős, Ortrud Oellermann, and I) adopted for Yousef Alavi. A graph G is *highly irregular* if for every vertex v of G, no two neighbors of v have the same degree. The three trees in Figure 1.6 of orders 4, 8 and 16 are therefore highly irregular. A paper [3] was then written with the title "Highly irregular graphs."

In [3] the following was observed.

Theorem 5.1 ([3]). *For every integer $n \geq 2$, except $n = 3, 5, 7$, there exists a highly irregular graph of order n.*

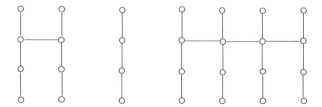

Fig. 1.6 Two highly irregular graphs

This does not mean that there are many highly irregular graphs, however.

Theorem 5.2 ([3]). *If $G(n)$ and $HI(n)$ denote the number of graphs of order n and the number of highly irregular graphs of order n, respectively, then*

$$\lim_{n \to \infty} \frac{HI(n)}{G(n)} = 0.$$

The *independence number* $\alpha(T)$ of a tree T is the maximum number of vertices in T, no two of which are adjacent. The independence numbers of highly irregular trees were investigated in [3] and the following result was stated.

Theorem 5.3 ([3]). *If T is a highly irregular tree of order $n \geq 2$, then $\alpha(T) \leq 12n/19$.*

It is known that there are highly irregular trees T of order n for which $\alpha(T) = 13n/21$, but no example of a highly irregular tree T of order n with $\alpha(T) > 13n/21$ has been found. Since $13n/21 \cong 0.62n$ and $12n/19 \cong 0.63n$, the upper bound stated in Theorem 5.3, if not sharp, is close to being sharp.

Problem 5.4. *For highly irregular trees T of order n, what is a sharp upper bound for $\alpha(T)$ in terms of n?*

After working on the paper [3], Erdős, Oellermann, and I continued to discuss the topic of irregular graphs and wrote a paper for the *College Mathematics Journal* titled "How to define an irregular graph" [10]. One idea was to define the *k-degree* $\deg_k v$ of a vertex v in a connected graph G as the number of vertices at distance k from v. Therefore, $\deg_1 v = \deg v$. A graph G is *distance-k irregular* if $\deg_k v \neq \deg_k u$ for every two distinct vertices u and v of G. That no graph is distance-1 irregular is a special case of the following observation.

Theorem 5.5 ([10]). *For each positive integer k, no graph is distance-k irregular.*

Proof. We already know that this is true for $k = 1$. Suppose, however, that there is a distance-k irregular graph of order n for some $k \geq 2$. Then for each integer $i \in \{0, 1, \ldots, n-1\}$, there is exactly one vertex of G whose k-degree is i. In particular, there is a vertex u such that $\deg_k(u) = 0$ and vertex v such that $\deg_k(v) = n-1$. This says that all vertices of $V(G) - \{v\}$ are at distance $n-1$ from v, including u, which is impossible. ∎

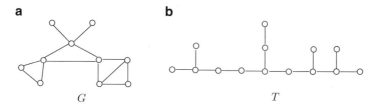

Fig. 1.7 k-Path irregular graphs for $k = 3, 4$. (**a**) G, (**b**) T

While this concept was not particularly interesting, a related one showed more promise. For a positive integer k, a connected graph G is k-path irregular (see [4]) if $\deg u \neq \deg v$ for every two vertices u and v connected by a path of length k in G. There are many 1-path irregular graphs, including the complete bipartite graphs $K_{s,t}$ with $s \neq t$. The 2-path irregular graphs are precisely the highly irregular graphs. The graph G in Figure 1.7(a) is 3-path irregular and the tree T in Figure 1.7(b) is 4-path irregular.

An open question in this topic is the following.

Problem 5.6. *For a given positive integer n, which sets S of positive integers have the property that there exists a connected graph G of order n that is k-path irregular if and only if $k \in S$?*

1.6 The Ascending Subgraph Decomposition Conjecture

Applying "irregular" to concepts has occurred in situations not concerning degrees of vertices. Decomposing a graph into subgraphs, each isomorphic to the same graph, has been a topic of interest for many years. Indeed, decomposing a complete graph into triangles is the concept of Steiner triple system, whose history goes back to Thomas Kirkman. I thought that an irregular version of this may be decomposing a graph into subgraphs no two of which are isomorphic. But every graph has this property – just require every two subgraphs to have different sizes. In fact, if G is a graph of size m, then $\binom{k+1}{2} \leq m \leq \binom{k+2}{2} - 1$ for a unique integer k and G can be decomposed into k subgraphs of different sizes. For every example I considered, I noticed that there was such a decomposition with another property, namely, for every graph G of size m with $\binom{k+1}{2} \leq m \leq \binom{k+2}{2} - 1$, there was a decomposition of G into k subgraphs $G_1, G_2, \ldots, G_k$ of sizes $m_1, m_2, \ldots, m_k$, respectively, where not only $m_1 < m_2 < \cdots < m_k$ but G_{i+1} contains a subgraph isomorphic to G_i for $i = 1, 2, \ldots, k - 1$. I had a meeting scheduled with a research group one afternoon and mentioned this possible conjecture to the group, which I called the *Ascending Subgraph Decomposition Conjecture*.

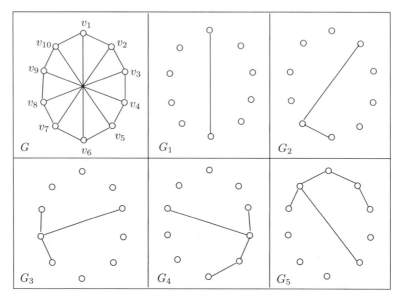

Fig. 1.8 An ascending subgraph decomposition of a graph

Conjecture 6.1. *Let G be a graph of size m, where $\binom{k+1}{2} \leq m \leq \binom{k+2}{2} - 1$. Then G can be decomposed into k subgraphs $G_1, G_2, \ldots, G_k$ of different sizes where G_{i+1} contains a subgraph isomorphic to G_i for $i = 1, 2, \ldots, k - 1$.*

Figure 1.8 shows an ascending subgraph decomposition of a 3-regular graph G of order 10 and size $15 = \binom{6}{2}$.

After a few minutes of discussing this conjecture with the research group, there was a knock on the door of our meeting room. It was Paul Erdős, who had arrived to give a lecture at Western Michigan University. He asked what we were working on. This conjecture was mentioned to him, but we had to stop discussing it with him as it was nearly time for his lecture, which we all attended. Near the end of his lecture, we were surprised to hear him mention this conjecture. Not only that, he offered \$5 for either a proof or a counterexample. As it turns out, I believe \$5 was too little as this conjecture has never been proved or disproved. Nevertheless, this led to a paper on the topic by Alavi, Boals, Erdős, Oellermann, and myself [2].

1.7 Panconnected Graphs

One idea that occurred to me in the 1970s was that of studying graphs having two vertices connected by paths of different lengths. A graph G of order n was defined to be *panconnected* if every two vertices of G are connected by paths of all possible lengths. More specifically, a graph G of order n is *panconnected* if for every two vertices u and v with distance $d(u, v)$ and every integer k with $d(u, v) \leq k \leq n - 1$, there exists a $u - v$ path of length k in G. Here, we will see an example of an open problem that was in existence for many years – but finally settled. I gave the topic of panconnected graphs to my doctoral student at the time, namely, Jim Williamson, to work on for his dissertation [27] on Hamiltonian-connected graphs (in which every two vertices are connected by a Hamiltonian path). Several well-known theorems occurring in Hamiltonian Graph Theory are the following. In a graph G, the notation $\sigma_2(G)$ denotes the minimum sum of the degrees of two nonadjacent vertices of G.

Theorem 7.1 ([8, p.152] [21, 23]). *Let G be a graph of order $n \geq 4$.*

(a) *If $\sigma_2(G) \geq n - 1$, then G has a Hamiltonian path.*
(b) *If $\sigma_2(G) \geq n$, then G is Hamiltonian.*
(c) *If $\sigma_2(G) \geq n + 1$, then G is Hamiltonian-connected.*

An immediate corollary of Theorem 7.1 is the following, where $\delta(G)$ denotes the minimum degree of a graph G.

Corollary 7.2. *Let G be a graph of order $n \geq 4$.*

(a) *If $\delta(G) \geq (n - 1)/2$, then G has a Hamiltonian path.*
(b) *If $\delta(G) \geq n/2$, then G is Hamiltonian.*
(c) *If $\delta(G) \geq (n + 1)/2$, then G is Hamiltonian-connected.*

Each of the results in Corollary 7.2 is sharp for if the lower bound is reduced by $1/2$ in each case, the result no longer holds. Williamson [28] proved the following result.

Theorem 7.3 ([28]). *If G is a graph of order $n \geq 4$ such that $\delta(G) \geq (n + 2)/2$, then G is panconnected.*

The lower bound in Theorem 7.3 is sharp in the sense that if this lower bound is reduced by $1/2$, then the result no longer holds [28]. Perhaps surprisingly, Williamson also showed that if G is a graph of order $n \geq 4$ such that $\sigma_2(G) \geq n + 2$, then G need not be panconnected. In the graph G of order 8 shown in Figure 1.9, $\sigma_2(G) = 10 = n + 2$ but G is not panconnected since, for example, $d(u, v) = 1$ but G contains no $u - v$ path of length 2.

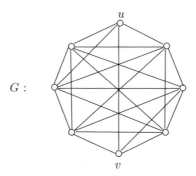

Fig. 1.9 A graph G of order n with $\sigma_2(G) = n + 2$ such that G is not panconnected

In fact, Williamson's research resulted in the following theorem.

Theorem 7.4 ([28]). *There exists no constant c such that if G is a graph of order $n \geq 4$ satisfying $\sigma_2(G) \geq n + c$, then G is panconnected.*

Theorem 7.4 is actually a consequence of the following result of Williamson.

Theorem 7.5 ([28]). *If G is a graph of order $n \geq 4$ such that $\sigma_2(G) \geq \frac{3n-2}{2}$, then G is panconnected.*

Williamson showed that if the lower bound in Theorem 7.5 is reduced by 1, then the result no longer holds. For a number of years, it remained an open question whether this result would hold if the lower bound in Theorem 7.5 were to be reduced by $1/2$. However, the situation was finally settled in [7], where Bi and Zhang proved the following.

Theorem 7.6 ([7]). *If G is a graph of order $n \geq 4$ such that $\sigma_2(G) \geq \frac{3n-3}{2}$, then G is panconnected.*

1.8   Rainbow Connection

In 2006 Ping Zhang and I were to be visited at Western Michigan University by Kathy McKeon and Garry Johns to work on a research problem – but we didn't have a problem to study. So I returned to my old standby: irregular something. I wondered if there might be an interesting version of paths of different lengths connecting each pair of vertices, but this returned me to panconnected graphs again.

The idea of locating monochromatically-colored subgraphs in edge-colored graphs is essentially what is done when studying Ramsey numbers. In recent years, problems concerning rainbow-colored subgraphs (in which no two edges are colored the same) have become increasingly popular.

For example, there have been problems dealing with the *rainbow Ramsey number* of two graphs F and H, which concerns edge colorings of a complete graph using an arbitrary number of colors that result in either a *monochromatic F* (in which all edges are colored the same) or a *a rainbow H* (in which no two edges are colored the same). This topic is surveyed in [14]. Another example (see [29]) concerns, for a prescribed complete graph, conditions under which proper edge colorings of complete graphs of sufficiently large order can be decomposed into rainbow copies of the prescribed complete graphs.

It occurred to me to consider assigning as few colors as possible to the edges of a graph G so that every two vertices of G are connected by a *rainbow path* (in which no two edges are colored the same). This concept could then be called *rainbow connection*, which appealed to me because this is the title of a song written by Paul Williams for Kermit the frog (Jim Henson) to sing in "The Muppet Movie."

Specifically, an edge-colored graph G is defined to be *rainbow connected* if every two vertices of G are connected by a rainbow path. The *rainbow connection number* rc(G) of a graph G is the minimum number of colors required of an edge coloring

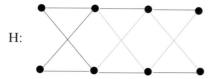

H:

Fig. 1.10 A graph with rainbow connection number 3

of G so that the resulting edge-colored graph is rainbow connected. The rainbow connection number of a (connected) graph G is necessarily at least as large as its diameter $diam(G)$. Since the diameter of the graph H in Figure 1.10 is 3, it follows that $rc(H) \geq 3$. However, because the 3-edge coloring of H shown in Figure 1.10 is a rainbow coloring, it follows that $rc(H) = 3$.

After an initial paper [12] on this topic, many other research papers in this area occurred. In fact, in 2012, the book *Rainbow Connections of Graphs* [19] by Xueliang Li and Yuefeng Sun was published.

In addition to the observations that (1) a graph G has $rc(G) = 1$ if and only if G is complete and (2) a nontrivial connected graph G of size m has $rc(G) = m$ if and only if G is a tree, the following formulas for the rainbow connection numbers of complete multipartite graphs were obtained.

Theorem 8.1 ([12]). *For integers s and t with $2 \leq s \leq t$,*

$$rc(K_{s,t}) = \min \left\{ \lceil \sqrt[s]{t} \rceil, 4 \right\}.$$

Theorem 8.2 ([12]). *For integers $n_1, n_2, \ldots, n_k$ where $k \geq 3$ and $1 \leq n_1 \leq n_2 \leq \cdots \leq n_k$ such that $s = \sum_{i=1}^{k-1} n_i$ and $t = n_k$,*

$$rc(K_{n_1,n_2,\ldots,n_k}) = \begin{cases} 1 & \text{if } n_k = 1 \\ 2 & \text{if } n_k \geq 2 \text{ and } s > t \\ \min\{ \lceil \sqrt[s]{t} \rceil, 3\} & \text{if } s \leq t. \end{cases}$$

Among the many unresolved questions in this area are the following Li, Personal communication (2014).

Problem 8.3. *Does there exist a constant $c \geq 0$ such that if G is a non-complete graph of order n with $\delta(G) \geq n/2 + c$, then $rc(G) = 2$? In particular, does $\delta(G) \geq n/2$ imply that $rc(G) = 2$?*

Problem 8.4. *Let k be a positive integer. Does there exist a constant $c \geq 0$ such that if G is a k-connected graph of order n, then $rc(G) \leq \lceil \frac{n}{k} \rceil + c$?*

Problem 8.4 is known to have an affirmative answer when $k = 1$ or $k = 2$ but the answer is unknown even when $k = 3$.

Acknowledgements I thank Xueliang Li and Tom Trotter for supplying useful information to me. I also thank Stephen Hedetniemi, one of the editors of this book, for numerous valuable suggestions.

References

1. Aigner, M., Triesch, E.: Irregular assignments of trees and forests. SIAM J. Discret. Math. **3**, 439–449 (1990)
2. Alavi, Y., Boals, A.J., Chartrand, G., Erdős, P., Oellermann, O.R.: The ascending subgraph decomposition problem. Congr. Numer. **58**, 7–14 (1987)
3. Alavi, Y., Chartrand, G., Chung, F.R.K., Erdős, P., Graham, R.L., Oellermann, O.R.: Highly irregular graphs. J. Graph Theory **11**, 235–249 (1987)
4. Alavi, Y., Boals, A.J., Chartrand, G., Erdős, P., Oellermann, O.R.: k-Path irregular graphs. Congr. Numer. **65**, 201–210 (1988)
5. Behzad, M., Chartrand, G.: No graph is perfect. Am. Math. Mon. **74**, 962–963 (1967)
6. Berge, C.: Some classes of perfect graphs. In: Six Papers on Graph Theory, pp. 1–21. Indian Statistical Institute, Calcutta (1963)
7. Bi, Z., Zhang, P.: On sharp lower bounds for panconnected, geodesic-pancyclic and path pancyclic graphs. J. Combin. Math. Combin. Comput. To appear
8. Chartrand, G., Zhang, P.: A First Course in Graph Theory. Dover, New York (2012)
9. Chartrand, G., Holbert, K.S., Oellermann, O.R., Swart, H.C.: F-Degrees in graphs. Ars Comb. **24**, 133–148 (1987)
10. Chartrand, G., Erdős, P., Oellermann, O.R.: How to define an irregular graph. Coll. Math. J. **19**, 36–42 (1988)
11. Chartrand, G., Jacobson, M.S., Lehel, J., Oellermann, O.R., Ruiz, S., Saba, F.: Irregular networks. Congr. Numer. **64**, 197–210 (1988)
12. Chartrand, G., Johns, G.L., McKeon, K.A., Zhang, P.: Rainbow connection in graphs. Math. Bohem. **133**, 85–98 (2008)
13. Ebert, G., Hemmeter, J., Lazebnik, F., Woldar, A.: Irregularity strengths for certain graphs. Congr. Numer. **71**, 39–52 (1990)
14. Fujita, S., Magnant, C., Ozeki, K.: Rainbow generalizations of Ramsey theory – a dynamic survey. Theory Appl. Graphs **0**, 1–28 (2014)
15. Gyárfás, A.: The irregularity strength of $K_{m,m}$ is 4 for odd m. Discret. Math. **71**, 273–274 (1988)
16. Kalkowski, M., Karoński, M., Pfender, F.: Vertex-coloring edge-weightings: towards the 1-2-3-conjecture J. Comb. Theory Ser. B **100**, 347–349 (2010)
17. Karoński, M., Łuczak, T., Thomason, A.: Edge weights and vertex colours. J. Comb. Theory Ser. B **91**, 151–157 (2004)
18. König, D.: Theorie der endlichen und unendliehen Graphen. Akademische Verlagsgesellschaft, Leipzig (1936)
19. Li, X.L., Sun, Y.F.: Rainbow Connections of Graphs. Springer, Boston (2012)
20. Nordhaus, E.A., Gaddum, J.W.: On complementary graphs. Am. Math. Mon. **63**, 175–177 (1956)
21. Ore, O.: Note on Hamilton circuits. Am. Math. Mon. **67**, 55 (1960)
22. Ore, O.: Theory of Graphs. American Mathematical Society Colloquium Publications, Providence (1962)
23. Ore, O.: Hamilton connected graphs. J. Math. Pures Appl. **42**, 21–27 (1963)
24. Petersen, J.: Die Theorie der regulären Graphen. Acta Math. **15**, 193–220 (1891)
25. Wells, D.: Which is the most beautiful? Math. Intell. **10**, 30–31 (1988)
26. Wells, D.: Are these the most beautiful? Math. Intell. **12**, 37–41 (1990)
27. Williamson, J.E.: On Hamiltonian-connected graphs. Ph.D. Dissertation, Western Michigan University (1973)
28. Williamson, J.E.: Panconnected graphs. II. Period. Math. Hung. **8**, 105–116 (1977)
29. Yuster, R.: Rainbow decomposition. Proc. Am. Math. Soc. **136**, 771–779 (2008)
30. Zhang, P.: Color-Induced Graph Colorings. Springer, New York (2015)

Chapter 2
Hamiltonian Extension

Ping Zhang

Abstract In the instructions accompanying William Hamilton's Icosian Game, it was written (by Hamilton) that every five consecutive vertices on a dodecahedron can be extended to produce a round trip on the dodecahedron that visits each vertex exactly once. This led to concepts for Hamiltonian graphs G dealing with (1) for any ordered list of k vertices in G, there exists a Hamiltonian cycle in G encountering these k vertices (not necessarily consecutively) in the given order and (2) determining the largest positive integer k for which any ordered list of k consecutive vertices in G lies on some Hamiltonian cycle in G as a path of order k. Whether G is Hamiltonian or not, there is a cyclic ordering of the vertices of G the sum of whose distances of consecutive vertices is minimum. These ideas are discussed in this chapter along with open questions dealing with them.

AMS classification subjects: 05C45

2.1 Introduction

What do Leonhard Euler, Kazimierz Kuratowski, Karl Menger, and Frank Ramsey have in common? While there are probably many possible answers to this question, the desired answer here is: they are four famous mathematicians after whom four famous theorems in graph theory are named, but who never worked in graph theory. They are not the only non-graph theorists whose names are closely associated with graph theory. One of the best known mathematicians belonging to this category is William Rowan Hamilton. This Irish mathematician and physicist was born in 1805. Thirty years later he was knighted, becoming Sir William Rowan Hamilton, for his accomplishments in physics. Among his accomplishments in mathematics, he is known for his creation of the quaternions, a 4-dimensional associative normed

P. Zhang (✉)

Department of Mathematics, College of Arts and Sciences, Western Michigan University, 1903 W Michigan Ave, Kalamazoo, MI 49008-5248, USA

e-mail: ping.zhang@wmich.edu

© Springer International Publishing Switzerland 2016

R. Gera et al. (eds.), *Graph Theory*, Problem Books in Mathematics,

DOI 10.1007/978-3-319-31940-7_2

division algebra over the real numbers. A quaternion is a number of the form $a +$ $b\mathbf{i} + c\mathbf{j} + d\mathbf{k}$ where $a, b, c,$ and d are real numbers and $\mathbf{i}^2 = \mathbf{j}^2 = \mathbf{k}^2 = \mathbf{ijk} = -1$. These numbers are an extension of the complex numbers and are noncommutative. In fact, the quaternions were the first example of a noncommutative division algebra.

One of Hamilton's last major discoveries was a noncommutative algebra he referred to as icosian calculus. This algebra was based on three symbols i, κ, and λ, all roots of unity, with $i^2 = 1$, $\kappa^3 = 1$, and $\lambda^5 = 1$, where $\lambda = i\kappa$. While this algebra is not commutative, it is associative. These elements generate a group isomorphic to the group of rotations of the regular dodecahedron. Hamilton saw that these symbols relate to journeys about a dodecahedron which led to his invention of a game called the *Icosian Game*. Hamilton was not only the inventor of this game, he was also involved in its commercialization and even wrote the instruction pamphlet that accompanied the game. A commercial version of this game was referred to as *Around the World*. Hamilton associated 20 cities with the 20 vertices of a dodecahedron where the names of the 20 cities began with the 20 consonants of the English alphabet (y being considered a vowel here). A dodecahedron together with the 20 cities Hamilton chose is shown in Figure 2.1. The goal then was to discover a round trip moving along the edges of the dodecahedron that visits each city exactly once.

Hamilton observed that it was possible to take a round trip about the dodecahedron, visiting each vertex exactly once. Hamilton actually envisioned this game as a two-person game, where the first player provides conditions that the second player was to follow, as one proceeds about the dodecahedron. In one version of this game

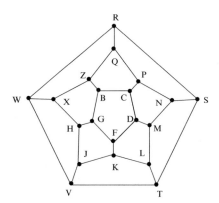

B. Brussels	H. Hanover	N. Naples	T. Toholsk
C. Canton	J. Jeddo	P. Paris	V. Vienna
D. Delhi	K. Kashmere	Q. Quebec	W. Washington
F. Frankfort	L. London	R. Rome	X. Xenia
G. Geneva	M. Moscow	S. Stockholm	Z. Zanzibar

Fig. 2.1 Hamilton's Icosian Game

described by Hamilton, there are 20 markers, numbered 1 to 20. The first player is to place markers $1, 2, 3, 4, 5$, in order, on five consecutive vertices of a dodecahedron. The second player is then to place the remaining markers $6, 7, \ldots, 20$, in order, on 15 consecutive unmarked vertices, such that markers 5 and 6, and 20 and 1 appear on consecutive vertices. This is the same as beginning with any path of order 5 on the graph of the dodecahedron. From his icosian calculus, Hamilton knew that this could always be done, no matter which five consecutive vertices are chosen first.

While Hamilton's Icosian Game is credited for being the origin of Hamiltonian cycles and Hamiltonian graphs, it is also known that Thomas Kirkman had earlier considered round trips on polyhedra passing through each vertex exactly once. Nevertheless, it is Hamilton's name that became associated with these cycles and graphs, not Kirkman's.

2.2   *k*-Ordered Hamiltonian Graphs

Although many graph theorists are well aware of this origin of Hamiltonian cycles, not many may be aware of the associated details. Indeed, Gary Chartrand once reported that he knew about Hamilton's *Around the World* game on a dodecahedron, but admitted that he had never read the details until the early 1990s. When he read what Hamilton had written in the instructions for his Icosian Game, two ideas occurred to him – one he thought was probably new, while about the other, he wasn't sure. His first idea led him to create the concept of *k*-ordered Hamiltonian graphs. We refer to the book [5] by Chartrand, Lesniak, and Zhang for graph theoretic notation and terminology not described here.

A Hamiltonian graph G is a *k-ordered Hamiltonian graph* if for every k vertices $v_1, v_2, \ldots, v_k$ of G, there exists a Hamiltonian cycle C of G encountering these k vertices (not necessarily consecutively) in the given order, somewhere on C. While it is obvious that every Hamiltonian graph is 3-ordered Hamiltonian, it is equally obvious that not every Hamiltonian graph is 4-ordered Hamiltonian (e.g., cycles). The graph of the dodecahedron, shown in Figure 2.1, is also not 4-ordered Hamiltonian, as there is no round trip visiting all 20 cities that passes through the cities Rome, Naples, Quebec, and Paris, in this order.

An obvious problem is to find sufficient conditions for a graph to be *k*-ordered Hamiltonian. Chartrand started working on this concept with the graduate student Michelle Schultz. Some time later, Chartrand received a letter from Joseph Gallian, who was running his well-known summer REU program and looking for a possible problem for one of his REU students. Chartrand suggested that Schultz and one of Gallian's students might work on this concept. Gallian chose to give the problem to Lenhard Ng. Schultz and Ng worked on the problem, which led to a paper [17].

For a graph G, the minimum sum of the degrees of two nonadjacent vertices of G is denoted by $\sigma_2(G)$, and the minimum degree of a (vertex in a) graph G is denoted by $\delta(G)$. In [17], it was proved that if G is a graph of order n and k is an integer with $3 \leq k \leq n$ such that $\sigma_2(G) \geq n + 2k - 6$, then G is k-ordered Hamiltonian.

Consequently, if $\delta(G) \geq \frac{n}{2} + k - 3$, then G is k-ordered Hamiltonian. Later, many others started investigating this concept, and the following improvements were obtained.

Theorem 2.1 ([14]). *Let G be a graph of order n and k an integer with $3 \leq k \leq n/2$. If $\sigma_2(G) \geq n + (3k - 9)/2$, then G is k-ordered Hamiltonian.*

Theorem 2.2 ([9]). *Let G be a graph of order $n \geq 11k - 3$ for some integer $k \geq 2$. If $\delta(G) \geq \lceil \frac{n}{2} \rceil + \lfloor \frac{k}{2} \rfloor - 1$, then G is k-ordered Hamiltonian.*

The lower bounds for $\sigma_2(G)$ and $\delta(G)$ presented in Theorems 2.1 and 2.2 that imply that a graph G is k-ordered Hamiltonian are sharp. In [17], it was shown that every k-ordered Hamiltonian graph is $(k-1)$-connected. Later, sufficient conditions were obtained on the connectivity of a graph, which imply that it is k-ordered Hamiltonian.

Theorem 2.3 ([7]). *Let G be a graph of order n such that $\sigma_2(G) \geq n$ and k an integer with $k \leq n/176$. If G is $\lfloor 3k/2 \rfloor$-connected, then G is k-ordered Hamiltonian.*

When only the minimum degree of a graph is considered, there is a slight improvement of Theorem 2.3.

Theorem 2.4 ([7]). *Let G be a graph of order n such that $\delta(G) \geq n/2$ and k an integer with $k \leq n/176$. If G is $3\lfloor k/2 \rfloor$-connected, then G is k-ordered Hamiltonian.*

Both lower bounds for the connectivity of a graph in Theorems 2.3 and 2.4 are best possible. A consequence of a result of Bollobás and Thomason[2] implies that every $22k$-connected graph G is k-*ordered* (i.e., for every k vertices $v_1, v_2, \ldots, v_k$ of G, there exists a cycle C of G encountering these k vertices in the given order, somewhere on C). This gave rise to the following question.

Problem 2.5 ([13]). *What is the least connectivity $f(k)$ for which every $f(k)$-connected graph is k-ordered?*

2.3 k-Path Hamiltonian Graphs

The second idea that occurred to Chartrand after reading Hamilton's introduction to the Icosian Game was one he thought may be known and concerned Hamiltonian extension. As Hamilton observed, beginning with any path P of order 5 on the graph H of the dodecahedron, P may be extended to a cycle containing every vertex of H. That is, for every path P of order 5 (or less) in H, there always exists a Hamiltonian cycle C of H such that P is a path on C. What Hamilton observed for paths of order 5 on the graph H does not hold for all paths of order 6. As is illustrated in Figure 2.2, the path of order 6 (drawn in bold edges) with initial vertex s cannot be extended to a Hamiltonian cycle on H, since the only way to reach y is through x, and then we cannot return to s. Hamilton never mentioned this however. Hamilton's observation led Chartrand to introduce a concept that is defined for every Hamiltonian graph.

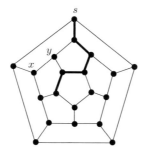

Fig. 2.2 The graph H of the dodecahedron

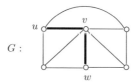

Fig. 2.3 A graph G with $he(G) = 2$

A Hamiltonian graph G of order $n \geq 3$ is called *k-path Hamiltonian*, for an integer k with $1 \leq k \leq n$, if for every path P of order k in G, there exists a Hamiltonian cycle C of G such that P lies on C. Certainly, every Hamiltonian graph is 1-path Hamiltonian. The largest integer k for which a Hamiltonian graph G is ℓ-path Hamiltonian for every integer ℓ with $1 \leq \ell \leq k$ is the *Hamiltonian extension number* $he(G)$. Therefore, $1 \leq he(G) \leq n$. If G is a Hamiltonian graph for which some automorphism maps any edge of G onto any other edge of G, then $he(G) \geq 2$. In fact, $he(G) = 2$ if and only if every edge of G lies on some Hamiltonian cycle of G but some path of order 3 in G does not lie on any Hamiltonian cycle of G. For example, the graph G of order 6 in Figure 2.3 has Hamiltonian extension number 2 since every edge of G lies on a Hamiltonian cycle, but the path (u, v, w) cannot be extended to a Hamiltonian cycle in G. Furthermore, $he(H) = 5$ for the graph H of the dodecahedron.

There is a rather curious question dealing with this topic that appears to have never been answered (or perhaps even asked).

Problem 3.1. *If G is a k-path Hamiltonian graph for some $k \geq 2$, is G also $(k-1)$-path Hamiltonian?*

If the question asked in Problem 3.1 has an affirmative answer, then the Hamiltonian extension number of a Hamiltonian graph G can then be defined as the largest positive integer k for which G is k-path Hamiltonian.

Observe that if G is a k-path Hamiltonian graph of order $n \geq 4$, where $2 \leq k \leq n$ such that $\delta(G) \geq k - 1$, then every path of order $k - 1$ lies on a path of order k

and so G is $(k-1)$-path Hamiltonian. It has been verified that the question stated in Problem 3.1 has an affirmative answer when $2 \le k \le 8$ or when $k \in \{n, n-1, n-2\}$.

All graphs G of order n for which $he(G) = n$ were characterized in a different context by Chartrand and Kronk [3]. They referred to a graph G of order $n \ge 3$ as *randomly Hamiltonian* if by beginning with any vertex v_1 of G, selecting any vertex v_2 adjacent to v_1, any vertex v_3 adjacent to v_2 not already selected, and so on, a Hamiltonian path $(v_1, v_2, v_3, \ldots, v_n)$ is always constructed with the additional property that v_n is adjacent to v_1. The randomly Hamiltonian graphs G of order $n \ge 3$ are precisely those graphs with $he(G) = n$. In particular, the result in [3] is then the following.

Theorem 3.2 ([3]). *A graph G of order $n \ge 3$ has Hamiltonian extension number n if and only if either G is the complete graph K_n, the n-cycle C_n, or when $n = 2r$ is even, the regular complete bipartite graph $K_{r,r}$.*

By Theorem 3.2, there are three graphs G of order $n \ge 3$ such that $he(G) = n$ if n is even and two such graphs if n is odd. There are no Hamiltonian graphs G of order n, however, for which $he(G) = n-1$ or $he(G) = n-2$ as the following result implies.

Proposition 3.3 ([6]). *If G is a Hamiltonian graph of order $n \ge 4$ for which $he(G) \ge n-2$, then $he(G) = n$.*

While no graph of order n has Hamiltonian extension number $n-1$ or $n-2$, a graph G of order $n \ge 4$ can have Hamiltonian extension number $n-3$. For example, $he(C_3 \,\square\, K_2) = n-3 = 3$. This is a consequence of the following result.

Proposition 3.4. *For the Cartesian product $C_n \,\square\, K_2$ of C_n ($n \ge 3$) and K_2,*

$$he(C_n \,\square\, K_2) = \begin{cases} 3 & \text{if } n \text{ is odd} \\ 4 & \text{if } n \text{ is even.} \end{cases}$$

Proof. Let $G = C_n \,\square\, K_2$, where the two copies of C_n in G are $(u_1, u_2, \ldots, u_n, u_1)$ and $(v_1, v_2, \ldots, v_n, v_1)$, and $u_i v_i$ is an edge of G for $1 \le i \le n$.

First, suppose that $n \ge 3$ is odd. It is obvious that $he(C_3 \,\square\, K_2) = 3$ and so assume that $n \ge 5$. Since G is Hamiltonian and every edge of G lies on a Hamiltonian cycle of G, we begin with a path P of order 3. By symmetry, we may assume, without loss of generality, that (i) $P = (u_1, u_2, u_3)$ or (ii) $P = (u_1, v_1, v_2)$. If (i) occurs, then P lies on the Hamiltonian cycle $(u_1, u_2, u_3, v_3, v_2, v_1, v_n, v_{n-1}, \ldots, v_4, u_4, u_5, \ldots, u_n, u_1)$, while if (ii) occurs, then P lies on the Hamiltonian cycle $(u_1, v_1, v_2, v_3, \ldots, v_n, u_n, u_{n-1}, \ldots, u_2, u_1)$. Thus, $he(G) \ge 3$. Since the path (u_1, v_1, v_2, u_2) of order 4 cannot be extended to a Hamiltonian cycle in G, it follows that $he(G) \le 3$ and so $he(G) = 3$ when n is odd.

When $n \ge 4$ is even, an argument similar to the one in the case where n is odd shows that every path of order 1, 2 or 3 can be extended to a Hamiltonian cycle in G. Next, we show that every path P of order 4 can be extended to a Hamiltonian cycle in G. By symmetry, there are five possible choices for P, namely, $Q_1 = (u_1, u_2, u_3, u_4)$,

$Q_2 = (u_1, u_2, u_3, v_3)$, $Q_3 = (u_1, u_2, v_2, v_3)$, $Q_4 = (u_1, u_2, v_2, v_1)$, and $Q_5 = (u_1, v_1, v_2, u_2)$. Each Q_i lies on a Hamiltonian cycle H_i for $1 \leq i \leq 5$ as follows:

$$H_1 = (u_1, u_2, u_3, u_4, u_5, \ldots, u_n, v_n, v_{n-1}, \ldots, v_1, u_1)$$

$$H_2 = (u_1, u_2, u_3, v_3, v_2, v_1, v_n, v_{n-1}, \ldots, v_4, u_4, u_5, \ldots, u_n, u_1)$$

$$H_3 = (u_1, u_2, v_2, v_3, u_3, u_4, v_4, v_5, u_5, u_6, \ldots, u_n, v_n, v_1, u_1)$$

$$H_4 = (u_1, u_2, v_2, v_1, v_n, v_{n-1}, \ldots, v_3, u_3, u_4, \ldots, u_n, u_1)$$

$$H_5 = (u_1, v_1, v_2, u_2, u_3, v_3, v_4, u_4, u_5, v_5, \ldots, u_{n-1}, v_{n-1}, v_n, u_n, u_1).$$

Thus, $\mathrm{he}(G) \geq 4$. Since the path $(u_1, u_2, u_3, v_3, v_4)$ of order 5 cannot be extended to a Hamiltonian cycle in G, it follows that $\mathrm{he}(G) \leq 4$, and so $\mathrm{he}(G) = 4$ when n is even. ∎

All five graphs G_1, G_2, G_3, G_4, G_5 shown in Figure 2.4 are 3-regular Hamiltonian graphs of order 20. Obviously, every vertex of G_1 lies on a Hamiltonian cycle of G_1, but the edge $u_1 v_1$ does not. Therefore, $\mathrm{he}(G_1) = 1$. On the other hand, every edge of G_2 lies on a Hamiltonian cycle of G_2, but the path (u_2, v_2, w_2) does not. That is, $\mathrm{he}(G_2) = 2$. In fact, for $i = 3, 4, 5$, every path of order i lies on a Hamiltonian cycle of G_i, but the path (u_3, v_3, w_3, x_3) lies on no Hamiltonian cycle of G_3, the path $(u_4, v_4, w_4, x_4, y_4)$ lies on no Hamiltonian cycle of G_4, and the path $(u_5, v_5, w_5, x_5, y_5, z_5)$ lies on no Hamiltonian cycle of G_5. Since the graph G_4 is $C_{10} \,\square\, K_2$, it follows that $\mathrm{he}(G_4) = 4$ from Proposition 3.4. The graph G_5, having Hamiltonian extension number 5, is due to Futaba Fujie (Personal communication 2015). Consequently, $\mathrm{he}(G_i) = i$ for $i = 1, 2, 3, 4, 5$.

There are 3-regular Hamiltonian graphs of order 20 having Hamiltonian extension number 1 other than the graph G_1 of Figure 2.4. For example, beginning with the cycle $(v_1, v_2, \ldots, v_{20}, v_1)$ of order 20 and adding the chords $v_1 v_{19}, v_2 v_{20}, v_9 v_{11}, v_{10} v_{12}$ and $v_i v_{21-i}$ for $i = 3, 4, \ldots, 8$ produces such a graph. This graph can then be generalized to produce a class of 3-regular Hamiltonian graphs of (even) order $n \geq 10$, all of which have Hamiltonian extension number 1. The graphs G_2 and G_3 are members of more general classes of 3-regular Hamiltonian graphs with Hamiltonian extension numbers 2 and 3, respectively. As we mentioned, the graph G_4 is a member of a more general class of 3-regular Hamiltonian graphs with Hamiltonian extension number 4, namely, the Cartesian products $C_n \,\square\, K_2$ of C_n and K_2 for all even integers $n \geq 4$. The graph H of the dodecahedron, having Hamiltonian extension number 5, can be redrawn as shown in Figure 2.5.

Several questions are suggested by these examples.

Problem 3.5. *Is there an infinite class of 3-regular Hamiltonian graphs having Hamiltonian extension number 5?*

Problem 3.6. *Is there a 3-regular Hamiltonian graph of order 20 having Hamiltonian extension number 6?*

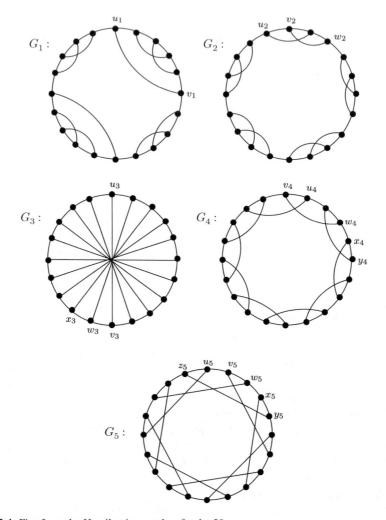

Fig. 2.4 Five 3-regular Hamiltonian graphs of order 20

Problem 3.7. *Let* $M = \max\{he(G) : G$ *is a 3-regular Hamiltonian graph}. Does* M *exist? If so, what is* M*?*

A number of results on Hamiltonian extension numbers were obtained in [6], including the following three theorems, where $\alpha(G)$ denotes the independence number of G.

Theorem 3.8 ([6]). *If* G *is a Hamiltonian complete* k*-partite graph of order* n *for some integer* $k \in \{3, 4, \ldots, n-1\}$, *then* $he(G) = n + 1 - 2\alpha(G)$.

Theorem 3.9 ([6]). *If* G *is a graph of order* $n \geq 3$ *and* $\delta(G) \geq n/2$, *then*

$$he(G) \geq 2\delta(G) - n + 1.$$

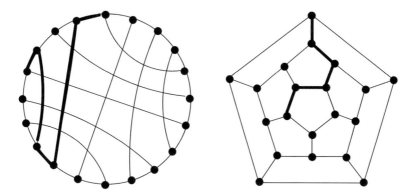

Fig. 2.5 The graph H of the dodecahedron

Theorem 3.10 ([6]). *If G is a graph of order $n \geq 4$ such that $\delta(G) \geq rn$ for some rational number r with $1/2 \leq r < 1$, then $he(G) \geq (2r-1)n+1$.*

Not only is the lower bound in Theorem 3.9 sharp, the lower bound presented in Theorem 3.10 for the Hamiltonian extension number of a graph is sharp for every rational number $r \in [1/2, 1)$. Of course, when $r = 1/2$ in Theorem 3.10, we have the well-known theorem of Dirac [8].

2.4   Hamiltonian Walks

While a graph can only be k-path Hamiltonian for some positive integer k if it is Hamiltonian, there is a Hamiltonian-related parameter, defined for all nontrivial connected graphs, Hamiltonian or not, that was introduced by Seymour Goodman and Stephen Hedetniemi [11, 12] in 1973.

Every connected graph G of order $n \geq 2$ and size m contains a closed spanning walk. Indeed, if every edge of a connected graph G is replaced by two parallel edges, then the resulting multigraph M of size $2m$ is Eulerian. Since an Eulerian circuit in M gives rise to a closed spanning walk in G in which each edge of G appears twice, it follows that G has a closed spanning walk of length $2m$. A *Hamiltonian walk* in G is a closed spanning walk of minimum length, and this length is denoted by $h(G)$. Therefore, $h(G) \leq 2m$. If G is a Hamiltonian graph of order n, then of course $h(G) = n$. Thus, $h(G)$ is only of interest to study when G is not Hamiltonian, in which case $h(G) \geq n + 1$. Goodman and Hedetniemi proved the following.

Theorem 4.1 ([12]). *If T is a nontrivial tree of order n, then $h(T) = 2(n-1)$.*

Proof. Since the size of a tree T of order n is $n - 1$, it follows by the remark above that $h(T) \leq 2(n-1)$. Let W be a Hamiltonian walk in T and let uv be an edge of T, where say u precedes v on W. Then uv is a bridge and so lies on W. We may therefore assume that W begins with u and is immediately followed by v. Since W terminates at u, the vertex u appears a second time on W, and this occurrence of u is immediately preceded by v. Thus the edge uv appears at least twice on W. Hence $h(T) \geq 2(n-1)$ and therefore $h(T) = 2(n-1)$. ∎

Since a Hamiltonian walk in a spanning tree T of a connected graph G is also a Hamiltonian walk in G, it follows that $h(G) \leq h(T) \leq 2(n-1)$ by Theorem 4.1. From this, the next result follows.

Corollary 4.2 ([12]). *For every nontrivial connected graph G of order n,*

$$n \leq h(G) \leq 2(n-1).$$

An upper bound for the length of a Hamiltonian walk in a connected graph G in terms of its order and $\sigma_2(G)$ was obtained by Jean-Claude Bermond [1].

Theorem 4.3. *If G is a connected graph of order $n \geq 3$, then $h(G) \leq 2n - \sigma_2(G)$.*

A well-known theorem by Oystein Ore [18] states that if G is a connected graph of order $n \geq 3$ such that $\sigma_2(G) \geq n$, then G is Hamiltonian. From this, Theorem 4.3 states that $h(G) \leq n$ and so $h(G) = n$, that is, G is Hamiltonian. Therefore, Theorem 4.3 is a generalization of Ore's theorem.

During the process of studying distance in graphs, Chartrand was led to the following concept. Let G be a connected graph of order $n \geq 2$ and let $s : v_1, v_2, \ldots, v_n, v_{n+1} = v_1$ be a cyclic ordering of the vertices of G. For the sequence s, the number $d(s)$ is defined by

$$d(s) = \sum_{i=1}^{n} d(v_i, v_{i+1}).$$

Since $d(v_i, v_{i+1}) \geq 1$ for $i = 1, 2, \ldots, n$, it follows that $d(s) \geq n$. The *Hamiltonian number* $h^*(G)$ is defined in [4] as min $\{d(s)\}$, where the minimum is taken over all cyclic orderings s of the vertices of G. Consequently, if G is a connected graph of order n, then $h^*(G) \geq n$ and $h^*(G) = n$ if and only if G is Hamiltonian. For the graph $G = K_{2,3}$ of Figure 2.6 and the two cyclic orderings $s_1 : v_1, v_2, v_3, v_4, v_5, v_1$

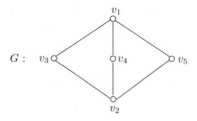

Fig. 2.6 A graph G with $h^*(G) = 6$

and $s_2 : v_1, v_3, v_2, v_4, v_5, v_1$ of the vertices of G, we have $d(s_1) = 8$ and $d(s_2) = 6$. Since G is a non-Hamiltonian graph of order 5 and $d(s_2) = 6$, it follows that $h^*(G) = 6$. In fact, for every cyclic ordering s of the vertices of G, either $d(s) = 6$ or $d(s) = 8$. It is not difficult to observe that not only is $h^*(G) = 6$, but $h(G) = 6$, as well. In fact, it was observed that $h^*(G) = h(G)$ for all graphs G that were initially considered. This led to a conjecture and then to the following theorem.

Theorem 4.4 ([4]). *If G is a nontrivial connected graph, then $h^*(G) = h(G)$.*

Proof. Since the result is obvious if the order n of G is 2, we may assume that $n \geq 3$. Let $s : v_1, v_2, \cdots, v_n, v_{n+1} = v_1$ be a cyclic ordering of the vertices of G for which $d(s) = h^*(G)$ and let $P^{(i)}$ be a $v_i - v_{i+1}$ geodesic in G. Then the walk that encounters the paths $P^{(1)}, P^{(2)}, \ldots, P^{(n)}$ in this order is a closed spanning walk of G having $h^*(G)$ as its length. Therefore, $h(G) \leq h^*(G)$.

Next, let $W = (u_0, u_1, \ldots, u_k)$ be a Hamiltonian walk in G, whose length $L(W)$ thus satisfies $h(G) = L(W) = k \geq n$. Let $v_i = u_{i+1}$ for $i = 1, 2$. For $3 \leq i \leq n$, let $v_i = u_{j_i}$, where j_i is the smallest positive integer such that $u_{j_i} \notin \{v_1, v_2, \ldots, v_{i-1}\}$. Then $s : v_1, v_2, \ldots, v_n, v_{n+1} = v_1$ is a cyclic ordering of the vertices of G. For $1 \leq i \leq n$, let W_i be the $v_i - v_{i+1}$ subwalk of W. Then

$$h^*(G) \leq d(s) = \sum_{i=1}^{n} d(v_i, v_{i+1}) \leq \sum_{i=1}^{n} L(W_i) = L(W) = h(G)$$

and so $h^*(G) = h(G)$. ∎

The fact that $h^*(G) = h(G)$ for every nontrivial connected graph G says that there is another way to look at the concept of Hamiltonian walks introduced by Goodman and Hedetniemi, which in turn suggests concepts and questions that would likely not have been considered otherwise. In particular, since the Hamiltonian number of a connected graph G of order $n \geq 2$ is the minimum value of $d(s)$ over all cyclic orderings s of the vertices of G, this makes one wonder if the values of $d(s)$ other than the minimum may provide information on the structure of G. One obvious concept, for example, is the *upper Hamiltonian number $h^+(G)$* of a connected graph G, defined by

$$h^+(G) = \max \{d(s)\},$$

where the maximum is taken over all cyclic orderings s of the vertices of G. Therefore, $h^+(G) = 8$ for the graph G of Figure 2.6. Observe that if $s : v_1, v_2, \ldots, v_n, v_{n+1} = v_1$ is any cyclic ordering of the vertices of a connected graph, then for each vertex v_i ($1 \leq i \leq n$), both $d(v_{i-1}, v_i) \leq e(v_i)$ and $d(v_i, v_{i+1}) \leq e(v_i)$, where $e(v_i)$ denotes the eccentricity of v_i (the distance from v_i to a vertex farthest from v_i). Therefore, if G is a connected graph of order $n \geq 3$ with $V(G) = \{v_1, v_2, \ldots, v_n\}$, then

$$h^+(G) \leq \sum_{i=1}^{n} e(v_i).$$

Since the eccentricity of a vertex in G is at most the diameter of G, we have the following upper bound for $h^+(G)$ in terms of the order and diameter of G.

Observation 4.5. *If G is a connected graph of order $n \geq 3$ and diameter d, then*

$$h^+(G) \leq nd.$$

The upper bound in Observation 4.5 for the upper Hamiltonian number of a connected graph has been shown to be sharp in [4]. For example, the graph $G = C_{2k+1} = (v_1, v_2, \ldots, v_{2k+1}, v_1)$, $k \geq 1$, has diameter k. For the sequence $s : v_1, v_{k+1}, v_{2k+1}, v_k, v_{2k}, \ldots, v_{k+2}, v_1$, we have $d(s) = nk = n \operatorname{diam}(G) = h^+(G)$.

For a connected graph G of diameter d and an integer p with $1 \leq p \leq d$, the *p*th *power G^p* of G is the graph with vertex set $V(G)$ for which $uv \in E(G^p)$ if and only if $1 \leq d_G(u, v) \leq p$. Thus, $\operatorname{diam}(G^p) = \lceil d/p \rceil$. In particular, for the graph $G = C_{2k+1}$ and the sequence s above, $d(s) = n\lceil k/p \rceil = n\lceil (n-1)/2p \rceil = h^+(G^p)$.

By definition, $h(G) \leq d(s) \leq h^+(G)$ for every cyclic ordering s of the vertices of G. The following result states that the Hamiltonian number and upper Hamiltonian number can be equal only for two classes of graphs.

Theorem 4.6 ([4]). *Let G be a nontrivial connected graph. Then $h(G) = h^+(G)$ if and only if G is either a complete graph or a star.*

By Theorem 4.6, only when a connected graph G is a complete graph or a star can $d(s)$ have a single value for each cyclic ordering of the vertices of G. More generally, we have the following concept, introduced by Král, Tong, and Zhu in [15]. For a connected graph G, the *Hamiltonian spectrum $\mathcal{H}(G)$* of G is defined by

$$\mathcal{H}(G) = \{d(s) : \ s \text{ is a cyclic ordering of the vertices of } G\}.$$

As an illustration, consider the Petersen graph P in Figure 2.7, whose vertices are labeled as shown in this figure. Since P is a non-Hamiltonian graph of order 10, $h(P) \geq 11$. On the other hand, let $s : x_1, x_2, \ldots, x_{11} = x_1$ be any cyclic ordering of

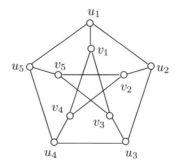

Fig. 2.7 The Petersen graph

the vertices of P. Since $diam(P) = 2$, it follows that $d(x_i, x_{i+1}) \leq 2$ for $1 \leq i \leq 10$. Hence $d(s) \leq 2 \cdot 10 = 20$ and so $h^+(P) \leq 20$. Therefore, $11 \leq h(P) \leq h^+(P) \leq 20$. In fact, $h(P) = 11$ and $h^+(P) = 20$. Consider the cyclic orderings s_i ($1 \leq i \leq 10$) as follows:

$$
\begin{aligned}
s_1 &: u_1, u_2, u_3, u_4, u_5, v_5, v_2, v_4, v_3, v_1, u_1 \\
s_2 &: u_1, u_2, u_3, u_4, u_5, v_5, v_2, v_3, v_4, v_1, u_1 \\
s_3 &: u_1, u_2, u_3, u_5, u_4, v_4, v_2, v_3, v_5, v_1, u_1 \\
s_4 &: u_1, u_3, u_5, u_2, u_4, v_4, v_2, v_5, v_3, v_1, u_1 \\
s_5 &: u_1, u_3, u_5, u_2, u_4, v_3, v_5, v_2, v_4, v_1, u_1 \\
s_6 &: u_1, u_3, u_5, u_2, u_4, v_5, v_2, v_4, v_3, v_1, u_1 \\
s_7 &: u_1, u_3, u_5, u_2, u_4, v_3, v_5, v_4, v_2, v_1, u_1 \\
s_8 &: u_1, u_3, u_5, u_2, v_2, u_4, v_3, v_4, v_5, v_1, u_1 \\
s_9 &: u_1, u_3, u_5, u_2, u_4, v_2, v_3, v_4, v_5, v_1, u_1 \\
s_{10} &: u_1, u_3, u_5, u_2, u_4, v_1, v_2, v_3, v_4, v_5, u_1.
\end{aligned}
$$

Since $d(s_i) = 10 + i$ for $1 \leq i \leq 10$, it follows for each integer k, with $11 \leq k \leq 20$, there exists a cyclic ordering s of the vertices of P such that $d(s) = k$. Therefore, $\mathcal{H}(P) = \{11, 12, \ldots, 20\}$.

By Theorem 4.6, if G is a complete graph or a star of order at least 2, then $h(G) = h^+(G)$. Hence, for every two cyclic orderings s and s' of the vertices of these two graphs, the numbers $d(s)$ and $d(s')$ are trivially of the same parity. The following theorem describes a more general class of graphs with this property.

Theorem 4.7 ([10]). *For every two cyclic orderings s and s' of the vertices of a nontrivial connected graph G, the numbers $d(s)$ and $d(s')$ are of the same parity if and only if G is complete or bipartite.*

Of course, every tree is a bipartite graph. The Hamiltonian spectrum of every tree was determined by Daphne Liu [16]. For a vertex v of a connected graph G, the *total distance* $td(v)$ of v is the sum of the distances from v to all vertices of G. The minimum total distance over all vertices of G is called the *median number* of G and is denoted by $med(G)$.

Theorem 4.8 ([16]). *For a tree T of order $n \geq 3$,*

$$\mathcal{H}(T) = \{2(n-1), 2n, 2(n+1), \ldots, 2med(T)\}.$$

Hence, for every tree T of order $n \geq 3$, an integer k is $d(s)$ for some cyclic ordering s of the vertices of T if and only if k is even with $2(n-1) \leq k \leq 2med(T)$.

A portion of the following result is an immediate consequence of Theorem 4.7.

Theorem 4.9 ([10]). *If G is a nontrivial connected graph that is neither complete nor bipartite, then there are not only two cyclic orderings s and s' of the vertices of G such that $d(s)$ and $d(s')$ are of opposite parity, there are cyclic orderings s and s' of the vertices of G such that $d(s) - d(s') = 1$.*

We are left with the following questions.

Problem 4.10. *For which finite sets S of positive integers, does there exist a nontrivial connected graph G whose Hamiltonian spectrum is S?*

Problem 4.11. *Let G be a connected bipartite graph of order n $\geq$ 3 with the property that an integer k is d(s) for some cyclic ordering s of the vertices of G if and only if k is an even integer with h(G) $\leq$ k $\leq$ h$^+$(G). Is it true that G is a tree?*

Problem 4.12. *The Petersen graph P has the property that $\mathcal{H}(P)$ is a set of ten consecutive integers. For every positive integer k, does there exist a connected graph G for which $\mathcal{H}(G)$ is a set of k consecutive integers?*

Acknowledgements I am grateful to Professor Gary Chartrand for suggesting the concept of Hamiltonian extension to me and kindly providing useful information on this topic. I also thank Stephen Hedetniemi, one of editors of this book, for numerous valuable suggestions.

References

1. Bermond, J.C.: On Hamiltonian walks. Congr. Numer. **15**, 41–51 (1976)
2. Bollobás, B., Thomason, A.: Highly linked graphs. Combinatorica **16**, 313–320 (1996)
3. Chartrand, G., Kronk, H.V.: Randomly traceable graphs. SIAM J. Appl. Math. **16**, 696–700 (1968)
4. Chartrand, G., Thomas, T., Saenpholphat, V., Zhang, P.: A new look at Hamiltonian walks. Bull. Inst. Comb. Appl. **42**, 37–52 (2004)
5. Chartrand, G., Lesniak, L., Zhang, P.: Graphs and Digraphs, 5th edn. Chapman and Hall/CRC, Boca Raton (2010)
6. Chartrand, G., Fujie, F., Zhang, P.: On an extension of an observation of Hamilton. J. Comb. Math. Comb. Comput. To appear.
7. Chen, G., Gould, R.J., Pfender, F.: New conditions for k-ordered Hamiltonian graphs. Ars Comb. **70**, 245–255 (2004)
8. Dirac, G.A.: Some theorems on abstract graphs. Proc. Lond. Math. Soc. **2**, 69–81 (1952)
9. Faudree, R.J., Gould, R.J., Kostochka, A.V., Lesniak, L., Schiermeyer, I., Saito, A.: Degree conditions for k-ordered Hamiltonian graphs. J. Graph Theory **42**, 199–210 (2003)
10. Fujie, F., Zhang, P.: Covering Walks in Graphs. Springer, New York (2014)
11. Goodman, S.E., Hedetniemi, S.T.: On Hamiltonian walks in graphs. In: Proceedings of the Fourth Southeastern Conference on Combinatorics, Graph Theory, and Computing (Florida Atlantic University, Boca Raton, 1973), pp. 335–342. Utilitas Mathematica, Winnipeg (1973)
12. Goodman, S.E., Hedetniemi, S.T.: On Hamiltonian walks in graphs. SIAM J. Comput. **3**, 214–221 (1974)
13. Gould, R.J.: Advances on the Hamiltonian problem – a survey. Graphs Comb. **19**, 7–52 (2003)
14. Kierstead, H.A., Sárközy, G.N., Selkow, S.M.: On k-ordered Hamiltonian graphs. J. Graph Theory **32**, 17–25 (1999)
15. Král, D., Tong, L.D., Zhu, X.: Upper Hamiltonian numbers and Hamiltonian spectra of graphs. Aust. J. Comb. **35**, 329–340 (2006)
16. Liu, D.: Hamiltonian spectrum for trees. Ars Comb. **99**, 415–419 (2011)
17. Ng, L., Schultz, M.: k-Ordered hamiltonian graphs. J. Graph Theory **24**, 45–57 (1997)
18. Ore, O.: Note on Hamilton circuits. Am. Math. Mon. **67**, 55 (1960)

Chapter 3
Conjectures on Cops and Robbers

Anthony Bonato

Abstract We consider some of the most important conjectures in the study of the game of Cops and Robbers and the cop number of a graph. The conjectures touch on diverse areas such as algorithmic, topological, and structural graph theory.

Mathematics Subject Classification 1991: 05C57, 05C85, 05C10

3.1   Introduction

The game of Cops and Robbers and its associated graph parameter, the cop number, have been studied for decades but are only now beginning to resonate more widely with graph theorists. One of the reasons for this owes itself to a challenging conjecture attributed to Henri Meyniel. Meyniel's conjecture, as it is now called, is arguably one of the deepest in the topic and will likely require new techniques to tackle. The conjecture has attracted the attention of the graph theory community and has helped revitalize the topic of Cops and Robbers. See Section 3.2.

As the game is not universally known, we define it here and provide some notation (it is customary to always begin a Cops and Robbers paper with the definition of cop number; regardless of best intentions, it is difficult to buck the trend). We consider only finite, undirected graphs in this paper, although we can play Cops and Robbers on infinite graphs or directed graphs in the natural way. Further, since the cop number is additive on connected components, we consider only *connected graphs*.

Supported by grants from NSERC.

A. Bonato (✉)
Department of Mathematics, Yeates School of Graduate Studies, Ryerson University,
350 Victoria Street, Toronto, ON, Canada M5B 2K3
e-mail: abonato@ryerson.ca

© Springer International Publishing Switzerland 2016   31
R. Gera et al. (eds.), *Graph Theory*, Problem Books in Mathematics,
DOI 10.1007/978-3-319-31940-7_3

We now formally define the game. Cops and Robbers is a game of perfect information; that is, each player is aware of all the moves of the other player. There are two players, with one player controlling a set of *cops* and the second controlling a single *robber*. The game is played over a sequence of discrete time-steps; a *round* of the game is a move by the cops together with the subsequent move by the robber. The cops and robber occupy vertices, and when a player is ready to move in a round, they must move to a neighboring vertex. The cops move first, followed by the robber; thereafter, the players move on alternate steps. Players can *pass* or remain on their own vertices. Observe that any subset of cops may move in a given round. The cops win if after some finite number of rounds, one of them can occupy the same vertex as the robber. This is called a *capture*. The robber wins if he can evade capture indefinitely. Note that the initial placement of the cops will not affect the outcome of the game, as the cops can expend finitely many moves to occupy a particular initial placement (the initial placement of the cops may, however, affect the length of the game).

Note that if a cop is placed at each vertex, then the cops are guaranteed to win. Therefore, the minimum number of cops required to win in a graph G is a well-defined positive integer, named the *cop number* of the graph G. The notation $c(G)$ is used for the cop number of a graph G. If $c(G) = k$, then G is *k-cop-win*. In the special case $k = 1$, G is *cop-win*.

For a familiar example, the cop number of the Petersen graph is 3. In a graph G, a set of vertices S is *dominating* if every vertex of G not in S is adjacent to some vertex in S. The *domination number* of a graph G is the minimum cardinality of a dominating set in G. Note that three cops are sufficient in the Petersen graph, as the domination number upper bounds the cop number. See Figure 3.1. This bound, however, is far from tight. For example, paths (or more generally, trees) have cop number 1.

There are now a number of conjectures that have arisen on Cops and Robbers, touching on many areas including algorithmic, topological, and structural graph theory. Some of these are more or less known. We will discuss these in the sections below.

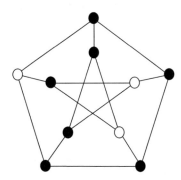

Fig. 3.1 The Petersen graph with white vertices dominating

When I am discussing Cops and Robbers with a newcomer, I am aware of the following, purely tongue-in-cheek principle:

Cops and Robbers Principle: Once you learn about Cops and Robbers, you are compelled to prove results about it.

The Cops and Robbers Principle, while itself is unverifiable, does seem fairly pervasive. One reason for this owes to the fact that the cop number, first defined in 1984, remains an unfamiliar parameter to many graph theorists. The cop number has limited connections (at least based on our current knowledge) to commonly studied graph parameters; this makes the field both challenging and fresh. The game is also simple to define and easy to play. You can even play it with some coins on a drawn graph with non-mathematicians.

Another reason why the Principle so often holds owes to the wealth of variations possible with the game. Almost every talk I give on the subject at a conference inspires the audience to spawn at least one (occasionally new) variation. This is not surprising as mathematicians have active imaginations, and Cops and Robbers definitely provides a fertile playground for the imaginative. One of my early mentors, the late lattice theorist Gunter Bruns, told the story of how he knew a mathematician who quit the field to become a poet. His reason for quitting was that he did not have enough imagination!

As a concrete example of this aspect of the Principle, at the 2014 SIAM Conference on Discrete Mathematics held in Minneapolis, colleagues Shannon Fitzpatrick and Margaret-Ellen Messenger suggested the new variant *Zombies and Survivors*. In this game, the zombies (cops) have minimal intelligence and always move directly toward the survivor (robber) along a shortest path (if there is more than one such path, then the zombies get to choose which one). We laughed at the following instance of the game. Consider a group of $\lfloor n/2 \rfloor - 2$ zombies on a cycle C_n, where $n \geq 4$. Place them on distinct, consecutive vertices, so they form a path of zombies. The survivor then chooses a vertex distance two from the "lead" zombie (that is the leaf of the zombie path). This placement of the zombies would result in zombies endlessly chasing the survivor in an orderly path. The survivor is forever just out of reach of the massive horde of hungry zombies! After learning about this variant, I told my colleagues they may be watching too many horror movies. In all seriousness, this variant speaks volumes about the broad appeal of the game.

The historical origin of the game is an interesting story in its own right. The game of Cops and Robbers was first considered by Quilliot [38] in his doctoral thesis. The game remained largely unknown at this time until it was considered independently by Nowakowski and Winkler [34]. According to Google Scholar, that five-page paper is the most cited of either author! Mathematics is no exception to the slogan "less is more."

Interestingly, both [34, 38] consider the game played with only one cop. In particular, they both focus on characterizing the cop-win graphs. The introduction of the cop number came a year later in 1984 with the important work of Aigner and Fromme [1].

Our book summarizes much of the research on Cops and Robbers up to 2011; see [12]. The interested reader is referred there for a broader background than provided here; see also the surveys [2, 8, 9, 27].

3.2   Meyniel's Conjecture

Graphs with cop number larger than one are not particulary well understood. The cop-win case is, on the other hand, well characterized as we describe next.

The *closed neighborhood* of a vertex x, written $N[x]$, is the set of vertices adjacent to x (including x itself). A vertex u is a *corner* if there is some vertex v such that $N[u] \subseteq N[v]$.

A graph is *dismantlable* if some sequence of deleting corners results in the graph K_1. For example, each tree is dismantlable: delete leaves repeatedly until a single vertex remains. The same approach holds with chordal graphs, which always contain at least two simplicial vertices (that is vertices whose neighbor sets are cliques). The following result characterizes cop-win graphs.

Theorem 1 ([34]). *A graph is cop-win if and only if it is dismantlable.*

The theorem provides a recursive structure to cop-win graphs, made explicit in the following sense. Observe that a graph is dismantlable if the vertices can be labeled by positive integers $\{1, 2, \ldots, n\}$, in such a way that for each $i < n$, the vertex i is a corner in the subgraph induced by $\{i, i + 1, \ldots, n\}$. This ordering of $V(G)$ is called a *cop-win ordering* (in the context of chordal graph theory, this is called an *elimination ordering*). See Figure 3.2 for a graph with vertices labeled by a cop-win ordering.

How big can the cop number be? First notice that for every positive integer n, there is a graph with cop number n. Hypercubes, written Q_n (where n is a nonnegative integer), are a family of graphs realizing every possible cop number (if we take Q_2 to be K_1). To see this, note that it was shown in [32] that for the hypercube Q_n of dimension n, $c(Q_n) = \lceil \frac{n+1}{2} \rceil$.

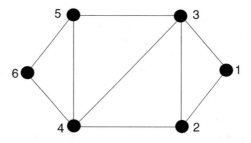

Fig. 3.2 A cop-win ordering of a cop-win graph

For a positive integer n, let $c(n)$ be the maximum cop number of a graph of order n (recall that we only consider connected graphs). *Meyniel's conjecture* states that there is a constant $d > 0$ such that for all positive integers n we have that

$$c(n) \leq d\sqrt{n}.$$

The conjecture was mentioned briefly in Frankl's paper [23] as a personal communication to him by Henri Meyniel in 1985 (see page 301 of [23] and reference [8] in that paper). As Meyniel has since passed away, we may never know his original motivation for the conjecture. Meyniel actually only published one short paper on Cops and Robbers, on a topic unrelated to the conjecture; see [32].

Meyniel's conjecture seemed to be largely unnoticed until recently. I may have been partly responsible for Meyniel's conjecture's rehabilitation. In 2006, I attended a small workshop organized by Geňa Hahn at the Bellairs Research Institute in Barbados, and I spoke about recent research on the cop number. The workshop was delightful, in no small part owing to the beautiful location. Jan Kratochvíl was there, and it appeared that the Cops and Robbers Principle was still in effect. He subsequently told Béla Bollobás about the parameter and conjecture, who then produced [7] (I am making this assumption based on the acknowledgment to Kratochvíl in that paper). Since then the interest in the conjecture has steadily grown. I also spoke at Bellairs about the capture time of a graph, which led to joint with Kratochvíl and others [13]. A play of the game with $c(G)$ cops is *optimal* if its length is the minimum over all possible plays for the cops, assuming the robber is trying to evade capture for as long as possible. There may be many optimal plays possible (for example on the path P_4 with four vertices, the cop may start on either of the two vertices in the center), but the length of an optimal game is an invariant of G. When $c(G)$ cops play on a graph G, we denote this invariant by capt(G) and refer to this as the *capture time* of G. In [13], the authors proved that if G is cop-win (that is has cop number 1) of order $n \geq 5$, then capt$(G) \leq n - 3$. By considering small-order cop-win graphs, the bound was improved to capt$(G) \leq n-4$ for $n \geq 7$ in [25]. Examples were given of planar cop-win graphs in both [13, 25] which prove that the bound of $n - 4$ is optimal. In addition to these works, capture time was studied in grids [33] and hypercubes [17].

For many years, the best known upper bound for general graphs was the one proved by Frankl [23].

Theorem 2 ([23]). *If n is a positive integer, then*

$$c(n) = O\left(n\frac{\log\log n}{\log n}\right).$$

I spoke about the cop number at the University of Waterloo in October 2007, to a group consisting mainly of theoretical computer scientists. My talk spurred a bright doctoral student Ehsan Chiniforooshan to consider improving on known upper bounds on the cop number. The Cops and Robbers Principle was again in full force that day! Chiniforooshan exploited similar ideas with retracts and proved the following bound, giving a modest improvement to Frankl's bound.

Theorem 3 ([18]). *If n is a positive integer, then*

$$c(n) = O\left(\frac{n}{\log n}\right).$$

At the time of writing this chapter in January 2015, the conjecture is still open. The best known upper bound was proved independently by three sets of authors. Interestingly, all of them use the probabilistic method in their proofs.

Theorem 4 ([24, 31, 41]). *If n is a positive integer, then*

$$c(n) = O\left(\frac{n}{2^{(1-o(1))\sqrt{\log n}}}\right).$$

To put Theorem 4 into perspective, even proving $c(n) = O(n^{1-\epsilon})$ for any given $\epsilon > 0$ remains open.

Prałat and Wormald in some recent work proved the conjecture for random graphs [36] and for random regular graphs [37], which gives us more evidence that the conjecture is true. I tend to believe the conjecture is true on good days; when I am in a bad mood, I imagine the universe contains some strange graph with cop number of larger order than $\sqrt{n}$.

There are graphs whose cop number is $\Theta(\sqrt{n})$; for example, consider the incidence graphs of finite projective planes. These graphs are of order $2(q^2 + q + 1)$, where q is a prime power, and have cop number $q + 1$. See Figure 3.3 for an example. The Cops and Robbers Principle was in effect when I described this graph family to the design theorist Andrea Burgess, which led to several other families with conjectured largest cop number; see [10].

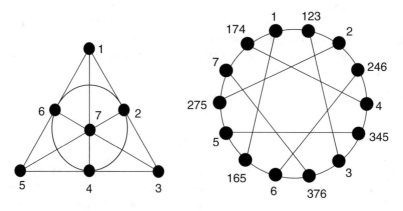

Fig. 3.3 The Fano plane and its incidence graph, the Heawood graph

Define m_k to be the minimum order of a connected graph G satisfying $c(G) \geq k$. Trivially, $m(1) = 1$ and $m(2) = 4$. The recent work [3, 4] establishes the fact that $m_3 = 10$. The unique isomorphism type of graph of order 10 with cop number 3 is the Petersen graph. It is easy to see that Meyniel's conjecture is equivalent to the property that

$$m_k = \Omega(k^2).$$

It might be fruitful to consider, therefore, the minimum orders of graph with a given cop number. We do not even know the exact value of m_4. The Petersen graph is the unique 3-regular graph of girth 5 of minimal order. A (k, g)-cage is a k-regular graph with girth g of minimal order. See [21] for a survey of cages. The Petersen graph is the unique $(3, 5)$-cage, and in general, cages exist for any pair $k \geq 2$ and $g > 3$. Aigner and Fromme [1] proved that graphs with girth 5 and degree k have cop number at least k; in particular, if G is a $(k, 5)$-cage, then $c(G) \geq k$. Let $n(k, g)$ denote the order of a (k, g)-cage. Is it true that a $(k, 5)$-cage is k-cop-win? It is natural to speculate whether $m_k = n(k, 5)$ for $k \geq 4$. It seems reasonable to expect that this is true at least for small values of k. It is known that $n(4, 5) = 19$, $n(5, 5) = 30$, $n(6, 5) = 40$, and $n(7, 5) = 50$. Do any of these cages attain the analogous m_k? More generally, we can ask the same question for large k: is m_k achieved by a $(k, 5)$-cage?

3.3 Graph Genus

With the analogy of the chromatic number in mind, what can be said on bounds on the cop number in planar graphs? This was settled early on by Aigner and Fromme [1].

Theorem 5 ([1]). *If G is a planar graph, then $c(G) \leq 3$.*

The idea of the proof of Theorem 5 is to increase the *cop territory*; that is, a set vertices S such that if the robber moved to S, then he would be caught. Hence, the number of vertices the robber can move to without being caught is eventually is reduced to the empty set, and so the robber is captured. While their proof is indeed elegant, it is not easy to follow. We wrote a proof which hopefully is easier to digest in Chapter 4 of [12] (based on ideas of Brian Alspach and Boting Yang).

The *genus* of a graph is the smallest integer n such that the graph can be drawn without edge crossings on a sphere with n handles. Note that a planar graph has genus 0. Less is known about the cop number of graphs with positive genus, and this provides our second major conjecture on the topic. The main conjecture in this area is due to Schroeder, and this conjecture I think deserves to be better known. In [40], Schroeder conjectured that if G is a graph of genus g, then $c(G) \leq g + 3$. Quilliot [39] proved the following.

Theorem 6 ([39]). *If G is a graph of genus g, then $c(G) \leq 2g + 3$.*

In the same paper where his conjecture was stated, Schroeder showed the following.

Theorem 7 ([40]). *If G is a graph of genus g, then*

$$c(G) \leq \left\lfloor \frac{3g}{2} \right\rfloor + 3.$$

Theorem 7 implies the following.

Corollary 8 ([40]). *If G is a graph that can be embedded on a torus, then $c(G) \leq 4$.*

We do not know much about the planar graphs with cop numbers 1, 2, or 3. As cop-win graphs have a dismantling structure, that might help to classify the planar cop-win graphs, but there is no success yet on that front.

3.4   Algorithms

We now describe a major conjecture on Cops and Robbers that was recently settled. Indeed, it is not every day that one of your postdocs comes into your office claiming to have proven a 20-year-old conjecture! We were lucky enough to have William Kinnersley as a postdoctoral fellow for two years starting in 2012. William came to Ryerson University having just completed his doctoral studies under Doug West's supervision, and he had a keen interest in the analysis of games played on graphs. While he had not worked much with Cops and Robbers before he came to Ryerson, the Cops and Robbers Principle was in effect, and he quickly delved into the topic.

EXPTIME is the class of decision problems solvable in exponential time. A decision problem is **EXPTIME**-complete if it is in **EXPTIME**, and for every problem in **EXPTIME**, there is a polynomial-time algorithm that transforms instances of one to instances of the other with the same answer. William proved that computing the cop number is **EXPTIME**-complete. Before going further to discuss this, let us formalize things and consider the following two graph decision problems.

k-COP NUMBER: Given a graph G and a positive integer k, is $c(G) \leq k$?

k-FIXED COP NUMBER: Let k be a fixed positive integer. Given a graph G, is $c(G) \leq k$?

The main difference between the two problems is that in k-COP NUMBER, the integer k may be a function of n and so grows with n. In k-FIXED COP NUMBER, k is fixed and not part of the input and so is independent of n.

The following result has been proved several times independently in the literature on the topic.

Theorem 9 ([6, 15, 28]). *The problem k-FIXED COP NUMBER is in **P**.*

If k is not fixed (and, hence, can be a function of n), then the problem becomes less tractable.

Theorem 10 ([22]). *The problem k-COP NUMBER is NP-hard.*

Theorem 10 is proved in Fomin et al. [22] by using a reduction from the following well-known **NP**-complete problem:

DOMINATION: Given a graph G and an integer $k \geq 2$, is there a dominating set in G of cardinality at most k?

Goldstein and Reingold [26] proved that it is **EXPTIME**-complete to compute the k-COP NUMBER problem assuming the initial position of the cops and robber is given as part of the input. They also conjectured in [26] that k-COP NUMBER is **EXPTIME**-complete. Kinnersley settled this conjecture in a recent tour de force [30], using a series of nontrivial reductions.

Note that Theorem 10 does not say that k-COP NUMBER is in **NP**; that is an open problem! There is little research on the optimal running times for polynomial-time algorithms to test if a graph has a small cop number such as 1, 2, or 3.

3.5 Variations

As you might expect, there are countless variations of the game of Cops and Robbers. Usually (though not always) such variations provide more complications than those found in the original game. One could play the game by giving the cops more power; in this direction, we studied the game of distance k Cops and Robbers [11, 15], where cops can capture the robber if it is within distance k. We could speed up the robber [24], allow the robber to capture a cop [16], or make the robber invisible [19, 20] (see Chapter 8 of [12] for more on these and other variants). We could also play on infinite graphs [14, 29], where many results from the finite landscape dramatically change.

We mention one variation in particular: the game of *Lazy Cops and Robbers*. This game is played in a similar fashion to Cops and Robbers, but only one cop may move at a time. Hence, Lazy Cops and Robbers is a game more akin to chess or checkers. The analogous parameter is the *lazy cop number*, written $c_L(G)$. Our knowledge of properties of the lazy cop number is limited, but in some cases its value is much larger than the classical cop number.

This game and parameter were first considered by Offner and Ojakian [35]. For hypercubes, it was proved in [32] that $c(Q_n) = \lceil \frac{n+1}{2} \rceil$. In contrast, the following holds for the lazy cop number.

$$2^{\lfloor \sqrt{n}/20 \rfloor} \leq c_L(Q_n) = O(2^n \log n / n^{3/2}). \tag{3.1}$$

A recent result of [5] improves the lower bound in (3.1).

Theorem 11 ([5]). *For all $\varepsilon > 0$, we have that*

$$c_L(Q_n) = \Omega\left(\frac{2^n}{n^{5/2+\varepsilon}}\right).$$

Thus, the upper and lower bounds on $c_L(Q_n)$ differ by only a polynomial factor. The proof uses the probabilistic method coupled with a potential function argument. It is an open problem to find the exact asymptotic order of $c_L(Q_n)$. The behavior of the lazy cop number on planar graphs or, more generally, graphs of higher genus is also not well understood.

Cops and Robbers represents the tip of the iceberg of what are called *vertex-pursuit games*, *graph searching*, or *good guys vs bad guys games* (the latter phrase was coined by Richard Nowakowski). A tough but fun problem in this general setting is on Firefighting in the infinite plane. Consider an infinite hexagonal grid. Every vertex is either on fire, clear, or protected. Initially, all vertices are clear. In the first round, fire breaks out on one vertex. In every round, a cop or *firefighter* protects one vertex which is not yet on fire. The fire spreads in the next round to all clear neighbors of the vertices already on fire. Once a vertex is on fire or is protected, it permanently remains in that state. Note that unlike Cops and Robbers, the firefighter does not play on the graph but can teleport anywhere it likes. Further, the fire the mindlessly spreads where it can.

Two firefighters can protect vertices so that the fire only burns two vertices in the hexagonal grid. It is not known if *one* firefighter can arrange things so the fire burns only *finitely* many vertices. In other words, can one firefighter build a wall containing the fire to a finite subgraph of the grid? It is conjectured that this is indeed impossible.

References

1. Aigner, M., Fromme, M., A game of cops and robbers. Discret. Appl. Math. **8**, 1–12 (1984)
2. Alspach, B.: Sweeping and searching in graphs: a brief survey. Matematiche **59**, 5–37 (2006)
3. Baird, W., Bonato, A.: Meyniel's conjecture on the cop number: a survey. J. Comb. **3**, 225–238 (2012)
4. Baird, W., Bonato, A., Beveridge, A., Codenotti, P., Maurer, A., McCauley, J., Valeva, S.: On the minimum order of k-cop-win graphs. Contrib. Discret. Math. **9**, 70–84 (2014)
5. Bal, D., Bonato, A., Kinnersley, W.B., Prałat, P.: Lazy cops and robbers played on hypercubes. Accepted to Comb. Probab. Comput. **24**, 829–837 (2015)
6. Berarducci, A., Intrigila, B.: On the cop number of a graph. Adv. Appl. Math. **14**, 389–403 (1993)
7. Bollobás, B., Kun, G., Leader, I.: Cops and robbers in a random graph. J. Comb. Theory Ser. B **103**, 226–236 (2013)
8. Bonato, A.: Catch me if you can: cops and robbers on graphs. In: Proceedings of the 6th International Conference on Mathematical and Computational Models (ICMCM'11) (2011)
9. Bonato, A.: WHAT IS . . . cop number? Not. Am. Math. Soc. **59**, 1100–1101 (2012)
10. Bonato, A., Burgess, A.: Cops and robbers on graphs based on designs. J. Comb. Des. **21**, 404–418 (2013)

11. Bonato, A., Chiniforooshan, E.: Pursuit and evasion from a distance: algorithms and bounds. In: Proceedings of ANALCO'09 (2009)
12. Bonato, A., Nowakowski, R.J.: The Game of Cops and Robbers on Graphs. American Mathematical Society, Providence (2011)
13. Bonato, A., Hahn, G., Golovach, P.A., Kratochvíl, J.: The capture time of a graph. Discret. Math. **309**, 5588–5595 (2009)
14. Bonato, A., Hahn, G., Tardif, C.: Large classes of infinite k-cop-win graphs. J. Graph Theory **65**, 234–242 (2010)
15. Bonato, A., Chiniforooshan, E., Prałat, P.: Cops and robbers from a distance. Theor. Comput. Sci. **411**, 3834–3844 (2010)
16. Bonato, A., Finbow, S., Gordinowicz, P., Haidar, A., Kinnersley, W.B., Mitsche, D., Prałat, P., Stacho, L.: The robber strikes back. In: Proceedings of the International Conference on Computational Intelligence, Cyber Security and Computational Models (ICC3) (2013)
17. Bonato, A., Gordinowicz, P., Kinnersley, W.B., Prałat, P.: The capture time of the hypercube. Electron. J. Comb. **20**(2), Paper #P24 (2013)
18. Chiniforooshan, E.: A better bound for the cop number of general graphs. J. Graph Theory **58**, 45 48 (2008)
19. Clarke, N.E., Connon, E.L.: Cops, robber, and alarms. Ars Comb. **81**, 283–296 (2006)
20. Clarke, N.E., Nowakowski, R.J.: Cops, robber and photo radar. Ars Comb. **56**, 97–103 (2000)
21. Exoo, G., Jajcay, R.: Dynamic cage survey. Electron. J. Comb. (2011). Dynamic Survey #16, revision #2
22. Fomin, F.V., Golovach, P.A., Kratochvl, J., Nisse, N., Suchan, K.: Pursuing a fast robber on a graph. Theor. Comput. Sci. **411**, 1167–1181 (2010)
23. Frankl, P.: Cops and robbers in graphs with large girth and Cayley graphs. Discret. Appl. Math. **17**, 301–305 (1987)
24. Frieze, A., Krivelevich, M., Loh, P.: Variations on cops and robbers. J. Graph Theory **69**, 383–402
25. Gavenčiak, T.: Cop-win graphs with maximal capture-time. Studentská vědecká a odbornáčinnost (SVOC) (2008)
26. Goldstein, A.S., Reingold, E.M.: The complexity of pursuit on a graph. Theor. Comput. Sci. **143**, 93–112 (1995)
27. Hahn, G.: Cops, robbers and graphs. Tatra Mt. Math. Publ. **36**, 163–176 (2007)
28. Hahn, G., MacGillivray, G.: A characterization of k-cop-win graphs and digraphs. Discret. Math. **306**, 2492–2497 (2006)
29. Hahn, G., Laviolette, F., Sauer, N., Woodrow, R.E.: On cop-win graphs. Discret. Math. **258**, 27–41 (2002)
30. Kinnersley, W.B.: Cops and robbers is **EXPTIME**-complete. Journal of Combinatorial Theory Series B **111**, 201–220 (2015)
31. Lu, L., Peng, X.: On Meyniel's conjecture of the cop number. J. Graph Theory **71**, 192–205 (2012)
32. Maamoun, M., Meyniel, H.: On a game of policemen and robber. Discret. Appl. Math. **17**, 307–309 (1987)
33. Merhrabian, A.: The capture time of grids. Discret. Math. **311**, 102–105 (2011)
34. Nowakowski, R.J., Winkler, P.: Vertex-t6o-vertex-pursuit in a graph. Discret. Math. **43**, 35–239 (1983)
35. Offner, D., Okajian, K.: Variations of cops and robber on the hypercube. Aust. J. Comb. **59**, 229–250 (2014)
36. Prałat, P., Wormald, N.: Meyniel's conjecture holds for random graphs. Random Structures and Algorithms **48**, 396–421 (2016)
37. Prałat, P., Wormald, N.: Meyniel's conjecture holds for random d-regular graphs. Preprint (2016)
38. Quilliot, A.: Jeux et pointes fixes sur les graphes. Thèse de 3ème cycle, Université de Paris VI, pp. 131–145 (1978)

39. Quilliot, A.: A short note about pursuit games played on a graph with a given genus. J. Comb. Theory Ser. B **38**, 89–92 (1985)
40. Schroeder, B.S.W.: The copnumber of a graph is bounded by $\lfloor \frac{3}{2} \mathrm{genus}(G) \rfloor + 3$. In: Categorical perspectives (Kent, OH, 1998). Trends in Mathematics, pp. 243–263. Birkhäuser, Boston (2001)
41. Scott, A., Sudakov, B.: A bound for the cops and robbers problem. SIAM J. Discret. Math. **25**, 1438–1442 (2011)

Chapter 4
On Some Open Questions for Ramsey and Folkman Numbers

Xiaodong Xu[*] and Stanisław P. Radziszowski[†]

Abstract We discuss some of our favorite open questions about Ramsey numbers and a related problem on edge Folkman numbers. For the classical two-color Ramsey numbers, we first focus on constructive bounds for the difference between consecutive Ramsey numbers. We present the history of progress on the Ramsey number $R(5, 5)$ and discuss the conjecture that it is equal to 43. For the multicolor Ramsey numbers, we focus on the growth of $R_r(k)$, in particular for $k = 3$. Two concrete conjectured cases, $R(3, 3, 3, 3) = 51$ and $R(3, 3, 4) = 30$, are discussed in some detail. For Folkman numbers, we present the history, recent developments, and potential future progress on $F_e(3, 3; 4)$, defined as the smallest number of vertices in any K_4-free graph which is not a union of two triangle-free graphs. Although several problems discussed in this paper are concerned with concrete cases and some involve significant computational approaches, there are interesting and important theoretical questions behind each of them.

AMS classification subjects: 05C55

[*]Supported by the National Natural Science Foundation (11361008) and the Guangxi Natural Science Foundation (2011GXNSFA018142).

[†]Work done while on sabbatical at the Gdańsk University of Technology, supported by a grant from the Polish National Science Centre grant 2011/02/A/ST6/00201. Support by the Institut Mittag-Leffler, Djursholm, Sweden, is gratefully acknowledged.

X. Xu
Guangxi Academy of Sciences, Nanning, Guangxi 530007, China
e-mail: xxdmaths@sina.com

S.P. Radziszowski (✉)
Department of Algorithms and System Modeling, Faculty of Electronics, Telecommunications and Informatics, Gdańsk University of Technology, 80-233 Gdańsk, Poland

Department of Computer Science, Rochester Institute of Technology, Rochester, NY 14623, USA
e-mail: spr@cs.rit.edu

© Springer International Publishing Switzerland 2016
R. Gera et al. (eds.), *Graph Theory*, Problem Books in Mathematics,
DOI 10.1007/978-3-319-31940-7_4

4.1   Introduction and Notation

In 2005, Arnold [4] wrote: *From the deductive mathematics point of view most of these results are not theorems, being only descriptions of several millions of particular observations. However, I hope that they are even more important than the formal deductions from the formal axioms, providing new points of view on difficult problems where no other approaches are that efficient.* The paper appeared in the *Journal of Mathematical Fluid Mechanics*, and it has not much to do with Ramsey theory. Yet the motivation of our paper is somewhat similar in that we may seem to focus much on concrete cases. Although several problems in this paper are concerned with concrete cases and some involve significant computational approaches, there are interesting and important theoretical questions behind each of them.

We obviously also try to look further for general results, but we do not want to skip observing what is happening with the basic small open cases. Understanding them better may lead to surprising general conclusions. For example, our work on an old construction for $R_4(6)$ and many unsuccessful attempts to prove that $\lim_{r \to \infty} R_r(3)^{\frac{1}{r}}$ is infinite led to interesting general connections between our methods and the Shannon capacity [78] discussed in Section 3.1.

The standard reference for Ramsey theory is a great book by Graham, Rothschild, and Spencer [40], *Ramsey Theory*. The subject first concerned mathematical logic, but over the years found its way into several areas of mathematics, computing, and other fields. For the discussion of numerous applications, see the survey paper by Rosta [66] and a very useful website by Gasarch [33]. There is also a colorful book by Soifer [72] on the history and results in Ramsey theory, followed by a collection of essays and technical papers based on presentations from the 2009 Ramsey theory workshop at DIMACS [73]. A regularly updated survey of the most recent results on the best-known bounds on various types of Ramsey numbers is maintained by the second author [63].

The most important operation involved in the concept of Ramsey and Folkman numbers is that of arrowing, which is defined as follows:

Definition 1.1 (Arrowing). *Graph F arrows graphs $G_1, \ldots, G_r$, written $F \to (G_1, \ldots, G_r)$, if and only if every r-coloring of the edges of F contains a monochromatic copy of G_{s_i} in color i, for some $1 \le i \le r$.*

The definition of the classical two-color *Ramsey numbers* can be stated in terms of the arrowing relation as $R(s, t) = \min\{n \mid K_n \to (K_s, K_t)\}$, with a straightforward generalization for more colors and noncomplete graphs. If all graphs G_i are the same G, we will use notation $R_r(G)$ for $R(G_1, \ldots, G_r)$, and if the graphs G_i are complete, we will write $s_i = |V(G_i)|$ instead of G_i. So, for example, $R(5, 5) = R(K_5, K_5)$, and $R_r(3)$ is the smallest n such that the r-color arrowing $K_n \to (K_3, \ldots, K_3)$ holds. The latter two cases are discussed in detail in Sections 2.2 and 4.3, respectively.

Ramsey proved a theorem which implies the following:

Theorem 1.2 (Ramsey 1930 [65]). *For $r \geq 1$ and all graphs $G_1, \ldots, G_r$, the Ramsey number $R(G_1, \ldots, G_r)$ exists.*

Any edge r-coloring witnessing $K_n \nrightarrow (G_1, \ldots, G_r)$ will be called a $(G_1, \ldots, G_r)$- or $(G_1, \ldots, G_r; n)$-coloring. Clearly, constructing any $(G_1, \ldots, G_r; n)$-coloring implies a lower bound $n < R(G_1, \ldots, G_r)$.

In the case of two colors, we will talk about (G, H)- and $(G, H; n)$-graphs, which are simply $(G, H; n)$-colorings of K_n, where the first color is interpreted as the graph, while the second as its complement. Let $\alpha(F)$ denote the independence number of graph F, which is the maximum number of vertices in any independent set of F. Note that $R(s, t)$ can be defined equivalently as the smallest integer n such that every graph F on n vertices contains K_s or has independence $\alpha(F) \geq t$. An (s, t)-graph G will be called *Ramsey critical* (for (s, t)) if it has $R(s, t) - 1$ vertices, i.e., it is an $(s, t; R(s, t) - 1)$-graph. $\delta(G)$ and $\Delta(G)$ will denote the minimum and maximum degree in G, respectively, and $K_n - e$ is the complete graph on n vertices with one edge removed. We will sometimes write $n(G) = |V(G)|$ for the number of vertices in G.

Next, we define the set of *edge Folkman graphs* by

$$\mathcal{F}_e(s, t; k) = \{F \mid F \rightarrow (s, t) \text{ and } K_k \nsubseteq F\}.$$

Then, the corresponding *edge Folkman number* $F_e(s, t; k)$ is the smallest order $n(F)$ of any graph F in $\mathcal{F}_e(s, t; k)$. Folkman proved that these graphs exist, as follows:

Theorem 1.3 (Folkman 1970 [28]). *If $k > \max(s, t)$, then $F_e(s, t; k)$ exists.*

Edge Folkman numbers have obvious generalizations to arrowing graphs other than complete graphs and to more colors as in $\mathcal{F}_e(s_1, \ldots, s_r; k)$ and $F_e(s_1, \ldots, s_r; k)$ [58]. One can also color vertices instead of edges, which leads to so-called *vertex Folkman numbers*. In general, much less is known about edge Folkman numbers than for more studied vertex Folkman numbers [18]. Here, however, we will discuss only the case of $F_e(3, 3; k)$, in Section 4.4. Finally, we note that there exists much more research activity related to Ramsey numbers than to Folkman numbers, though the latter area is attracting significantly more attention in recent years.

The problem of deciding whether a graph F arrows triangles, that is, whether $G \rightarrow (3, 3)$, is of particular interest in Ramsey theory. This is **coNP**-complete, and it appeared in the classical complexity text by Garey and Johnson in 1979 [32]. Some related Ramsey graph coloring problems are **NP**-hard or lie even higher in the polynomial hierarchy. For example, Burr [8, 9] showed that arrowing $(3, 3)$ is **coNP**-complete together with other results about arrowing, and Schaefer [68] showed that for general graphs F, G, and H, $F \rightarrow (G, H)$ is Π_2^P-complete.

4.2   Two-Color Ramsey Numbers

4.2.1   *Difference and Connectivity*

The estimates of the difference between consecutive (in various meanings) Ramsey numbers are difficult. What we know, most of the time, implies much weaker bounds in comparison to what we expect to hold.

Problem 2.1.1 (Erdős-Sós 1980 [12, 20]). *Let* $\Delta_k = R(3,k) - R(3, k-1)$. *Is it true that*

$$\Delta_k \overset{k}{\to} \infty ? \quad (\Delta_k/k) \overset{k}{\to} 0 ?$$

Only easy bounds $3 \le \Delta_k \le k$ are known. The upper bound k is obvious since the maximum degree of $(3, k)$-graphs is at most $k - 1$. The lower bound of 3 looks misleadingly simple, but it is not trivial (Theorems 2.1.2 and 2.1.3 imply it). It was argued in [36] that better understanding of Δ_k may come from the study of $R(K_3, K_k - e)$ relative to $R(K_3, K_k) = R(3, k)$, since

$$\Delta_k = \big(R(K_3, K_k) - R(K_3, K_k - e)\big) + \big(R(K_3, K_k - e) - R(K_3, K_{k-1})\big).$$

Recent progress on what we know for small cases is significant [35, 36]; however some very simple-looking questions still remain open. For example, we do not even know for certain whether $R(K_3, K_k - e) - R(K_3, K_{k-1})$ is positive for all large k.

The following three theorems were proved by constructive methods as parts of Theorems 2 and 3 in [79, 81] and Theorem 9 in [80]:

Theorem 2.1.2 ([81]). *Given a (k, s)-graph G and a (k, t)-graph H, for some $k \ge 3$ and $s, t \ge 2$, if both G and H contain an induced subgraph isomorphic to some K_{k-1}-free graph M, then $R(k, s + t - 1) \ge n(G) + n(H) + n(M) + 1$.*

Theorem 2.1.3 ([81]). *If $2 \le s \le t$ and $k \ge 3$, then*

$$R(k, s + t - 1) \ge R(k, s) + R(k, t) + \begin{cases} k - 3, & \text{if } s = 2; \\ k - 2, & \text{if } s \ge 3. \end{cases}$$

The first inequality of Theorem 2.1.3 for $s = 2$, $R(k, t + 1) \ge R(k, t) + 2k - 3$, was proved by Burr et al. in 1989 [10].

Theorem 2.1.4 ([80]). *If $k \ge 2, s \ge 5$, then $R(2k - 1, s) \ge 4R(k, s - 1) - 3$.*

We think that the progress on constructive lower bounds illustrated in Theorems 2.1.2–2.1.4 is quite representative for the area, but it seems slow. It is much slower than it was once anticipated by Erdős, Faudree, Schelp, and Rousseau. In 1980, Paul Erdős wrote in [20], page 11 (using r for our R): *Faudree, Schelp, Rousseau and I needed recently a lemma stating*

$$\lim_{n\to\infty} \frac{r(n+1,n) - r(n,n)}{n} = \infty \tag{a}$$

We could prove (a) without much difficulty, but could not prove that $r(n+1,n) - r(n,n)$ increases faster than any polynomial of n. We of course expect

$$\lim_{n\to\infty} \frac{r(n+1,n)}{r(n,n)} = C^{\frac{1}{2}}, \tag{b}$$

where $C = \lim_{n\to\infty} r(n,n)^{1/n}$.

Based on the above theorems and considerations in [82], the best-known lower bound estimate for the difference in (a) seems to be barely $\Omega(n)$. Asking others, including collaborators of Erdős, did not lead us to any proof of this result, leaving however some possibility that Erdős knew it. In summary, we think that it is reasonable to consider (a) to be only a conjecture.

Beveridge and Pikhurko in [5], using Theorem 2.1.3, proved that the connectivity of any $(k, s; R(k, s) - 1)$-graph, i.e., Ramsey-critical (k, s)-graph, is at least $k - 1$ for all $k, s \geq 3$. In Theorem 8 of [82], we increased this bound on connectivity to k for $k \geq 5$, and then we obtained further results about which Ramsey-critical graphs must be Hamiltonian.

Theorem 2.1.5 ([82]). *If $k \geq 5$ and $s \geq 3$, then the connectivity of any $(k, s; R(k, s) - 1)$-graph is at least k. Furthermore, if $k \geq s - 1 \geq 1$ and $k \geq 3$, except $(k, s) = (3, 2)$, then any $(k, s; R(k, s) - 1)$-graph is Hamiltonian.*

In particular, all diagonal Ramsey-critical (k, k)-graphs are Hamiltonian for every $k \geq 3$. It remains an open question for which k and s, when $3 \leq k < s - 1$, Ramsey-critical (k, s)-graphs are still Hamiltonian. We believe that the answer is positive at least in the cases when $s - k$ is small.

Conjecture 2.1.6. *For all $k \geq 2$, there exists a Ramsey-critical $(k + 1, k)$-graph with maximum degree at least $R(k + 1, k)/2 - 1$.*

This conjecture seems weak, but we still have no idea how to prove or disprove it. Many would even readily agree with an intuition that any Ramsey-critical $(k+1, k)$-graph G satisfies the bound $\Delta(G) \geq |V(G)|/2$. On the other hand, we clearly have $\Delta(G) < R(k, k)$. Putting it together with the classical bound $R(k + 1, k + 1) \leq 2R(k + 1, k)$, we propose the next conjecture.

Conjecture 2.1.7. *$R(k + 1, k) \leq 2R(k, k)$ and $R(k + 1, k + 1) \leq 4R(k, k)$.*

By the comments above, a yes answer to Conjecture 2.1.6 implies a yes for Conjecture 2.1.7. We note that a very similar inequality, $R(k + 1, k + 1) \leq 4R(k + 1, k - 1) + 2$, was proved by Walker [75] in 1968. There are straightforward generalizations of these thoughts to other close-to-diagonal cases and to more than two colors, but we stop short of proposing them as conjectures.

4.2.2   On the Ramsey Number $R(5, 5)$

What is the largest number of vertices in any K_5-free graph with independence number less than 5? The answer is $R(5, 5) - 1$. The values of $R(s, t)$ are known for all s and t with $s + t < 10$ [63], so in this sense $R(5, 5)$ is the smallest open case in Ramsey theory.

The progress of knowledge about lower and upper bounds on $R(5, 5)$ first spanned more than three decades; then it apparently stopped in 1997. What we know now is almost the same as 17 years ago, while a significant gap between bounds remains unchanged. The effort required to lower the upper bound on $R(5, 5)$ from 50 down to 49 was very significant, but still 49 is quite far from the best-known lower bound of 43, which was obtained by Exoo in 1989 [23].

Theorem 2.2.1 ([23]). $43 \leq R(5, 5) \leq 49$ *[57].*

The history of bounds on $R(5, 5)$ is presented in Table 4.1. None of the results in references listed until 1973 depended in a significant way on computer algorithms. All of the later items involved at least some computational components to the degree that their full verification by hand seems infeasible. Note that Table 4.1 stops the listing in 1997. It is not the case that people did not try since then. We are aware of several such attempts, but it seems that none of them were finally published. The constructions allegedly improving on the lower bound of 43, which we have seen, each contained an error. A few attempts to improve the upper bound tried to derive some properties of, say, $(5, 5; 45)$-graphs; however we are not aware of any recognized and significant results in this direction.

In 1997, McKay and the second author [57] posed the following conjecture:

Conjecture 2.2.2. $R(5, 5) = 43$, *and the number of $(5, 5; 42)$-graphs is precisely* 656.

Table 4.1 The history of bounds on $R(5, 5)$, based on [57]. (LP refers to linear programming techniques)

Year	Reference	Lower	Upper	Comments
1965	Abbott [1]	38		Quadratic residues in $\mathcal{Z}_{37}$
1965	Kalbfleisch [44]		59	Pointer to a future paper
1967	Giraud [34]		58	Combinatorics, LP
1968	Walker [75]		57	Combinatorics, LP
1971	Walker [76]		55	Combinatorics, LP
1973	Irving [43]	42		Sum-free sets
1989	Exoo [23]	43		Simulated annealing
1992	McKay-Radziszowski [54]		53	$(4, 4)$-graph enumeration, LP
1994	McKay-Radziszowski [55]		52	LP, computation
1995	McKay-Radziszowski [56]		50	Implication of $R(4, 5) = 25$
1997	McKay-Radziszowski [57]		49	

The authors of [57] provided some strong evidence for its correctness. Of particular strength seems to be the fact that a few distinct methods to generate $(5, 5; 42)$-graphs ended up in the same final set of 656 graphs. Three hundred twenty-eight of these graphs, with the number of edges ranging from 423 to 430, are posted at a website by McKay [53]; the other 328 on at least 431 edges are their complements. All of the 656 graphs have the minimum degree 19 and maximum degree 22. The automorphism groups of these graphs are surprisingly small; none has order larger than 2, or more precisely 232 are involutions without fixed points, and the remaining 424 groups are trivial. This is somewhat against an intuition that complete sets of extreme graphs for typical Ramsey cases should contain some graph with a larger automorphism group. We note, however, that graphs with more symmetries in general are easier to find, and thus we think that any such $(5, 5; 42)$-graph would have been already found if it existed.

In, McKay and Lieby (Research School of Computer Science, Australian National University, personal communication, 2014) provided the following new evidence for Conjecture 2.2.2, which required a computational effort of about 9 CPU years. Define the distance between two graphs on n vertices to be k if their largest common induced subgraph has $n - k$ vertices. McKay and Lieby report that any new $(5, 5; 42)$-graph H would have to be in distance at least 6 from every graph in the set of 656 known $(5, 5; 42)$-graphs.

Some improvement of the upper bound in Theorem 2.2.1 might be possible, but we consider that lowering it even just by 1 would be a great accomplishment.

Although the authors of this work share their ideas on most problems presented herein, there is an exception in our positions on the so-called almost regular Ramsey graphs and in consequence on Conjecture 2.2.2. A graph G is *almost regular* if $\Delta(G) - \delta(G) \leq 1$. The following Conjecture 2.2.3 on almost regular Ramsey graphs was proposed by the first author, who explored it with Zehui Shao and Linqiang Pan in 2008. Shao's computational work in this direction appears in his thesis [70], but otherwise was not published.

Conjecture 2.2.3. *For all positive s and t and every $1 \leq n < R(s, t)$, there exists an almost regular $(s, t; n)$-graph.*

Needless to say, no counterexample to Conjecture 2.2.3 is known. However, since none of the 656 known $(5, 5; 42)$-graphs are almost regular, if Conjecture 2.2.2 holds, then Conjecture 2.2.3 is false. The first author supports Conjecture 2.2.3, but not Conjecture 2.2.2, while the second author supports Conjecture 2.2.2 and thus not Conjecture 2.2.3, unless the latter is restated only for sufficiently large s and t.

4.2.3 Constructive Lower Bounds for $R(3, k)$

In 1995, Kim [47] obtained a breakthrough result establishing the asymptotics of $R(3, k)$ up to a multiplicative constant, when he raised the lower bound to match the upper bound.

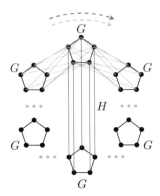

Fig. 4.1 Construction of a $(3, 9; 30)$-graph H using C_5 as a $(3, 3; 5)$-graph G, for $k = 2$

Theorem 2.3.1 ([47]). $R(3, k) = \Theta(k^2/\log k)$.

Recently, in independent work by Bohman and Keevash [7] and by Fiz, Griffiths, and Morris [27], impressive further progress has been obtained in closing on the actual constants of Theorem 2.3.1.

Theorem 2.3.2 ([7, 27]). $\left(\frac{1}{4} + o(1)\right)k^2/\log k \leq R(3, k) \leq (1 + o(1))k^2/\log k$ *[71].*

The progress on asymptotic lower bounds for $R(3, k)$ was obtained by the probabilistic method [7, 27, 47, 74], which often yields very weak bounds for concrete small cases. The upper bound of Theorem 2.3.2 is implicit in [71]. The best specific constructions are usually obtained by insight, computations, and ad hoc means. We lack general constructions which give both clear structure of the graphs and good Ramsey lower bounds. One of the best-known and elegant constructions is a recursive method by Chung, Cleve, and Dagum from 1993 [14]. We present an instance of it in Figure 4.1.

Let G be a triangle-free graph on n vertices with independence $\alpha(G) = k$, i.e., G is a $(3, k + 1; n)$-graph. Consider graph H, called a *fibration* of G, formed by six disjoint copies of G with two types of edges joining them (see Figure 4.1), as described in [14]. Chung et al. proved that their construction produces a $(3, 4k + 1; 6n)$-graph H, which easily gives $R(3, 4k + 1) \geq 6R(3, k + 1) - 5$. By solving the recurrence, one obtains the asymptotic lower bound $R(3, k) = \Omega(k^{\log 6/\log 4}) \approx \Omega(k^{1.29})$.

Other explicit constructions for $R(3, k)$ leading to a better lower bound $\Omega(k^{3/2})$ were presented by Alon in 1994 [2] and Codenotti, Pudlák, and Resta in 2000 [15]. In 2010, Kostochka, Pudlák, and Rödl [48] improved further known constructive lower bounds for $R(k, n)$ for fixed $4 \leq k \leq 6$, but their results still lagged behind those obtained by the probabilistic method. For example, with $k = 4$, the probabilistic K_4-free process used by Bohman yields $R(4, n) = \Omega(n^{5/2}/\log^2 n)$ [6], while the constructive approach of [48] gives only $R(4, n) = \Omega(n^{8/5})$.

Challenge 2.3.3. *Design a recursive lower bound construction of* $(3, k; n)$-*graphs for* $R(3, k)$, *with the number of vertices* n *larger than* $\Omega(k^{3/2})$.

4.3 Multicolor Ramsey Numbers

Using elementary methods in 1955, Greenwood and Gleason [41] established that for the multicolor Ramsey numbers, for all $k_i \geq 2$ and $r \geq 2$, we have

$$R(k_1, \ldots, k_r) \leq 2 - r + \sum_{i=1}^{r} R(k_1, \ldots, k_{i-1}, k_i - 1, k_{i+1}, \ldots, k_r), \qquad (4.1)$$

with strict inequality if the right-hand side of (4.1) is even and the sum has an even term. The bound (4.1) reduces to the classical two-color upper bound for $r = 2$. There are only two known non-trivial multicolor cases ($r \geq 3$ and $k_i \geq 3$), for the parameters (3,3,3,3) and (3,3,4), where this bound was improved. On the other hand, very likely the bound (4.1) is never tight for $r \geq 3$, except for (3,3,3). We will discuss the special cases of $R_4(3) = R(3, 3, 3, 3)$ and $R(3, 3, 4)$ in Subsections 4.3.2 and 4.3.3, respectively, in more detail.

4.3.1 Constructions and Limits

In 1973, Chung [11] proved constructively that $R_r(3) \geq 3R_{r-1}(3) + R_{r-3}(3) - 3$, and in 1983 Chung and Grinstead [13] showed that the limit

$$L = \lim_{r \to \infty} R_r(3)^{\frac{1}{r}} \qquad (4.2)$$

exists, though it may be infinite.

One of the most successful techniques for deriving lower bounds on $R_r(3)$ is constructions based on Schur partitions and closely related cyclic and linear colorings. A Schur partition of the integers from 1 to n, $[1, n]$, is a partition into sum-free sets. The *Schur number* $s(r)$ is the maximum n for which there exists a Schur partition of $[1, n]$ into r sets. A simple argument gives $s(r) + 2 \leq R_r(3)$.

In an early work, Abbott [1] showed that $s(r) > 89^{r/4 - c \log r} > 3.07^r$. After much more effort, the exact values of $s(r)$ have been found for $1 \leq r \leq 4$. What we know now about Schur numbers $s(r)$ provides the best-known lower bound of 3.199 for L, which is implied by the lower bound of 536 on $s(6)$. This was obtained by Fredricksen and Sweet in 2000 [31]. In Table 4.2, which summarizes the best-known bounds on $R_r(3)$, three lower bounds for $5 \leq r \leq 7$ are implied by constructions of partitions for Schur numbers. For additional results and comments on constructive lower bounds on $R_r(3)$ and general $R(k_1, \cdots, k_r)$, see [80].

Table 4.2 Bounds and values of $R_r(3)$

r	Value or bounds	References
2	6	Folklore
3	17	Greenwood-Gleason 1955 [41]
4	51–62	Chung 1973 [11] – Fettes-Kramer-R 2004 [26]
5	162–307	Exoo 1994 [25] – bound (4.1)
6	538–1838	Fredricksen-Sweet 2000 [31] – bound (4.1)
7	1682–12861	Fredricksen-Sweet 2000 [31] – bound (4.1)

Recently, we improved a construction from [80] to one which permits us to double the number of colors in a special way, as stated in the next theorem.

Theorem 3.1.1 ([78]). *For integers $k, n, m, s \geq 2$, let G be a $(k, \ldots, k; s)$-coloring with n colors containing an induced subcoloring of K_m using less than n colors. Then*

$$R_{2n}(k) \geq s^2 + m(R_n(k-1, k, \ldots, k) - 1) + 1.$$

The *Shannon capacity* $c(G)$ of a noisy channel modeled by graph G is defined as $\lim_{n\to\infty} \alpha(G^n)^{1/n}$, where $\alpha(G^n)$ is the independence number of the n-th power of graph G using the strong product of graphs. We proved in [78] that the construction in the proof of Theorem 3.1.1 with $k = 3$ implies the following:

Theorem 3.1.2 ([78]). *The supremum of the Shannon capacity over all graphs with independence number 2 cannot be achieved by any finite graph power.*

We also generalized Theorem 3.1.2 to graphs with bounded independence number. The link between Shannon capacity and multicolor Ramsey numbers was first studied by Erdős, McEliece, and Taylor [22] in 1971, but it was not much exploited afterward. As we showed in [78], the limits involved in the definition of $c(G)$ and L can be linked via constructions as in Theorem 3.1.1. We note that at least three different graph products are used in the work in this area: strong product in the definition of $c(G)$ [69], Cartesian product [1], and the so-called composition used by us in [78, 80]. Each of these products is useful in a different way. We now propose two conjectures related to Theorem 3.1.2.

Conjecture 3.1.3. *For each $k \geq 3$, there does not exist any finite graph G with independence number equal to $k - 1$ such that $c(G) = \lim_{n\to\infty} R_n(k)^{1/n}$.*

Conjecture 3.1.4. *There exists a positive integer k such that $\lim_{n\to\infty} R_n(k)^{1/n} = \infty$.*

The limit $\lim_{n\to\infty} R_n(k)^{1/n}$ exists for each $k \geq 3$ by an argument similar to that in the proof for $k = 3$ [13]. What remains open is for which k this limit is infinite. The second of these conjectures seems a little easier, if it is true. If Conjecture 3.1.3 is false, then $L_k = \lim_{n\to\infty} R_n(k)^{\frac{1}{n}}$ is finite, and actually we have $L_k \leq |V(G)|$ where G is a counterexample graph. Hence, a proof of Conjecture 3.1.4 would imply a proof for Conjecture 3.1.3. We note that Conjecture 3.1.3 is not true for infinite graphs. This was not considered in [22], but one could prove it using the same methods as in [22].

The known values and bounds on $R_r(3)$ for small r are listed in Table 4.2 below. The first open case for $r = 4$ is perhaps the most studied specific multicolor Ramsey number, and we give more details about it in the next subsection. The lower bounds in Table 4.2 for $5 \leq r \leq 7$ were obtained by constructions of Schur colorings.

4.3.2 On the Ramsey Number $R(3, 3, 3, 3)$

The best-known bounds on $R_4(3) = R(3, 3, 3, 3)$ are given in the next theorem, after which we conjecture that the actual value is 51.

Theorem 3.2.1 ([11]). $51 \leq R(3, 3, 3, 3) \leq 62$ *[26]*.

Conjecture 3.2.2. $R(3, 3, 3, 3) = 51$.

We first overview the history of the upper bounds. The bound of 66 follows from (4.1) and $R_3(3) = 17$ [41]. The result $R(3, 3, 3, 3) \leq 65$ appeared first in a 1973 paper by Whitehead [77], although he gives credit for part of the proof to Folkman. Notes by Folkman were printed posthumously in 1974 [29]. No progress was made on lowering further the upper bound until Sánchez-Flores [67] gave a computer-free proof that $R(3, 3, 3, 3) \leq 64$. In his 1995 article, Sánchez-Flores proved some properties of 3- and 4-colorings of K_n without monochromatic triangles and then used them to derive the new upper bound. In 1994, Kramer [49] gave a series of talks at a graph theory seminar at Iowa State University to show that $R(3, 3, 3, 3) \leq 62$. These talks led to a 116-page-long unpublished manuscript [49], which provided the spark to develop the algorithms for the computational verification of this result in [26]. We consider it feasible to decide whether $R(3, 3, 3, 3) \leq 61$ with the techniques similar to those in [26]; however we also consider that going down to 60 or less would necessarily require a significantly new insight.

Between 1955 and 1973, the best-known lower bound was moving from 42 to 51 as listed in Table 4.3. In her 1973 article, Chung took an incidence matrix for one of the two proper 3-colorings of K_{16} and constructed from it the incidence matrix corresponding to a good 4-coloring of K_{50}, thereby establishing $R(3, 3, 3, 3) > 50$. Actually, this is a special case of the general construction by Chung for any number of colors, mentioned at the opening of Subsection 3.1. To date, it gives the best-known lower bound for four colors. Many other nonisomorphic proper 4-colorings of K_{50} were obtained by the second author, though all of them had a structure very similar to the one constructed by Chung, in that all of them have significantly less edges in one of the colors. We summarize all these developments in Table 4.3.

We are aware of several attempts to use heuristic algorithms for the lower bound, which had a hard time to produce correct constructions for the number of vertices well below 50. Actually, we consider that designing a general heuristic method

Table 4.3 History of bounds on $R_4(3)$, based on [26]

Year	References	Lower	Upper
1955	Greenwood-Gleason [41]	42	66
1967	False rumors	[66]	
1971	Golomb, Baumert [38]	46	
1973	Whitehead [77]	50	65
1973	Chung [11], Porter cf. [11]	51	
1974	Folkman [29]		65
1995	Sánchez-Flores [67]		64
1995	Kramer (no computer) [49]		62
2004	Fettes-Kramer-R (computer) [26]		62

which can come close to match or perhaps even beat Chung's bound is an interesting challenge for the computationally oriented approach. There exists a very large number of 4-colorings of K_n without monochromatic triangles for n equal to 49 or slightly less, yet the standard heuristic search techniques somehow fail to find them. Understanding why this is happening could give new insights on how to design better general search techniques.

4.3.3   On the Ramsey Number $R(3, 3, 4)$

In the multicolor case, when only complete graphs are avoided, the only known nontrivial value of such type of Ramsey number is $R(3, 3, 3) = 17$ [41]. The only other case whose evaluation does not look hopeless is $R(3, 3, 4)$, which currently is known to be equal to 30 or 31. The lower bound $30 \leq R(3, 3, 4)$ was obtained by Kalbfleisch in 1966 [45], while the best-known upper bound $R(3, 3, 4) \leq 31$ by Piwakowski and the second author [60] in 1998. The same authors obtained some further constraints on the final outcome in 2001 [61]. We are not aware of any further progress on this case since then. Perhaps it is time to attack it again.

Conjecture 3.3.1 ([60, 61]). $R(3, 3, 4) = 30$.

It is known that if $R(3, 3, 4) = 31$, then any witness $(3, 3, 4; 30)$-coloring must be very special. The known results of [44, 45, 60, 61], all obtained with the help of computer algorithms, are summarized in the next three theorems. For edge coloring C of K_n, the set $C[k]$ consists of the edges in color k.

Theorem 3.3.2 ([44, 60]). $30 \leq R(3, 3, 4) \leq 31$, and $R(3, 3, 4) = 31$ if and only if there exists a $(3, 3, 4; 30)$-coloring C such that every triangle $T \subset C[3]$ has a vertex $x \in T$ with $\deg_{C[3]}(x) = 13$. Furthermore, C has at least 14 vertices v such that $\deg_{C[1]}(v) = \deg_{C[2]}(v) = 8$ and $\deg_{C[3]}(v) = 13$.

Theorem 3.3.3 ([61]). $R(3, 3, 4) = 31$ if and only if there exists a $(3, 3, 4; 30)$-coloring C such that every triangle $T \subseteq C[3]$ has at least two vertices $x, y \in T$ with $\deg_{C[3]}(x) = \deg_{C[3]}(y) = 13$.

Theorem 3.3.4 ([61]). $R(3, 3, 4) = 31$ *if and only if there exists a* $(3, 3, 4; 30)$-*coloring C such that every edge in the third color has at least one endpoint x with* $\deg_{C[3]}(x) = 13$. *Furthermore, C has at least 25 vertices v such that* $\deg_{C[1]}(v) = \deg_{C[2]}(v) = 8$ *and* $\deg_{C[3]}(v) = 13$.

Further elimination of all vertices of degree at least 14 in the third color, on triangles in the third color, is perhaps within the reach of feasible computations. Unfortunately, we don't know of any approach which likely could be efficient enough to proceed similarly as in [60, 61] for the remaining cases (including a 13-regular graph in the third color).

If you like this type of problem and wish to attack $R(3, 3, 4)$, we would recommend to try first a somewhat similar case of $R_3(K_4 - e)$. This is almost certainly easier than $R(3, 3, 4)$, but still difficult enough to pose a serious computational challenge. The best-known bounds are [24] $28 \le R_3(K_4 - e) = R(K_4 - e, K_4 - e, K_4 - e) \le 30$ [59]. With some new approach and a lot of good luck, it might be even possible to solve this case without the help of intensive computations.

Finally, we note that the Ramsey numbers of the form $R(3, 3, k)$ are special since their asymptotics are known up to a poly-log factor. A surprising result by Alon and Rödl from 2005 [3] implies that $R(3, 3, k) = \Theta(k^3 \text{poly-log} k)$. They actually prove a more general result that for every fixed number of colors $r \ge 2$, when we avoid triangles in the first $r - 1$ colors and K_k in color r, we have $R(3, \ldots, 3, k) = \Theta(k^r \text{poly-log} k)$.

Note. Recently (March 2015), Codish, Frank, Itzhakov, and Miller posted an `arXiv` manuscript [16] reporting on a very significant progress of work toward the Ramsey number $R(3, 3, 4)$. Namely, they apply a SAT solver to prove that if any $(3, 3, 4; 30)$-coloring exists, then it must be 8-regular in the first two colors and 13-regular in the third. Furthermore, they anticipate that full analysis of all such colorings will be completed, and thus the exact value of $R(3, 3, 4)$ will be known soon.

4.4 Edge Folkman Numbers

In 1967, Erdős and Hajnal [21] posed a problem asking for a construction of a K_6-free graph G whose every coloring of the edges with two colors contains a monochromatic triangle. The proposers also expected (but did not prove) that for every number of colors r, there is a K_4-free graph G whose every coloring of the edges with r colors contains a monochromatic triangle. The latter for $r = 2$ reduces to the question: Does there exist a K_4-free graph that is not a union of two triangle-free graphs? In 1970, Folkman [28] proved a general result implying that such graphs exist, but far from providing their effective construction. We recommend Chapter 27 in a book by Soifer [72] for an earlier, alternate, and complementary perspective on problems discussed in this section.

Table 4.4 Edge Folkman numbers $F_e(3, 3; k)$

k	$F_e(3, 3; k)$	Graphs	References
≥ 7	6	K_6	Folklore
6	8	$C_5 + K_3$	Graham 1968 [39]
5	15	659 graphs	P-R-U 1999 [62]
4	19–786	See Table 4.5	2007 [64], 2014 [50]

Table 4.5 History of the edge Folkman number $F_e(3, 3; 4)$

Year	Lower/upper bounds	Who/what
1967	Any?	Erdős-Hajnal [21]
1970	Exist	Folkman [28]
1972	10	Lin [51]
1975	10^{10}?	Erdős offers $100 for proof [19]
1986	8×10^{11}	Frankl-Rödl [30]
1988	3×10^9	Spencer [74]
1999	16	Piwakowski-R-Urbański, implicit in [62]
2007	19	R-X [64]
2008	9697	Lu [52]
2008	941	Dudek-Rödl [17]
2012	786	Lange-R-X [50]
2012	100?	Graham offers $100 for proof

Using notation from Section 4.1, we wish to understand the structure of the graphs in the set $\mathcal{F}_e(s, t; k)$ and in particular those with the smallest number of vertices which define the value of the corresponding Folkman number $F_e(s, t; k)$. Much work has been done for the general cases, but here we concentrate mainly on the simplest-looking but already difficult case of arrowing triangles, namely, for $s = t = 3$.

The state of knowledge about the cases $F_e(3, 3; k)$ is summarized in Table 4.4. It is easy to see that $k > R(s, t)$ implies $F_e(s, t; k) = R(s, t)$, which gives the first row. Graham [39] found that $C_5 + K_3 \rightarrow (3, 3)$, which solved the first question by Erdős and Hajnal, and it gives the second row with $k = 6$. The next entry for $k = 5$, after numerous papers on this case, was finally completed in 1999 by Piwakowski, Urbański, and the second author [62] who used significant help of computer algorithms. The case of $k = 4$ is the hardest and still open. The known bounds are stated in Theorem 4.1 below. We expect that any further improvements to these bounds will be very hard to obtain. We discuss $F_e(3, 3; 4)$ in more detail in the remainder of this section.

Theorem 4.1 ([64]). $19 \leq F_e(3, 3; 4) \leq 786$ *[50]*.

The history of events and progress on $F_e(3, 3; 4)$ is summarized in Table 4.5 starting with Erdős and Hajnal's [21] original question. The positive answer follows from a theorem by Folkman [28] proved in 1970, which when instantiated to two colors produces a very large upper bound for $F_e(3, 3; 4)$. In 1975, Erdős [19] offered $100 (or 300 Swiss francs) for deciding if $F_e(3, 3; 4) < 10^{10}$, which later resulted to be remarkably close to what can be obtained by using probabilistic methods. This question remained open for over 10 years. Frankl and Rödl [30] nearly met Erdős' request in 1986 when they showed that $F_e(3, 3; 4) < 7.02 \times 10^{11}$. In 1988, Spencer [74], using probabilistic techniques, proved the existence of a Folkman graph of order 3×10^9 (after an erratum by Hovey), without explicitly constructing it. The main idea of these probabilistic proofs [30, 74] is quite simple. Any K_4-free graph G such that $G \to (3, 3)$ proves the bound $F_e(3, 3; 4) \leq |V(G)|$. How to find such a G? First, take randomly a graph F from the set $G(n, p)$ of all graphs on n vertices with edge probability p, and then remove one edge from every K_4 in F. The resulting graph G is clearly K_4-free and so has some probability of being the graph we need. The difficult part is showing that this probability is positive for certain values of n and p.

In 2008, Lu [52] showed that $F_e(3, 3; 4) \leq 9697$ by constructing a family of K_4-free circulant graphs and showing that some such graphs arrow $(3, 3)$ using spectral analysis. Dudek and Rödl [17] developed a strategy to construct new Folkman graphs by approximating the maximum cut of a related graph and used it to improve the upper bound to 941. Lange and the authors [50] improved this bound first to 860 and then further to 786 with the MAX-CUT semidefinite programming relaxation as in the Goemans-Williamson algorithm. The results of [50] were obtained by 2012, though its publication year is 2014. During the 2012 SIAM Conference on Discrete Mathematics in Halifax, Nova Scotia, Ronald Graham announced a $100 award for determining if $F_e(3, 3; 4) < 100$.

Conjecture 4.2. $50 \leq F_e(3, 3; 4) \leq 94$.

At the end of Chapter 27 of *The Mathematical Coloring Book* by Soifer [72], it is stated that a double prize of $500 was offered by the second author of this paper for proving the bounds $50 \leq F_e(3, 3; 4) \leq 127$. These bounds are much stronger than the best-known bounds in Theorem 4.1, but note that we are lowering further the upper bound in Conjecture 4.2 because of Conjecture 4.4 and comments after it.

Next, we give more details on the upper bounds obtained in recent years. Building on other methods, Dudek and Rödl [17] showed how to construct a graph H_G from graph G, such that the maximum cut size of H_G determines whether or not $G \to (3, 3)$. The vertices of H_G are the edges of G, so $|V(H_G)| = |E(G)|$. For $e_1, e_2 \in V(H_G)$, if edges $\{e_1, e_2, e_3\}$ form a triangle in G, then $\{e_1, e_2\}$ is an edge in H_G. Let $t(G)$ denote the number of triangles in G, so $|E(H_G)| = 3t(G)$. Let $MC(H)$ denote the MAX-CUT size of graph H.

Theorem 4.3 (Dudek-Rödl 2008 [17]). $G \to (3, 3)$ *if and only if* $MC(H_G) < 2t(G)$.

The intuition behind Theorem 4.3 is as follows. Any coloring of the edges G can be seen as a partition of the vertices in H_G, with two colors giving a bipartition of $V(H_G)$. If a triangle in G is not monochromatic, then its edges are in both parts. If we treat this bipartition as a cut, then the size of the cut counts each triangle twice for the two edges that cross it. Since there is only one triangle in a graph that contains two given edges, this counts the number of non-monochromatic triangles. Therefore, if there exists a cut of size $2t(G)$, then it defines an edge 2-coloring of G without monochromatic triangles. However, if $MC(H_G) < 2t(G)$, then in each coloring all three edges of some triangle are in one part, and thus $G \to (3, 3)$.

A benefit of converting the problem of arrowing $(3, 3)$ to MAX-CUT is that the latter is well known and has been studied extensively in computer science and mathematics. The related decision problem MAX-CUT(H, k) asks whether $MC(H) \geq k$. MAX-CUT is **NP**-hard, and its decision problem was one of Karp's 21 **NP**-complete problems [46].

The Goemans-Williamson MAX-CUT approximation algorithm [37] is a polynomial-time algorithm that relaxes the problem to a semidefinite program (SDP). It involves the first use of SDP in combinatorial approximation and has inspired a variety of other successful algorithms. This randomized algorithm returns a cut with expected size at least 0.878 of the optimal value. However, in our case, all that is needed is the solution to the SDP, as it gives an upper bound on $MC(H)$. Another often effective method approximates MAX-CUT using the minimum eigenvalue, or one can combine a partial exhaustive search with one of the approximation methods [17, 50].

Define graphs $G_{n,r}$ on vertices $\mathbb{Z}_n$ with an edge connecting x and y if and only if $x - y = \alpha^r$ for some nonzero $\alpha \in \mathbb{Z}_n$. If the graph $G_{n,r}$ is K_4-free, then it may be a good candidate for a witness to the upper bound of n. Using the minimum eigenvalue method, Dudek and Rödl [17] found that the graph $G_{941,5}$ is a witness of $F_e(3, 3; 4) \leq 941$. A reduction of the same graph led to a better bound 860 [50], and some modifications of graphs considered by Lu [52] produced the best-to-date bound of 786 [50].

A puzzling question about triangle arrowing is however for a much smaller graph, namely, for $G_{127,3}$. This graph was used by Hill and Irving [42] in 1982 to establish the bound $128 \leq R(4, 4, 4)$. About 10 years ago, Exoo proposed to consider this graph for triangle arrowing. Since then, Exoo, us, and many others tried to decide whether $G_{127,3}$ forces a monochromatic triangle if its edges are colored with two colors. As far as we are aware, all are to no avail. Nevertheless, all failed attempts build up more evidence for the positive answer to the following:

Conjecture 4.4. *Exoo,* $G_{127,3} \to (3, 3)$.

Exoo suggested that even a 94-vertex-induced subgraph of $G_{127,3}$, obtained by removing from it three disjoint independent sets of order 11, may still work. If true, this would imply $F_e(3, 3; 4) \leq 94$.

One of the approaches for verifying the conjecture is by reducing $\{G \mid G \not\to (3, 3)\}$ to the 3-SAT problem. We map the edges $E(G)$ to the variables of $\phi_G \in$ 3-SAT, and for each (edge)-triangle xyz in $E(G)$, we add to ϕ_G two clauses $(x + y + z) \wedge (\bar{x} + \bar{y} + \bar{z})$. One can easily see that $G \not\to (3, 3)$ if and only if ϕ_G is satisfiable. Conjecture 4.4 above is equivalent to the unsatisfiability of ϕ_G for $G = G_{127,3}$. In this case, the formula ϕ_G has 2667 variables and 19558 3-clauses, two for each of the

9779 triangles. In all, this is considered of only moderate size for the state-of-the-art SAT solvers. Still, all of several attempts to decide this ϕ_G by us and others failed.

The lower bound on $F_e(3, 3; 4)$ is a challenge as well, as it is quite surprising that only 19 is the best known. Even an improvement to $20 \leq F_e(3, 3; 4)$ would be a good progress. Lin [51] obtained a lower bound of 10 in 1972 without the help of a computer. All 659 graphs on 15 vertices witnessing $F_e(3, 3; 5) = 15$ [62] contain K_4, thus giving the bound $16 \leq F_e(3, 3; 4)$. In 2007, the authors gave a computer-free proof of $18 \leq F_e(3, 3; 4)$ and improved the lower bound further to 19 with the help of computations [64]. Any proof or computational technique improving further the lower bound of 19 very likely will be of significant interest.

We also wish to mention another interesting open problem about a related Folkman number $F_e(K_4 - e, K_4 - e; 4)$. Note that clearly we have $F_e(3, 3; 4) \leq F_e(K_4 - e, K_4 - e; 4)$. As commented by Lu [52] in his work on $F_e(3, 3; 4)$, he also obtained as a side result the bound $F_e(K_4 - e, K_4 - e; 4) \leq 30193$. The gap here between the known lower and upper bounds is much larger than that for $F_e(3, 3; 4)$, so it should in principle be more feasible to make progress here.

Acknowledgements We are grateful to the anonymous reviewers whose comments helped us to improve the presentation of this survey.

References

1. Abbott, H.L.: Some problems in combinatorial analysis. Ph.D. thesis, Department of Mathematical and Statistical Sciences, University of Alberta, Edmonton (1965)
2. Alon, N.: Explicit Ramsey graphs and orthonormal labelings. Electron. J. Comb. **1**, #R12, 8 (1994). http://www.combinatorics.org
3. Alon, N., Rödl, V.: Sharp bounds for some multicolor Ramsey numbers. Combinatorica **24**, 125–141 (2005)
4. Arnold, V.: Number-theoretical turbulence in Fermat-Euler arithmetics and large young diagrams geometry statistics. J. Math. Fluid Mech. **7**, S4–S50 (2005)
5. Beveridge, A., Pikhurko, O.: On the connectivity of extremal Ramsey graphs. Aust. J. Comb. **41**, 57–61 (2008)
6. Bohman, T.: The triangle-free process. Adv. Math. **221**, 1653–1677 (2009)
7. Bohman, T., Keevash, P.: Dynamic concentration of the triangle-free process. Preprint (2013). http://arxiv.org/abs/1302.5963
8. Burr, S.A.: Determining generalized Ramsey numbers is NP-hard. Ars Comb. **17**, 21–25 (1984)
9. Burr, S.A.: On the computational complexity of Ramsey-type problems, in mathematics of Ramsey theory. Algoritm. Comb. **5**, 46–52 (1990)
10. Burr, S.A., Erdős, P., Faudree, R.J., Schelp, R.H.: On the difference between consecutive Ramsey numbers. Utilitas Math. **35**, 115–118 (1989)
11. Chung, F.R.K.: On the Ramsey numbers $N(3, 3, \ldots, 3; 2)$. Discret. Math. **5**, 317–321 (1973)
12. Chung, F.R.K., Graham, R.L.: Erdős on Graphs, His Legacy of Unsolved Problems. A K Peters, Wellesley (1998)
13. Chung, F.R.K., Grinstead, C.: A survey of bounds for classical Ramsey numbers. J. Graph Theory **7**, 25–37 (1983)
14. Chung, F.R.K., Cleve, R., Dagum, P.: A note on constructive lower bounds for the Ramsey numbers $R(3, t)$. J. Comb. Theory Ser. B **57**, 150–155 (1993)

15. Codenotti, B., Pudlák, P., Resta, G.: Some structural properties of low-rank matrices related to computational complexity. Theor. Comput. Sci. **235**, 89–107 (2000)
16. Codish, M., Frank, M., Itzhakov, A., Miller, A.: Solving graph coloring problems with abstraction and symmetry. Preprint, http://arxiv.org/abs/1409.5189,v3 (2015)
17. Dudek, A., Rödl, V.: On the Folkman number $f(2, 3, 4)$. Exp. Math. **17**, 63–67 (2008)
18. Dudek, A., Rödl, V.: On K_s-free subgraphs in K_{s+k}-free graphs and vertex Folkman numbers. Combinatorica **31**, 39–53 (2011)
19. Erdős, P.: Problems and results on finite and infinite graphs. Recent Advances in Graph Theory (Proceedings of the Second Czechoslovak Symposium, Prague, 1974), pp. 183–192. Academia, Prague (1975)
20. Erdős, P.: Some new problems and results in graph theory and other branches of combinatorial mathematics. In: Combinatorics and Graph Theory (Calcutta 1980). Lecture Notes in Mathematics, vol. 885, pp. 9–17. Springer, Berlin (1981)
21. Erdős, P., Hajnal, A.: Research problem 2-5. J. Comb. Theory. **2**, 104 (1967)
22. Erdős, P., McEliece, R.J., Taylor, H.: Ramsey bounds for graph products. Pac. J. Math. **37**, 45–46 (1971)
23. Exoo, G.: A lower bound for $R(5, 5)$. J. Graph Theory **13**, 97–98 (1989)
24. Exoo, G.: Three color Ramsey number of $K_4 - e$. Discret. Math. **89**, 301–305 (1991)
25. Exoo, G.: A lower bound for Schur numbers and multicolor Ramsey numbers of K_3. Electron. J. Comb. **1**, #R8, 3 (1994). http://www.combinatorics.org
26. Fettes, S., Kramer, R., Radziszowski, S.: An upper bound of 62 on the classical Ramsey number $R(3, 3, 3, 3)$. Ars Comb. **LXXII**, 41–63 (2004)
27. Fiz Pontiveros, G., Griffiths, S., Morris, R.: The triangle-free process and $R(3, k)$. Preprint, http://arxiv.org/abs/1302.6279 (2013)
28. Folkman, J.: Graphs with monochromatic complete subgraphs in every edge coloring. SIAM J. Appl. Math. **18**, 19–24 (1970)
29. Folkman, J.: Notes on the Ramsey number $N(3, 3, 3, 3)$. J. Comb. Theory Ser. A **16**, 371–379 (1974)
30. Frankl, P., Rödl, V.: Large triangle-free subgraphs in graphs without K_4. Graphs Comb. **2**, 135–144 (1986)
31. Fredricksen, H., Sweet, M.M.: Symmetric sum-free partitions and lower bounds for Schur numbers. Electron. J. Comb. **7**, #R32, 9 pp. (2000). http://www.emis.ams.org/journals/EJC/Volume_7/PDF/v7i1r32.pdf
32. Garey, M.R., Johnson, D.S.: Computers and Intractability: a Guide to the Theory of NP-Completeness. W.H. Freeman, New York (1979)
33. Gasarch, W.: Applications of Ramsey theory to computer science, collection of pointers to papers (2009/2012). http://www.cs.umd.edu/~gasarch/ramsey/ramsey.html
34. Giraud, G.R.: Une majoration du nombre de Ramsey binaire-bicolore en (5,5). C.R. Acad. Sci. Paris **265**, 809–811 (1967)
35. Goedgebeur, J., Radziszowski, S.: New computational upper bounds for Ramsey numbers $R(3, k)$. Electron. J. Comb. **20**(1), #P30, 28 (2013). http://www.combinatorics.org
36. Goedgebeur, J., Radziszowski, S.: The Ramsey number $R(3, K_{10}-e)$ and computational bounds for $R(3, G)$. Electron. J. Comb. **20**(4), #P19, 25 pp. (2013). http://www.combinatorics.org
37. Goemans, M., Williamson, D.: Improved approximation algorithms for maximum cut and satisfiability problems using semidefinite programming. J. ACM **42**, 1115–1145 (1995)
38. Golomb, S.W., Baumert, L.D.: Backtrack programming. J. Assoc. Comput. Mach. **12**, 516–524 (1965)
39. Graham, R.L.: On edgewise 2-colored graphs with monochromatic triangles and containing no complete Hexagon. J. Comb. Theory **4**, 300 (1968)
40. Graham, R.L., Rothschild, B.L., Spencer, J.H.: Ramsey Theory. Wiley, New York (1990)
41. Greenwood, R.E., Gleason, A.M.: Combinatorial relations and chromatic graphs. Can. J. Math. **7**, 1–7 (1955)
42. Hill, R., Irving, R.W.: On group partitions associated with lower bounds for symmetric Ramsey numbers. Eur. J. Comb. **3**, 35–50 (1982)

43. Irving, R.W.: Contributions to Ramsey theory. Ph.D. thesis, University of Glasgow (1973)
44. Kalbfleisch, J.G.: Construction of special edge-chromatic graphs. Can. Math. Bull. **8**, 575–584 (1965)
45. Kalbfleisch, J.G.: Chromatic graphs and Ramsey's theorem. Ph.D. thesis, University of Waterloo (1966)
46. Karp, R.M.: Reducibility among combinatorial problems. In: Miller, R.E., Thatcher, J.W. (eds.) Complexity of Computer Computations, pp. 85–103. Plenum, New York (1972)
47. Kim, J.H.: The Ramsey number $R(3, k)$ has order of magnitude $t^2/\log t$. Random Struct. Algoritm. **7**, 173–207 (1995)
48. Kostochka, A., Pudlák, P., Rödl, V.: Some constructive bounds on Ramsey numbers. J. Comb. Theory Ser. B **100**, 439–445 (2010)
49. Kramer, R.L.: The classical Ramsey number $R(3, 3, 3, 3; 2)$ is no greater than 62. Preprint, Iowa State University (1994)
50. Lange, A.R., Radziszowski, S.P., Xu, X.: Use of MAX-CUT for Ramsey arrowing of triangles. J. Comb. Math. Comb. Comput. **88**, 61–71 (2014)
51. Lin, S.: On Ramsey numbers and K_r-coloring of graphs. J. Comb. Theory Ser. B **12**, 82–92 (1972)
52. Lu, L.: Explicit construction of small Folkman graphs. SIAM J. Discret. Math. **21**, 1053–1060 (2008)
53. McKay, B.D.: Ramsey graphs, Research School of Computer Science, Australian National University. http://cs.anu.edu.au/people/bdm/data/ramsey.html
54. McKay, B.D., Radziszowski, S.P.: A new upper bound for the Ramsey number $R(5, 5)$. Aust. J. Comb. **5**, 13–20 (1992)
55. McKay, B.D., Radziszowski, S.P.: Linear programming in some Ramsey problems. J. Comb. Theory Ser. B **61**, 125–132 (1994)
56. McKay, B.D., Radziszowski, S.P.: $R(4, 5) = 25$. J. Graph Theory **19**, 309–322 (1995)
57. McKay, B.D., Radziszowski, S.P.: Subgraph counting identities and Ramsey numbers. J. Comb. Theory Ser. B **69**, 193–209 (1997)
58. Nešetřil, J., Rödl, V.: The Ramsey property for graphs with forbidden complete subgraphs. J. Comb. Theory Ser. B **20**, 243–249 (1976)
59. Piwakowski, K.: A new upper bound for $R_3(K_4 − e)$. Congr. Numer. **128**, 135–141 (1997)
60. Piwakowski, K., Radziszowski, S.: $30 \leq R(3, 3, 4) \leq 31$. J. Comb. Math. Comb. Comput. **27**, 135–141 (1998)
61. Piwakowski, K., Radziszowski, S.: Towards the exact value of the Ramsey number $R(3, 3, 4)$. Congr. Numer. **148**, 161–167 (2001)
62. Piwakowski, K., Radziszowski, S., Urbański, S.: Computation of the Folkman number $F_e(3, 3; 5)$. J. Graph Theory **32**, 41–49 (1999)
63. Radziszowski, S.: Small Ramsey numbers. Electron. J. Comb. Dyn. Surv. **DS1**, revision #14, 94 (2014). http://www.combinatorics.org/
64. Radziszowski, S., Xu, X.: On the most wanted Folkman graph. Geombinatorics **XVI**(4), 367–381 (2007)
65. Ramsey, F.P.: On a problem of formal logic. Proc. Lond. Math. Soc. **30**, 264–286 (1930)
66. Rosta, V.: Ramsey theory applications. Electron. J. Comb. Dyn. Surv. **DS13**, 43 (2004). http://www.combinatorics.org
67. Sánchez-Flores, A.T.: An improved upper bound for Ramsey number $N(3, 3, 3, 3; 2)$. Discret. Math. **140**, 281–286 (1995)
68. Schaefer, M.: Graph Ramsey theory and the polynomial hierarchy. J. Comput. Syst. Sci. **62**, 290–322 (2001)
69. Shannon, C.E.: The zero error capacity of a noisy channel. Inst. Radio Eng. Trans. Inf. Theory **IT-2**, 8–19 (1956)
70. Shao, Z.: Construction and computation on graphs in Ramsey theory. Ph.D. thesis, Huazhong University of Science and Technology, Wuhan (2008)
71. Shearer, J.B.: A note on the independence number of triangle-free graphs. Discret. Math. **46**, 83–87 (1983)

72. Soifer, A.: The Mathematical Coloring Book, Mathematics of Coloring and the Colorful Life of Its Creators. Springer, New York (2009)
73. Soifer, A.: Ramsey Theory: Yesterday, Today and Tomorrow. Progress in Mathematics, vol. 285. Springer, New York (2011)
74. Spencer, J.: Three hundred million points suffice. J. Comb. Theory Ser. A **49**, 210–217 (1988). Also see erratum by M. Hovey in vol. **50**, 323
75. Walker, K.: Dichromatic graphs and Ramsey numbers. J. Comb. Theory **5**, 238–243 (1968)
76. Walker, K.: An upper bound for the Ramsey number $M(5, 4)$. J. Comb. Theory **11**, 1–10 (1971)
77. Whitehead, E.G.: The Ramsey number $N(3, 3, 3; 2)$. Discret. Math. **4**, 389–396 (1973)
78. Xu, X., Radziszowski, S.P.: Bounds on Shannon capacity and Ramsey numbers from product of graphs. IEEE Trans. Inf. Theory **59**(8), 4767–4770 (2013)
79. Xu, X., Radziszowski, S.P.: A small step forwards on the Erdős-Sós problem concerning the Ramsey numbers $R(3, k)$, to appear in *Discrete Applied Mathematics*. Preprint (2015). http://arxiv.org/abs/1507.01133
80. Xu, X., Xie, Z., Exoo, G., Radziszowski, S.P.: Constructive lower bounds on classical multicolor Ramsey numbers. Electron. J. Comb. **11**(1), #R35, 24 (2004). http://www.combinatorics.org
81. Xu, X., Xie, Z., Radziszowski, S.P.: A constructive approach for the lower bounds on the Ramsey numbers $R(s, t)$. J. Graph Theory **47**, 231–239 (2004)
82. Xu, X., Shao, Z., Radziszowski, S.P.: More constructive lower bounds on classical Ramsey numbers. SIAM J. Discret. Math. **25**, 394–400 (2011)

Chapter 5
All My Favorite Conjectures Are Critical

Teresa W. Haynes

Abstract My favorite graph theory conjectures involve the effects of edge removal on the diameter of a graph and the effects of edge addition on the domination and total domination numbers of a graph. Loosely speaking, "criticality" means that the value of the parameter in question always changes under the graph modification. This chapter presents five conjectures concerning criticality, namely, a conjecture by Sumner and Blitch on the criticality of domination upon edge addition, a conjecture by Murty and Simon on the criticality of diameter upon edge removal, and three conjectures on the criticality of total domination upon edge addition. These last three conjectures involving total domination are closely related, and surprisingly, a solution to one of them would provide a solution to the Murty-Simon Conjecture on diameter.

With acknowledgments and thanks to my special critical partners: Michael Henning, Lucas van der Merwe, and Anders Yeo.

5.1 Introduction

When studying a graph parameter such as the diameter or domination number, it is sometimes important to know the parameter's behavior when changes are made to the graph. The effects of modifying a graph by removing or adding an edge or by removing a vertex are of particular interest in many applications of graph theory. For example, in network design, fault tolerance is the ability of a network to withstand the failure of a link (removing an edge) or a node (removing a vertex). On the other hand, networks can be reinforced by adding links (edges). My favorite graph

T.W. Haynes (✉)
Department of Mathematics and Statistics, East Tennessee State University,
Johnson City, TN 37614, USA

Department of Mathematics, University of Johannesburg, Auckland Park,
Johannesburg, South Africa
e-mail: haynes@etsu.edu

© Springer International Publishing Switzerland 2016
R. Gera et al. (eds.), *Graph Theory*, Problem Books in Mathematics,
DOI 10.1007/978-3-319-31940-7_5

63

theory conjectures arise from questions involving the effects of edge removal on the diameter of a graph and the effects of edge addition on the domination and total domination numbers of a graph. Informally, the concept of "criticality" with respect to a given graph parameter generally means that the value of the parameter always changes under the graph modification.

In this chapter, I will discuss five such "critical" conjectures. The first of which by Sumner and Blitch [35] concerns the effects of edge addition on the domination number of a graph. Although this conjecture was settled in 1999, it remains one of my favorite conjectures. The other four conjectures are still open and are closely related as we shall see. One of them is a conjecture by Murty and Simon [5, 33] about the effects of edge removal on the diameter of a graph, and the remaining three concern the effects of edge addition on the total domination number of a graph. Surprisingly, a solution to one of the conjectures involving total domination would provide a solution to the Murty-Simon Conjecture on diameter.

Before presenting the conjectures, we give some basic terminology and a brief background. The maximum distance between any two vertices in a graph $G = (V, E)$ is the *diameter* of G, denoted diam(G). A set S of vertices is a *dominating set* of a graph G if every vertex in $V \setminus S$ is adjacent to a vertex in S. The *domination number* $\gamma(G)$ is the minimum cardinality of any dominating set of G. A set of vertices S is a *total dominating set* of a graph G if every vertex in V is adjacent to a vertex in S, and the *total domination number* $\gamma_t(G)$ is the minimum cardinality of any total dominating set of G. Distance in graphs, diameter in particular, is well studied in the literature. Domination and total domination are as well. Total domination in graphs was introduced by Cockayne, Dawes, and Hedetniemi [6]. For more details, the reader is referred to Henning and Yeo's [27] superb book on total domination.

Change alone is unchanging. Heraclitus

To state the obvious, a graph parameter will either change or remain the same when the graph is modified by adding/removing an edge or removing a vertex. The question is does the parameter change or remain unchanged? I was first introduced to these types of questions by my Ph.D. advisor, Professor Robert Brigham, when I was a graduate student at the University of Central Florida. Professor Brigham is a gifted teacher and was a wonderful mentor to me. At the very beginning of my Ph.D. research, he instructed me to go to the library (yes, those were the days before we could search databases like MathSciNet) and browse through articles to see what appealed to me. He asked me to select one publication per week to report to him. I was like a kid in a candy store looking through the papers. Eager to please, I dutifully reported to him a different paper each week. After a few weeks of this and what I am sure seemed like no end in sight to Professor Brigham, sounding exasperated he said, "Teresa, just choose something. It doesn't matter what. If you study spiders long enough, you'll learn to like them." I did not know how to reply because, although it should have been obvious, I had not realized that the goal of looking through the papers was that on my own I would choose one. In hindsight, I am not sure what I was thinking; I guess I must have thought Professor

Brigham and I would choose the topic together or that during one of my reporting sessions, he would suddenly announce, "This is the one." I felt a little embarrassed for not recognizing what was expected of me, so I immediately chose a topic. Fortunately, I had just finished reading a manuscript that Brigham and Professor Ronald Dutton [7] had recently submitted for publication. It was on the minimum number of edges in domination insensitive graphs. A graph is *domination insensitive* if its domination number does not change upon the removal of any edge. Not only was this paper the most recent I had read, but the one I found most interesting of all the ones I had reported. Hence, I selected domination insensitive graphs as my research topic, and the good news is that, unlike spiders, it did not have to grow on me. Following Brigham's advice, for my Ph.D. research, I generalized domination insensitive graphs by studying domination k-insensitive graphs, that is, those graphs whose domination number remains unchanged when any $k \geq 2$ edges are removed. This proved to be quite a difficult topic and sparked my lifelong interest in such questions.

In the final year of my graduate studies, Professors Brigham and Dutton took Julie Carrington, also a student at the time, and me to the *19th Southeastern International Conference on Combinatorics, Graph Theory, and Computing* in Baton Rouge, Louisiana. It was at this conference I had the privilege of meeting Professor Frank Harary. Professor Brigham introduced me to him on Wednesday night at the conference banquet. I felt so honored and excited to meet him, as I was acquainted with his graph theory book and had heard many refer to him as the "father of modern-day graph theory." The next day at the conference, Professor Harary approached me and started talking. He was friendly and seemed to be genuinely interested in the research I was doing. He asked me the topic of my dissertation and I told him. He smiled and said it was interesting. Then he went to his room to rest. A few hours later, I heard someone calling my name and saw Professor Harary walking toward me. And in what became a life-changing moment for me, he handed me a piece of paper on which he had handwritten six problems (see Figure 5.1 for a copy of the handwritten letter). The problems were stated in terms of the domination number because Harary knew I was studying it, but he said to pick your favorite graph parameter and plug it into these six problems. The six problems he suggested were

Changing and Unchanging Domination

Characterize the graphs G for which

1. $\gamma(G - v) \neq \gamma(G)$ for all $v \in V$.
2. $\gamma(G - e) \neq \gamma(G)$ for all $e \in E$.
3. $\gamma(G + e) \neq \gamma(G)$ for all $e \in E(\overline{G})$.
4. $\gamma(G - v) = \gamma(G)$ for all $v \in V$.
5. $\gamma(G - e) = \gamma(G)$ for all $e \in E$.
6. $\gamma(G + e) = \gamma(G)$ for all $e \in E(\overline{G})$.

In essence, Harary's problems are to characterize graphs based on whether a given graph parameter changes or remains the same when the graph is modified by

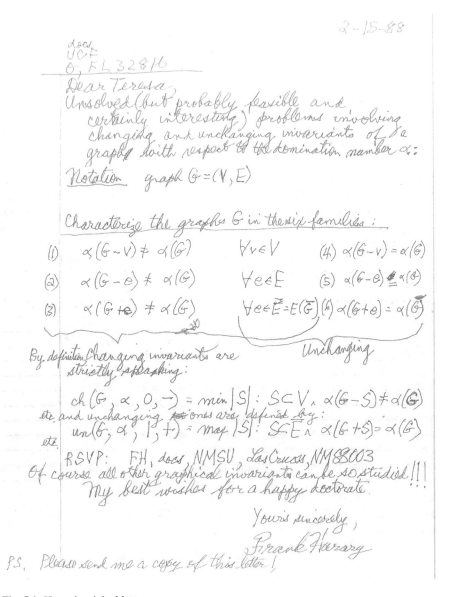

Fig. 5.1 Harary's original letter

one of three operations, namely, removing an edge, adding an edge, or removing a vertex. (It is worth mentioning that graphs can be modified in other ways, e.g., subdividing an edge or contracting an edge, but these modifications are not the subject of the conjectures considered here.) Note that the "unchanging" graphs from Problem 5 on Harary's list are precisely the domination insensitive graphs defined by Dutton and Brigham [7]. Further, Problems 1, 2, and 3 define "critical graphs." I found these problems extremely interesting and exciting when I first saw them, and my interest in them has not waned.

This encounter also began several years of research collaboration with Professor Harary. I feel very fortunate to have had this opportunity as he was a great mentor and teacher. He encouraged me, shared many research ideas with me, and helped me hone my writing skills. In addition, we became good friends. One of the most valuable lessons that he taught me was that there is no such thing as revising a paper too many times. When I was visiting him at New Mexico State University, I walked into his office one morning, and there he sat revising one of his papers. Although there was nothing strange about his doing this, what was strange was that the paper he was marking up with his red pen had already been published! I asked him what good would it do to mark up his copy of the actual journal article, especially since no one would ever see it. He just smiled and said that the paper needed it and he would see it. Those who knew Harary will attest that working with him could be a bit trying at times, but I always appreciated his personality. He once told me, jokingly, that I held the distinction of being his only female coauthor who had not thrown something at him. Harary was definitely unique and creative; I often thought of him as a fountain flowing with new research ideas. Professors Stephen Hedetniemi and Gary Chartrand are two others with whom I have had the privilege of working who naturally "ask the right questions" and "flow with many new ideas."

5.2 The Sumner-Blitch Conjecture

History is always changing. Aung San Suu Kyi

Although the 1983 Sumner-Blitch Conjecture is a history, as it was settled in 1999, it was and still is one of my favorite "critical conjectures." A graph G is *domination k-edge critical* or just *k-critical*, if $\gamma(G) = k$ and the addition of any edge decreases the domination number. Since adding an edge cannot increase the domination number, these graphs are precisely the "changing" graphs defined by Harary's Problem 3. For more details, see Chapter 5 of [20] and Chapter 16 of [21].

It is straightforward to characterize the k-critical graphs for $k \leq 2$. Vacuously, a graph is 1-critical if and only if it is the complete graph. The 2-critical graphs are characterized as follows. A *nontrivial star* is the complete bipartite graph $K_{1,m}$ for $m \geq 1$.

Theorem 1 ([35]). *A graph G is 2-critical if and only if its complement $\overline{G}$ is the union of nontrivial stars.*

Although the k-critical graphs for $k \leq 2$ are easy to characterize, the level of difficulty increases significantly for $k \geq 3$, and to date these graphs have not been characterized. The Sumner-Blitch Conjecture deals with a property of 3-critical

Fig. 5.2 The smallest connected, 3-critical graph

Fig. 5.3 A 3-critical graph

graphs. The graph shown in Figure 5.2 is the smallest (in terms of order) connected, 3-critical graph [35]. Another example of a 3-critical graph is given in Figure 5.3.

The *independent domination number* $i(G)$ is the minimum cardinality of an independent dominating set of G. Sumner and Blitch [35] conjectured that for any k-critical graph G, $\gamma(G) = i(G) = k$. In 1994 Ao [1] disproved the conjecture for $k = 4$, and a couple of years later Ao, Cockayne, MacGillivray, and Mynhardt [2] disproved it for $k \geq 4$ by giving an elegant construction of a counterexample. However, for the case of $k = 3$, the conjecture remained open and is restated as follows:

Conjecture 1 ([35] Sumner-Blitch Conjecture, 1983). *If G is a 3-critical graph, then $\gamma(G) = i(G) = 3$.*

In [36] Sumner stated that although the stronger conjecture for all k-critical graphs had been disproven for $k \geq 4$ in [1, 2], he still believed that Conjecture 1 for $k = 3$ is true. His belief was based mainly on an extensive computer search that failed to find a counterexample. The results from this search had at one point led him to believe that Conjecture 1 was too weak and that in fact, every vertex of a 3-critical graph belongs to an independent dominating set of cardinality 3. However, at the time of writing Chapter 16 of [21], Sumner was aware of counterexamples to this stronger form of the conjecture.

Conjecture 1 was listed in [16, 20, 32] as a major outstanding conjecture and attracted attention of researchers worldwide. Sumner was not alone in believing that the conjecture was true. Support for its validity mounted as it was proved in the affirmative for subfamilies of 3-critical graphs. For instance, it was shown in [35] that Conjecture 1 is true for a 3-critical graph G with any of the following properties: G has a leaf, G has a cutvertex, or G has diameter 3. Further, Favaron, Tian, and Zhang [9] showed that if a 3-critical graph G has minimum degree $\delta(G) \geq 2$ and independence number $\delta(G) + 2$, then Conjecture 1 holds. Hence, computer data [35] and the subsequent research supported the validity of Conjecture 1. I was among the researchers that believed Conjecture 1 was true and personally spent much time trying to prove it.

When the Sumner-Blitch Conjecture was about 15 years old, Professor C. M. (Kieka) Mynhardt and I were co-advisors for Lucas van der Merwe, a Ph.D. student at the University of South Africa. Mynhardt and I had both been interested in Conjecture 1 for several years. The main topic we gave Lucas for his Ph.D. research was to characterize the total domination critical graphs. There will be more on these graphs when we discuss the Murty-Simon Conjecture in Section 5.3. As a side dish to his main research, we also asked Lucas to work on Conjecture 1. We told him to use it as something to consider during the dry periods of his research on total domination. The Sumner-Blitch Conjecture intrigued Lucas as it had many of us, so whenever he hit a sticking point in his research or needed a break from his main topic, he turned his attention to this conjecture.

When Mynhardt and I gave the Sumner-Blitch Conjecture to Lucas, we knew it was a very difficult problem and did not fully expect a solution. And, if he did manage to settle the conjecture, we expected it to be in the affirmative. Were we ever surprised! Before revealing the end of the story, I want to say a little more about Conjecture 1. While it was a testy little conjecture, it was also very alluring. My attempts to prove it seemed to get tantalizingly close, that is, the proof would work for every case except for possibly one, seemingly insignificant, tiny one. One always felt "if I could just account for that one missing edge," then it would be proven. I know, it seems that at this point, I should have considered looking for a counterexample, but I had fallen into the trap of believing that the conjecture was true.

Surprise is the greatest gift which life can grant us. Boris Pasternak

Time passed and Lucas' research on total domination critical graphs was nearing completion. Professor Mynhardt and I agreed that his results were sufficient for a dissertation, so we encouraged him to begin putting his research in the final form required by the University of South Africa. It is no secret (sorry, Lucas) that Lucas' least favorite part of doing research is the writing process. Hence, I was not surprised when Lucas kept coming up with excuses to postpone the writing. In an effort to prompt Lucas to finish his dissertation, I called him to say that he really needed to use the upcoming weekend to write. I was going out of town to visit my grandmother and wanted him to have made substantial progress on it by my return on Monday. He said, "I've been working on the Sumner-Blitch Conjecture and think I'm close, so I'd like to continue working on it through the weekend." Thinking that Lucas was stalling and procrastinating writing, I told him that I'd been "close" to a solution of the Sumner-Blitch Conjecture for years and that he really needed to write his thesis. I reminded him that he would have plenty of time to work on the conjecture after his graduation. He pleaded with me saying, "But I'm so close, if I can find one more edge, I'll be finished." Remembering my attempts to prove the conjecture, I assured him that I too had thought I was within "one edge" of proving the conjecture many times. At this point, he started to bargain saying, "If you'll just let me work on it this weekend, I promise that I will start writing my dissertation first thing Monday morning." Finally, I thought what can two more days matter, so I agreed and left to go to my grandmother's home. This was before cell phones were commonplace, and I did not have one. I was surprised when Lucas called me on Saturday on my grandmother's home phone saying that he had solved the conjecture. I must admit that I did not take him too seriously, probably because there had been times when I thought I had proven it and realized my mistake later. Not expecting much, I told him I would call him when I returned home. Sunday evening I gave him a call, and he proceeded to describe a beautiful counterexample to the Sumner-Blitch Conjecture! The minute I heard the counterexample, I knew he had it. I was so impressed by Lucas' counterexample, excited that Conjecture 1 had been settled, and shocked that the conjecture was not true; emotions overwhelmed me. Although it was the middle of the night at Mynhardt's home in South Africa, we called her. She was so excited that she did not sleep that night either.

Writing the paper to present the counterexample to this 16-year-old conjecture was one of the fastest projects I have ever completed. In fact, the entire writing and review process took only two weeks from start to acceptance, and it was published shortly thereafter. Using notes and drafts from our previous work on the conjecture, Mynhardt and I were prepared and able to do the actual writing of the paper [41] in just a couple of days. We wanted to get the result out quickly, so we chose to submit it to the *Bulletin of the Institute of Combinatorics and Its Applications*, which is known for its rapid turnaround time. We included a cover letter with the paper submission that explained the significance of the result and asked that the review of our paper be expedited. It was published a couple of weeks later.

Van der Merwe's counterexample not only shows that $\gamma(G)$ is not necessarily equal to $i(G)$ for 3-critical graphs G; it constructs 3-critical graphs G with $i(G) = k$ for every integer $k \geq 3$. Note for $k = 3$, the constructed graph G has $\gamma(G) = i(G) = 3$, so, in order to have a counterexample, we must have $k \geq 4$. Hence, the smallest counterexample to the Sumner-Blitch Conjecture provided by this construction is the graph for $k = 4$, which has 56 vertices. It is still unknown whether there exists a smaller counterexample. Congratulations again, Lucas.

We conclude this section by giving the construction of the counterexample first presented in [38] and later in [41]. Construct G_k for $k \geq 3$ as follows: Begin with a factorization of the complete graph K_{2k} with vertex set $v_1, v_2, \ldots, v_{2k}$ into the 1-factors $F_1, F_2, \ldots, F_{2k-1}$. Let the vertices of each F_i be labeled $v_{i,1}, v_{i,2}, \ldots, v_{i,2k}$ such that $v_{i,j}v_{i,p} \in E(F_i)$ if and only if $v_j v_p$ is an edge of the 1-factor F_i. Then add the edges $v_{i,j}v_{h,p}$ for all $i \neq h$ and $j \neq p$, that is, add edges such that vertex $v_{i,j}$ is adjacent to every vertex in every other factor F_h, where $h \neq i$, except the vertex $v_{h,j}$ for $1 \leq i \leq 2k - 1$ and $1 \leq j \leq 2k$. Assume, without loss of generality, that $v_{1,1}v_{1,2}$ is an edge in F_1. It is straightforward to check that no two vertices dominate G_k and that $\{v_{1,1}, v_{2,1}, v_{2,2}\}$ is a dominating set of G_k, so $\gamma(G_k) = 3$. There are only two possibilities to check to show that G_k is 3-critical, namely, adding an edge $v_{i,a}v_{i,b}$ in F_i or adding an edge $v_{i,a}v_{h,a}$ for $i \neq h$. If $v_{i,a}v_{i,b}$ is not an edge in G_k, then $v_{j,a}v_{j,b}$ is in some F_j, where $j \neq i$. Hence, $\{v_{i,a}, v_{j,b}\}$ is a dominating set of $G_k + v_{i,a}v_{i,b}$.

Further, if $v_{i,a}v_{i,b} \in E(F_i)$, then $\{v_{i,a}, v_{h,b}\}$ is a dominating set of $G_k + v_{i,a}v_{h,a}$ for $h \neq i$. In both cases, it follows that G_k is 3-critical. To see that $i(G_k) = k$, observe that since $v_{i,j}$ is adjacent to every vertex except $v_{h,j}$ in F_h for every $h \neq i$, it follows that an independent dominating set containing vertices from different 1-factors must have at least $2k - 1$ vertices (one from each F_i). On the other hand, an independent dominating set contained in a single F_i needs only k vertices (one from each adjacent pair). Moreover, no fewer than k vertices form an independent dominating set of G_k, so $i(G) = k$. Hence, the graph G_k is 3-critical and has $i(G_k) = k$.

5.3 The Murty-Simon Conjecture

In this section, we will consider another "critical conjecture." This long-standing conjecture proposes an upper bound on the number of edges in a graph of diameter 2 having the property that the removal of any edge increases its diameter. Such graphs are called *diameter-2-critical* graphs. Clearly, removing an edge cannot decrease the diameter, so the diameter-critical graphs are the "changing" graphs defined by Problem 5 (substituting diameter for domination) on Harary's list in Figure 5.1.

Murty and Simon (see [5, 33]) independently made the following conjecture:

Conjecture 2 (Murty-Simon Conjecture). *If G is a diameter-2-critical graph with n vertices and m edges, then $m \leq \lfloor n^2/4 \rfloor$, with equality if and only if G is the complete bipartite graph $K_{\lfloor \frac{n}{2} \rfloor, \lceil \frac{n}{2} \rceil}$.*

This conjecture is credited to Murty and Simon in 1970s. However, according to Füredi [10], Erdős attributed it to the work of Ore in the 1960s. In either case, the conjecture is approximately a half-century old.

The Murty-Simon Conjecture, as we shall refer to it, has been proven for some families of graphs. Mantel's [30] result (a special case of a classic result of Turán [37]) proves the Murty-Simon Conjecture for triangle-free graphs. Bounding the number of triangles in a diameter-2-critical graph, Caccetta and Häggkvist [5] proved that the conjecture holds for graphs of order n which have $O(n^{3-\epsilon})$ triangles with a certain property. Hence, it would seem that looking at graphs having a large number of triangles would be a place to start the search for a counterexample to the conjecture. However, Plesník [34] constructed families of diameter-2-critical graphs with the property that each edge belongs to at least one triangle of the graph, and Madden [29] constructed families of diameter-2-critical graphs with $\Theta(n^3)$ triangles. None of the graphs in these families have enough edges to be a counterexample to the Murty-Simon Conjecture, and their constructions seem to indicate that this approach is unlikely to lead to one. See also [11–14].

In 1975 Plesník [33] proved that $m \leq 3n(n - 1)/8$ for any diameter-2-critical graph with n vertices and m edges. In 1979 Caccetta and Häggkvist [5] proved that $m < .27n^2$ for such graphs. In 1984 Xu [44] published a "proof" of the conjecture, which he later retracted after discovering a mistake in it. In 1987 Fan [8] proved the bound of the conjecture for $n \leq 24$ and for $n = 26$. For diameter-2-critical graphs of order $n \geq 25$, he showed that $m < n^2/4 + (n^2 - 16.2n + 56)/320 < .2532n^2$.

Perhaps Füredi's [10] astounding asymptotic result from 1992 is the most noteworthy contribution to date on the conjecture. He proved that the conjecture is true for large n, that is, for $n > n_0$ where n_0 is a tower of 2s of height about 10^{14}. However, even this striking result does not put the conjecture to rest. As remarked by Madden [29], "n_0 is an inconceivably (and inconveniently) large number: it is a tower of 2's of height approximately 10^{14}. Since, for practical purposes, we are usually interested in graphs which are smaller than this, further investigation is warranted." The Murty-Simon Conjecture has been studied by several other authors; see, for example, [4, 28, 31] and elsewhere.

Although many impressive partial results have been obtained, the conjecture remains open for general n. On the other hand, a recent discovery gives a new way to look at the Murty-Simon Conjecture from the point of view of total domination, a seemingly unrelated parameter. This is the topic of our next section.

5.4   The Equivalent Total Domination Edge-Critical Conjecture

As we have seen, several attempts have been made to solve the Murty-Simon Conjecture by attacking it head-on, that is, from the viewpoint of diameter-2-critical graphs, and valuable partial results have been obtained using this direct method. Nonetheless, in more than 50 years since the conjecture was first posed, it remains unsettled.

I look at this game from a different perspective. Pitbull

Recently, however, Hanson and Wang [15] observed a relationship that essentially equates the Murty-Simon Conjecture with a conjecture involving total domination in the complement graphs. Hence the problem can now be approached from a new perspective, by coming at it "through the back door." This surprising connection, linking two seemingly disparate parameters, has given a breath of fresh air into this problem.

I first heard of this fascinating connection from Lucas van der Merwe. He called to tell me about it and suggested that we use it to attack the Murty-Simon Conjecture from the total domination standpoint. He also shared several very good ideas that he had for approaching it from this new perspective. We invited Professor Michael Henning to join us on this adventure. Given our lifelong interests in domination, of course, the three of us were excited by this unpredictable association with total domination and eager to tackle the Murty-Simon Conjecture from this angle.

This pivotal link to total domination has resulted in a flurry of newfound interest in the problem and allowed significant progress to be made on the conjecture. In the remainder of this section, we will discuss the link and progress made using it.

5.4.1 The Relationship: Diameter-2-Critical and 3_t-Critical

In this section, we present a third "critical" conjecture involving total domination edge critical graphs. We will see that a solution to this conjecture is, in fact, a solution to the Murty-Simon Conjecture.

As mentioned in Section 5.2, the investigation of total domination edge critical graphs was the topic of van der Merwe's doctoral research [38, 40, 42, 43]. A graph G is *total domination edge critical* if $\gamma_t(G + e) < \gamma_t(G)$ for every edge $e \in E(\overline{G}) \neq \emptyset$. Further if $\gamma_t(G) = k$, then we say that G is a k_t-*critical graph*. Therefore, a k_t-critical graph has total domination number k, and the addition of any edge decreases the total domination number. Note that here we are adding an edge to a graph, whereas the criticality in the Murty-Simon Conjecture refers to removing an edge from a graph. Since adding an edge cannot increase the total domination number, the k_t-critical graphs are the "changing" graphs in Harary's Problem 3, where total domination number is the selected parameter.

It is shown in [40] that the addition of an edge to a graph can change the total domination number by at most two. Total domination edge critical graphs G with the property that $\gamma_t(G) = k$ and $\gamma_t(G + e) = k - 2$ for every edge $e \in E(\overline{G})$ are called k_t-*supercritical graphs*.

Theorem 2 ([40]). *For any edge $e \in E(\overline{G})$, $\gamma_t(G) - 2 \leq \gamma_t(G + e) \leq \gamma_t(G)$.*

> *It's useful to go out of this world and see it from the perspective of another one.* Terry Pratchett

We are now ready to discuss Hanson and Wang's result, which gives the key association between diameter-2-critical graphs and k_t-critical graphs. The proof to this unexpected result is short and simple, so we include a slightly modified version of the proof in [15] here. First we make an observation.

Since any pair of vertices at distance 3 or more apart in G forms a total dominating set of $\overline{G}$, it follows that if diam$(G) \geq 3$, then $\gamma_t(\overline{G}) = 2$. Moreover, if $S = \{u, v\}$ is a total dominating set of a graph G, then in $\overline{G}$, u and v are not adjacent and have no common neighbors, that is, diam$(\overline{G}) \geq 3$. Hence, we have the following result:

Observation 3. *A graph G has $\gamma_t(G) = 2$ if and only if* diam$(\overline{G}) \geq 3$.

Theorem 4 ([15]). *A graph is diameter-2-critical if and only if its complement is 3_t-critical or 4_t-supercritical.*

Proof. Let G be a diameter-2-critical graph. Observation 3 implies that $\gamma_t(\overline{G}) \geq 3$. Moreover, diam$(G - uv) \geq 3$ for any edge $uv \in E(G)$. Again, by Observation 3, we have that $\gamma_t(\overline{G} + uv) = 2$. It follows from Theorem 2 that $\overline{G}$ is 3_t-critical or 4_t-supercritical.

Assume that $\overline{G}$ is 3_t-critical or 4_t-supercritical. Then $\gamma_t(\overline{G} + uv) = 2$ for any $uv \in E(G)$. Observation 3 implies that diam$(G - uv) \geq 3$ for any edge $uv \in E(G)$. If diam$(G) \geq 3$, then $\gamma_t(\overline{G}) = 2$, a contradiction. Hence, diam$(G) \leq 2$.

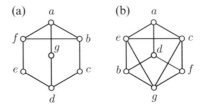

Fig. 5.4 A diameter-2-critical graph G and its 3_t-critical complement $\overline{G}$. (a) G, (b) $\overline{G}$

If $\operatorname{diam}(G) = 1$, then G is complete and $E(\overline{G})$ is empty, contradicting that $\gamma_t(\overline{G}) \in \{3, 4\}$. Thus, we have that $\operatorname{diam}(G) = 2$, and so, G is a diameter-2-critical graph. $\square$

For example, the self-complementary 5-cycle is both diameter-2-critical and 3_t-critical. Figure 5.4 shows a diameter-2-critical graph G and its complement $\overline{G}$, a 3_t-critical graph.

A result from van der Merwe's dissertation [38, 39] characterized the 4_t-supercritical graphs, supporting the Murty-Simon Conjecture, though we were unaware of the connection at the time.

Theorem 5 ([39]). *A graph G is 4_t-supercritical if and only if G is the disjoint union of two nontrivial complete graphs.*

The reverse side also has a reverse side. Japanese Proverb

Note that the complement of a 4_t-supercritical graph is a complete bipartite graph, and the number of edges in a complete bipartite graph is maximized when the partite sets differ in cardinality by at most one. Thus, Conjecture 2 (the Murty-Simon Conjecture) holds for diameter-2-critical graphs whose complements are 4_t-supercritical, and a subset of these graphs are the extremal graphs of the conjecture. Noting also that for a graph G of order n, $|E(G)| < \left\lfloor \frac{n^2}{4} \right\rfloor$ if and only if $|E(\overline{G})| > \left\lceil \frac{n(n-2)}{4} \right\rceil$, it follows from Theorem 4 and Theorem 5 that proving the Murty-Simon Conjecture is equivalent to proving the following conjecture:

Conjecture 3. *If G is a 3_t-critical graph with order n and size m, then $m > \left\lceil \frac{n(n-2)}{4} \right\rceil$.*

Conjecture 3 arose from Theorem 4 in 2003. Proving this conjecture, and consequently the Murty-Simon Conjecture, requires determining the minimum number of edges in a 3_t-critical graph on n vertices. When Mynhardt and I gave total domination edge critical graphs to van der Merwe as a research topic, we started out with the lofty goal of asking him to characterize the k_t-critical graphs for all $k \geq 3$. It did not take us long to realize that we needed to limit our scope. Noting that $k = 2$ is the smallest k for which ordinary domination can be k-critical and that these graphs were easily characterized in [35], Mynhardt and I thought that we would ask him to consider the smallest k for which a graph could be k_t-critical, that is, $k = 3$.

Hoping to jump start his research by giving him success with some low-lying fruits, we narrowed the question to characterizing the 3_t-critical graphs. Of course, we were unaware that we were asking Lucas to prove the Murty-Simon Conjecture and more. Lucas did a beautiful dissertation [38] and made significant progress toward his goal, including determining several useful properties of 3_t-critical graphs. He characterized an infinite family of 3_t-critical graphs (the graphs with a cutvertex), solving a more general problem, and hence, Conjectures 2 and 3, for this family. But he was not able to obtain a characterization of all 3_t-critical graphs. Now that we know the relationship to the Murty-Simon Conjecture, it is not surprising that the problem of characterizing the 3_t-critical graphs proved to be unattainable.

5.4.2 Progress on Conjecture 3

Prior to the knowledge that 3_t-critical graphs were related to the diameter-2-critical graphs of the Murty-Simon Conjecture, most of the work on 3_t-critical graphs focused on either trying to characterize them or obtaining bounds on their graphical parameters, such as diameter. It was not until Hanson and Wang [15] discovered the link that any attempt was made to determine a lower bound on their size or to study the extremal (edge-minimal) graphs. This linkage has turned attention to these types of questions concerning 3_t-critical graphs as well as breathed new life into attacking the Murty-Simon Conjecture by allowing it to be approached from the point of view of domination. Our initial excitement about the connection has escalated as it has been used to make significant progress on this elusive problem. We summarize the progress in this section.

After van der Merwe contacted me about Hanson and Wang's result, we organized a summer workshop to focus on the problem. Michael Henning and Anders Yeo visited Tennessee to work with us. We concentrated solely on the Murty-Simon Conjecture via Conjecture 3 and worked very hard during this workshop and subsequent meetings. It was at the first gathering that we proved Conjecture 3 for 3_t-critical graphs having diameter 3.

One of the useful properties of 3_t-critical graphs established by van der Merwe [38] was the following bounds on their diameter:

Theorem 6 ([38, 40]). *If G is a 3_t-critical graph, then $2 \leq \operatorname{diam}(G) \leq 3$.*

By Theorem 6, every 3_t-critical graph has diameter 2 or 3. Hanson and Wang [15] proved the following result:

Theorem 7 ([15]). *If G is a 3_t-critical graph of diameter 3, order n, and size m, then $m \geq \left\lceil \frac{n(n-2)}{4} \right\rceil$.*

Note that in order to prove that Conjecture 3 holds for 3_t-critical graphs of diameter 3, strict inequality is needed in Theorem 7. Hence, the first task we set out to accomplish in our workshop was to prove strictness of their bound. We thought that it would be easy to show, but it turned out to be much more difficult than we anticipated. To prove strictness, we needed to be able to add one more edge to the edge count for 3_t-critical graphs of diameter 3. While the bound in Theorem 7 has a straightforward, short proof (only about a page), an extensive amount of work was required to find this one additional edge. The counting arguments to prove the following were lengthy and detailed:

Theorem 8 ([22]). *If G is a 3_t-critical graph of diameter 3, order n, and size m, then $m > \left\lceil \frac{n(n-2)}{4} \right\rceil$.*

Thus, by Theorem 8, the Murty-Simon Conjecture is proven for the graphs whose complements have diameter 3. Hence, the problem is reduced to determining the minimum number of edges in 3_t-critical graphs of diameter 2.

Several years after our initial workshop, we [26] were able to prove the following result for 3_t-critical graphs with sufficiently small minimum degree:

Theorem 9 ([26]). *If G is a 3_t-critical graph of order n, size m, and minimum degree $\delta(G) \geq 1$, then the following hold:*

(a) *If $\delta \leq 0.3\,n$, then $m > \lceil n(n-2)/4 \rceil$.*
(b) *If $n \geq 2000$ and $\delta \leq 0.321\,n$, then $m > \lceil n(n-2)/4 \rceil$.*

As before, we used the link between diameter-2-critical graphs and 3_t-critical graphs to prove the Murty-Simon Conjecture for diameter-2-critical graphs with sufficiently large maximum degree $\Delta(G)$ by restating Theorem 9 in its equivalent form as follows. Note that $\delta(G) = n - 1 - \Delta(\overline{G})$.

Theorem 10 ([26]). *Let G be a diameter-2-critical graph of order n, size m, and maximum degree $\Delta(G)$. Then the following hold:*

(a) *If $\Delta \geq 0.7\,n$, then $m < \lfloor n^2/4 \rfloor$.*
(b) *If $n \geq 2000$ and $\Delta \geq 0.6787\,n$, then $m < \lfloor n^2/4 \rfloor$.*

Recall that Theorem 8 settles Conjecture 3 for 3_t-critical graphs of diameter 3, leaving the conjecture open only for the 3_t-critical graphs of diameter 2. Restricting their attention to these graphs, subsets of authors from Haynes, Henning, van der Merwe, and Yeo proved Conjecture 3 for several families of 3_t-critical graphs and, hence, Conjecture 2 for their complements. The graph classes for which Conjecture 3 is known to hold are summarized with references in Table 5.1. More details on these results can be found in the respective references and in the survey paper [25] on the Murty-Simon Conjecture.

A *claw* is an induced $K_{1,3}$, that is, a star with three edges. The *house graph* is a 5-cycle with a chord, and the *diamond* is a 4-cycle with a chord. A graph is said to be H-free if it has no induced subgraph H. A graph G of order n is called *k-vertex-connected* (or simply *k-connected*) if $n \geq k + 1$ and deletion of any $k - 1$ or fewer vertices leaves a connected graph. We say that G is a graph of connectivity-k to mean that G is k-connected and G has a cutset of cardinality k.

Table 5.1 Graph classes for which Conjecture 3 holds

Diameter-3 graphs	[22]
Connectivity-1 graphs	[22, 38]
Connectivity-2 graphs	[24]
Connectivity-3 graphs	[24]
Claw-free graphs	[23]
C_4-free graphs	[17]
Diamond-free graphs	[18]
House-free graphs	[19]

By Theorem 9, Conjecture 3 holds if $\delta(G) \leq 0.3\,n$. As summarized in [25], any counterexample to Conjecture 3 would have the following properties:

Theorem 11. *If G is a 3_t-critical graph of order n that is a counterexample to Conjecture 3, then the following hold:*

(a) $\mathrm{diam}(G) = 2$.
(b) $\delta(G) > 0.3\,n$.
(c) *G is 4-connected.*
(d) *G has a claw.*
(e) *G has an induced C_4.*
(f) *G has an induced diamond.*
(g) *G has an induced house.*

Thus, the Murty-Simon Conjecture has been verified for a number of infinite families of graphs, namely, the complements of the graphs listed in Table 5.1. Many of the arguments used to prove the results presented in this section involved intricate counting methods. For example, one of the first approaches suggested by van der Merwe to count the number of edges in a 3_t-critical graph is to partition its vertex set into cliques. Notice that Conjecture 3 is true for a 3_t-critical graph G of order n if V can be partitioned into two cliques such that at least one of the cliques has more than $\lceil n/2 \rceil$ vertices or the cliques are balanced with $\lfloor n/2 \rfloor$ and $\lceil n/2 \rceil$ vertices, respectively, and G has at least one additional edge. Clearly, this type of partition does not always exist, so for the purpose of counting edges, we considered what we called pseudo-cliques. A set S of k vertices forms a *pseudo-clique* if we can uniquely associate $\binom{k}{2}$ edges of G with S. If S is a pseudo-clique and e is an edge of G associated with a missing edge in S, then we call e a *pseudo-edge* associated with the pseudo-clique S. Thus, our aim in several of the proofs was to partition the vertices of G into two pseudo-cliques (where any edge is a pseudo-edge for at most one of the pseudo-cliques) and show that the pseudo-cliques have the properties to yield the desired edge count.

Unfortunately, the techniques we used in each subproblem collapsed when applied to the general problem. For example, the counting technique used to prove the conjecture in the claw-free case does not apply to graphs having a claw.

Similarly, different proof methods used for the other graph structures failed in the more general case. Thus, although we have partial solutions, Conjecture 3, and, hence, the Murty-Simon Conjecture, is still unsolved. The fact that the Murty-Simon Conjecture remains open is a testimony to its difficulty and that there is still much work to do. On the other hand, the significant partial results attained by using the association to total domination are evidence that this new approach is promising. Further, we believe that we have only skimmed the surface from the vantage point of total domination, but in doing so have begun to understand it better and to develop proof techniques that we can apply in future work. I would love to see the Murty-Simon Conjecture and its equivalents settled and plan to continue diligently working with my colleagues toward this goal. In fact, as we shall see in the next section, our new aim is to prove an even stronger result than Conjecture 3 for the open portion of the problem, that is, for the subfamily of 3_t-critical graphs having diameter 2.

5.5   Two New Conjectures

As we have seen, the Murty-Simon Conjecture is true for diameter-2-critical graphs whose complements have diameter 3 and that it remains to be settled only for the diameter-2-critical graphs whose complements have diameter 2. Hence, proving the Murty-Simon Conjecture is equivalent to proving Conjecture 3 for 3_t-critical graphs of diameter 2. Recently, Balbuena, Hansberg, Henning, and I [3] joined forces to focus on this subfamily of 3_t-critical graphs. We believe that in fact Conjecture 3 can be strengthened for this class of graphs. Our stronger conjectures are the subject of this section.

Less is only more where more is no good. Frank Lloyd Wright

Noticing that the 3_t-critical graphs of order n and diameter 2 seem to have many more edges than the conjectured lower bound $\lfloor (n(n-2)/4 \rfloor$ of Conjecture 3, we posed the following stronger conjecture in [3]:

Conjecture 4. *If G is a 3_t-critical graph with order n, size m, and* $\operatorname{diam}(G) = 2$, *then $m \geq \lfloor (n^2 - 4)/4 \rfloor$.*

It was proved in [3] that Conjecture 4 holds for a family $\mathcal{H}$ of graphs defined as follows. A pair of nonadjacent vertices, say u and v, is called a *dominating pair* of G if $\{u, v\}$ dominates G. Let $\mathcal{H}$ be the subset of 3_t-critical graphs of diameter 2 for which every pair of nonadjacent vertices is a dominating pair. Family $\mathcal{H}$ is important because we [3] believe that almost all extremal (edge-minimal) 3_t-critical graphs of diameter 2 belong to this family.

A *bull graph* consists of a triangle with two disjoint pendant edges as illustrated in Figure 5.5. A graph is *bull-free* if it has no bull as an induced subgraph. It turns out that family $\mathcal{H}$ is precisely the family of bull-free, 3_t-critical graphs.

In order to characterize the extremal graphs, a subfamily of $\mathcal{H}$ was defined in [3]. Let $C_5(n_1, n_2, n_3, n_4, n_5)$ denote the graph that can be obtained from a 5-cycle

Fig. 5.5 The bull graph

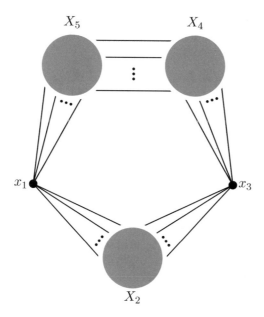

Fig. 5.6 A graph in the family $\mathcal{H}_n$

$x_1x_2x_3x_4x_5x_1$ by replacing each vertex x_i, $1 \leq i \leq 5$, with a nonempty clique X_i, where $|X_i| = n_i \geq 1$, and adding all edges between X_i and X_{i+1}, where addition is taken modulo 5. For $n \geq 5$, let $\mathcal{H}_n$ be the subfamily of $\mathcal{H}$ defined by $\mathcal{H}_n = \{C_5(n_1, n_2, n_3, n_4, n_5) \mid n_1 = n_3 = 1, n_2 = \lfloor \frac{n-3}{2} \rfloor$ or $n_2 = \lceil \frac{n-3}{2} \rceil$, and $n = 2 + n_2 + n_4 + n_5\}$. As illustrated in [3], Figure 5.6 gives a schematic of graphs in the family $\mathcal{H}_n$, where in this diagram each X_i represents the clique replacing x_i, all edges exist between the vertices of X_4 and X_5, the vertex x_1 dominates $X_2 \cup X_5$, and the vertex x_3 dominates $X_2 \cup X_4$.

The following result shows that Conjecture 4 holds for the graphs in $\mathcal{H}$ and that the graphs $\mathcal{H}_n$ are, in fact, the extremal ones in this family:

Theorem 12 ([3]). *Let $G \in \mathcal{H}$ have order n and size m. Then, $m \geq \lfloor (n^2 - 4)/4 \rfloor$, with equality if and only if $G \in \mathcal{H}_n$.*

As an immediate consequence of Theorem 12, Conjecture 4, and, hence, Conjecture 3, holds for the graphs in family $\mathcal{H}$. In other words, the Murty-Simon Conjecture is true for the diameter-2-critical graphs whose complements are $\mathcal{H}$.

Recognizing that it was not necessary for every pair of nonadjacent vertices to be a dominating pair (as required in family $\mathcal{H}$) to prove Conjecture 4, we [3] relaxed the condition to only requiring "many dominating pairs" as follows:

Theorem 13 ([3]). *Let G be a 3_t-critical graph of order $n > 5$, of size m, and of diameter 2. If G has fewer than $\lceil n^2/(4(n-5)) \rceil$ non-dominating pairs, then $m > \lceil n(n-2)/4 \rceil$.*

The second conjecture posed in [3] predicts that the 3_t-critical graphs of sufficiently large order that achieve equality in the lower bound of Conjecture 4 belong to the family $\mathcal{H}_n$.

Conjecture 5. *Let G be a 3_t-critical graph of diameter 2, of size m, and of sufficiently large order n. Then, $m \geq \lfloor (n^2 - 4)/4 \rfloor$, with equality if and only if $G \in \mathcal{H}_n$.*

In closing, we remark that proving Conjecture 4 would prove both Conjecture 3 and the Murty-Simon Conjecture. Also, we believe that further study of the graphs in family $\mathcal{H}$ could shed some light on resolving these conjectures. We remain excited about the connection between diameter-2-critical and 3_t-critical graphs and are convinced that continuing to approach the Murty-Simon Conjecture from the standpoint of total domination shows promise. Recalling that I fell into the trap of believing the Sumner-Blitch Conjecture and so wanted to prove it that I did not search for a counterexample, it seems prudent to question whether history is repeating itself. While one must remain open to the possibility of a counterexample, the evidence seems to point in the other direction. In particular, Füredi's [10] result proving the Murty-Simon Conjecture for graphs of very large order offers strong support for its validity. Also, the restrictions on any potential counterexample seem to reinforce this. For more information on the Murty-Simon Conjecture and its equivalents, the reader is referred to the survey paper [25].

Acknowledgements First I thank the editors for their initial conception of this volume and their subsequent hard work to complete it. I also want to thank Wyatt Desormeaux and the anonymous reviewers of this chapter for their many helpful comments.

References

1. Ao, S.: Independent domination critical graphs. Master's thesis, University of Victoria (1994)
2. Ao, S., Cockayne, E.J., MacGillivray, G., Mynhardt, C.M.: Domination critical graphs with higher independent domination numbers. J. Graph Theory **22**, 9–14 (1996)
3. Balbuena, C., Hansberg, A., Haynes, T.W., Henning, M.A.: On total domination edge critical graphs with total domination number three and with many dominating pairs. To appear in Graphs Comb. **31**, 1163–1176 (2015)
4. Bondy, J.A., Murty, U.S.R.: Extremal graphs of diameter two with prescribed minimum degree. Stud. Sci. Math. Hung. **7**, 239–241 (1972)
5. Caccetta, L., Häggkvist, R.: On diameter critical graphs. Discret. Math. **28**(3), 223–229 (1979)

6. Cockayne, E., Dawes, R., Hedetniemi, S.: Total domination in graphs. Networks **10**, 211–219 (1980)
7. Dutton, R., Brigham, R.C.: An extremal problem for the edge domination insensitive graphs. Discret. Appl. Math. **20**, 113–125 (1988)
8. Fan, G.: On diameter 2-critical graphs. Discret. Math. **67**, 235–240 (1987)
9. Favaron, O., Tian, F., Zhang, L.: Independence and hamiltonicity in 3-critical graphs. J. Graph Theory **25**, 173–184 (1997)
10. Füredi, Z.: The maximum number of edges in a minimal graph of diameter 2. J. Graph Theory **16**, 81–98 (1992)
11. Gliviak, F.: On certain classes of graphs of diameter two without superfluous edges. Acta F.R.N. Univ. Comen. Math. **21**, 39–48 (1968)
12. Gliviak, F.: On the impossibility to construct diametrically critical graphs by extensions. Arch. Math. (Brno) **11**(3), 131–137 (1975)
13. Gliviak, F.: On certain edge-critical graphs of a given diameter. Matematický časopis **25**(3), 249–263 (1975)
14. Gliviak, F., Kyš, P., Plesník, J.: On the extension of graphs with a given diameter without superfluous edges. Matematický časopis **19**, 92–101 (1969)
15. Hanson, D., Wang, P.: A note on extremal total domination edge critical graphs. Util. Math. **63**, 89–96 (2003)
16. Haynes, T.W.: Domination in graphs: a brief overview. J. Comb. Math. Comb. Comput. **24**, 225–237 (1997)
17. Haynes, T.W., Henning, M.A.: A characterization of diameter-2-critical graphs with no antihole of length four. Cent. Eur. J. Math. **10**(3), 1125–1132 (2012)
18. Haynes, T.W., Henning, M.A.: A characterization of diameter-2-critical graphs whose complements are diamond-free. Discret. Appl. Math. **160**, 1979–1985 (2012)
19. Haynes, T.W., Henning, M.A.: A characterization of P_5-free, diameter-2-critical graphs. Discret. Appl. Math. **169**, 135–139 (2014)
20. Haynes, T.W., Hedetniemi, S.T., Slater, P.J.: Fundamentals of Domination in Graphs. Marcel Dekker, New York (1998)
21. Haynes, T.W., Hedetniemi, S.T., Slater, P.J. (ed.), Domination in Graphs: Advanced Topics. Marcel Dekker, New York (1998)
22. Haynes, T.W., Henning, M.A., van der Merwe L.C., Yeo, A.: On a conjecture of Murty and Simon on diameter-2-critical graphs. Discret. Math. **311**, 1918–1924 (2011)
23. Haynes, T.W., Henning, M.A., Yeo, A.: A proof of a conjecture on diameter two critical graphs whose complements are claw-free. Discret. Optim. **8**, 495–501 (2011)
24. Haynes, T.W., Henning, M.A., Yeo, A.: On a conjecture of Murty and Simon on diameter two critical graphs II. Discret. Math. **312**, 315–323 (2012)
25. Haynes, T.W., Henning, M.A., van der Merwe, L.C., Yeo, A.: Progress on the Murty-Simon conjecture on diameter-2 critical graphs: a survey. J. Comb. Optim. (2013). doi:10.1007/s10878-013-9651-7
26. Haynes, T.W., Henning, M.A., van der Merwe, L.C., Yeo, A.: A maximum degree theorem for diameter-2-critical graphs. Cent. Eur. J. Math. **12**, 1882–1889 (2014)
27. Henning, M.A., Yeo, A.: Total Domination In Graphs. Springer Monographs in Mathematics. Springer, New York (2013). ISBN:978-1-4614-6524-9 (Print) 978-1-4614-6525-6 (Online)
28. Krishnamoorthy, V., Nandakumar, R.: A class of counterexamples to a conjecture on diameter critical graphs. Combinatorics and Graph Theory. Lecture Notes in Mathematics, pp. 297–300. Springer, Berlin/Heidelberg (1981)
29. Madden, J.: Going critical: an investigation of diameter-critical graphs. Master's thesis, The University of Victoria Columbia (1999)
30. Mantel, W.: Wiskundige Opgaven **10**, 60–61 (1906)
31. Murty, U.S.R.: On critical graphs of diameter 2. Math. Mag. **41**, 138–140 (1968)
32. Mynhardt, C.M.: On two conjectures concerning 3-domination critical graphs. Congr. Numer. **135**, 119–138 (1998)
33. Plesník, J.: Critical graphs of given diameter. Acta F.R.N Univ. Comen. Math. **30**, 71–93 (1975)

34. Plesník, J.: On minimal graphs of diameter 2 with every edge in a 3-cycle. Math. Slovaca **36**, 145–149 (1986)
35. Sumner, D.P., Blitch, P.: Domination critical graphs. J. Comb. Theory B **34**, 65–76 (1983)
36. Sumner, D.P., Wojcicka, E.: Graphs critical with respect to the domination number. In: Haynes, T.W., et al. (eds.) Domination in Graphs: Advanced Topics, pp. 439–468. Marcel Dekker, Inc, New York (1998)
37. Turán, P.: Eine Extremalaufgabe aus der Graphentheorie. Mat. Fiz. Lapok **48**, 436–452 (1941)
38. van der Merwe, L.C.: Total domination edge critical graphs. Ph.D. Dissertation, University of South Africa (2000)
39. van der Merwe, L.C., Mynhardt, C.M., Haynes, T.W.: Criticality index of total domination. Congr. Numer. **131**, 67–73 (1998)
40. van der Merwe, L.C., Mynhardt, C.M., Haynes, T.W.: Total domination edge critical graphs. Util. Math. **54**, 229–240 (1998)
41. van der Merwe, L.C., Mynhardt, C.M., Haynes, T.W.: 3-domination critical graphs with arbitrary independent domination numbers. Bull. Inst. Comb. Appl. **27**, 85–88 (1999)
42. van der Merwe, L.C., Mynhardt, C.M., Haynes, T.W.: Total domination edge critical graphs with maximum diameter. Discuss. Math. Graph Theory **21**, 187–205 (2001)
43. van der Merwe, L.C., Mynhardt, C.M., Haynes, T.W.: Total domination edge critical graphs with minimum diameter. Ars Comb. **66**, 79–96 (2003)
44. Xu, J.: A proof of a conjecture of Simon and Murty (in Chinese). J. Math. Res. Exp. **4**, 85–86 (1984)

Chapter 6
Efficient Local Representations of Graphs

Edward Scheinerman

Abstract Informally, an *efficient local representation* of a graph G is a scheme in which we assign short labels (representable by a "small" number of bits, hence *efficient*) to G's vertices so that we can determine if two vertices are adjacent simply by examining the labels assigned to the pair of vertices (hence *local*). For some classes of graphs (such as planar graphs), one can devise local representations, but for others (such as bipartite graphs), this is not possible.

We present a conjecture due to Muller [22] and to Kannan, Naor, and Rudich [15] that distinguishes those hereditary classes of graphs (closed under induced subgraphs) for which an efficient local representation is feasible from those for which it is not.

Notation All graphs in this chapter are *simple*: their edges are undirected, and they have neither loops nor multiple edges. For a graph $G = (V, E)$, the number of vertices is nearly always denoted by the letter n. The notation $v \sim w$ indicates that vertices v and w are adjacent, i.e., $vw \in E$. For a positive integer n, we write $[n]$ for the set $\{1, 2, \ldots, n\}$. We write $\lg n$ for the base-2 logarithm of n. We also write $\log n$ but that is invariably wrapped in big-oh notation, so the base is irrelevant.

6.1 Seeking an Efficient Data Structure for Graphs

The *efficient local representation of graphs* problem is due to Muller [22] and to Kannan, Naor, and Rudich [15].

Here's the overarching question: *How do we efficiently represent a graph in a computer?*

Perhaps the simplest method is via an adjacency matrix; the memory to store this matrix uses $\Theta(n^2)$ bits, but this may be wasteful if the graph does not have

E. Scheinerman (✉)
Department of Applied Mathematics and Statistics, Johns Hopkins University,
Baltimore, MD 21218, USA
e-mail: ers@jhu.edu

© Springer International Publishing Switzerland 2016
R. Gera et al. (eds.), *Graph Theory*, Problem Books in Mathematics,
DOI 10.1007/978-3-319-31940-7_6

83

too many edges. In the latter case, adjacency lists may be a better option. See, for example, [8, 12], or [31] for a discussion of data structures one may use to represent a graph.

If we think of a graph as modeling relations between its vertices, then determining if two vertices are adjacent "should" only depend on some shared property of the two nodes under consideration.

To illustrate our thoughts, we begin by discussing *interval representations* of graphs.

6.1.1   Interval Representations of Graphs

Let G be a graph. We say that G has an *interval representation* if we can assign to each vertex $v \in V(G)$ a real interval J_v so that for distinct vertices v and w we have

$$v \sim w \iff J_v \cap J_w \neq \varnothing.$$

For example, let G be the path graph with $1 \sim 2 \sim 3 \sim 4$. The following assignment

$$1 \mapsto [1, 3], \quad 2 \mapsto [2, 5], \quad 3 \mapsto [4, 7], \quad \text{and} \quad 4 \mapsto [6, 8].$$

gives an interval representation for G. See [11] and [19]. Interval representations of graphs are a special case of intersection representations [7, 12, 21, 24, 25].

This representation provides a terrific way to store a graph in a computer. For each vertex, we only need to hold two numbers: the left and right end points of its interval. To check if two vertices are adjacent, we just do some quick checks on four numbers.

How much storage space does such a representation consume? We might be concerned that we may need a great deal of precision to specify the intervals' end points. However, it's not hard to show that if G has an interval representation, then we can find a representation in which the end points are distinct values[1] in $[2n]$.

This implies that for each vertex of the graph, we hold a scant $O(\log n)$ bits of information, and we can test adjacency simply by examining the information attached to just the two vertices of interest.

Informally, this is what we mean by an *efficient local representation* of a graph. To each vertex v we attach a "short" [hence *efficient*] label $\ell(v)$, and adjacency

---

[1]Here's why: Closed intervals are compact. Therefore, given a finite collection of intervals, there is a positive ε such that the sizes of the gaps between nonintersecting pairs of these intervals are all greater than ε. This means we can enlarge intervals by moving left end points to the left and right end points to the right by amounts less than $\varepsilon/2$ and not create any additional intersections. In this way, we may modify the representation so that all $2n$ end points of the intervals are distinct. To determine if two intervals intersect, one only needs to know the relative order of the four end points. Therefore, we may reassign the end points to be distinct values in $\{1, 2, \ldots, 2n\}$ so long as we preserve their order.

between vertices v and w can be tested via a calculation whose inputs are just $\ell(v)$ and $\ell(w)$ [hence *local* as we do not consider labels on any other vertices nor do we reference a global data structure such as an adjacency matrix].

Sadly, interval representations are not the ideal we seek because not all graphs admit such a representation. A moment's doodling shows that the four cycle C_4 has no such representation.

Graphs that have interval representations are called *interval graphs*, and for that family, an efficient local representation is available. What about other families of graphs?

6.1.2   Additional Examples

Trees

Let T be a tree with n vertices. Arbitrarily select a vertex of T to be called the *root* and give the edges directions so that on any path to the root, all edges point toward the root. Thus every vertex (other than the root) has a unique parent.

We now attach labels to the vertices of T. Each label $\ell(v)$ is an ordered pair of integers (a_v, b_v) drawn from the set $[n]$. The a-values are assigned arbitrarily; the only requirement is that they be distinct. The b-labels depend on the direction of the edge. If the edge from v to w is oriented $v \rightarrow w$, then $b_v = a_w$. That is, the second element of v's label is the first element of v's parent. Since the root r does not have a parent, its label is simply (a_r, a_r). See Figure 6.1.

We now observe that the vertex labels are $2 \lg n$ bits in length, and testing vertices for adjacency only requires examining the labels of the two vertices:

$$v \sim w \iff a_v = b_w \text{ or } a_w = b_v. \tag{6.1}$$

This efficient local representation method easily extends from trees to acyclic graphs (forests). Indeed, the scheme works for graphs that are not trees. For example, consider an n-cycle and label its vertices, in order, as follows:

$$(1, 2) \quad (2, 3) \quad (3, 4) \quad \cdots \quad (n-1, n) \quad (n, 1).$$

Test (6.1) applies here as well.

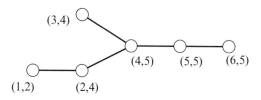

Fig. 6.1   An efficient local representation for a tree. Each label is an ordered pair (a, b). Observe that the a-values are distinct. The vertex labeled $(5, 5)$ is the root and the b-label of every other vertex is the a-label of its parent

Planar Graphs

We can extend this technique to planar graphs. The minimum degree of planar graph is at most 5. Let G be a planar graph with n vertices and assume that $V(G) = [n]$. Here is how we label $V(G)$.

Let v be a vertex of minimum degree. The label we assign to v is a tuple of the form $\ell(v) = (v; v_1, v_2, v_3, v_4, v_5)$ where the v_i's are v's neighbors (or 0s if v has fewer than 5 neighbors). Now delete v from G and repeat. Since $G - v$ is planar, we select a vertex, say w, of minimum degree in $G - v$ and assign to it the label $\ell(w) = (w; w_1, w_2, w_3, w_4, w_5)$ where the w_i's are (up to) five of its neighbors in $G - v$. Delete w and continue until the entire graph is consumed. This labeling uses $6 \lg n$ bits per vertex.

Once this labeling is created, we can test the adjacency of any two vertices x and y just by examining their labels $\ell(x) = (x; x_1, \ldots, x_5)$ and $\ell(y) = (y; y_1, \ldots, y_5)$: we simply check if $x = y_i$ or $y = x_i$ for some i.

k-Degenerate Graphs

The representation method we presented for planar graphs relies on an idea that we can generalize. For a positive integer k, a graph G is called k-degenerate if $\delta(H) \leq k$ for all subgraphs $H \subseteq G$. We write $\mathscr{D}_k$ for the set of k-degenerate graphs. Note that trees are 1-degenerate and planar graphs are 5-degenerate.

The labeling method we used for planar graphs readily extends to graphs in $\mathscr{D}_k$. Let G be an n-vertex graph in $\mathscr{D}_k$. Without loss of generality we assume $V(G) = [n]$. Choose a minimum degree vertex v of G, and let $\ell(v) = (v; v_1, v_2, \ldots, v_k)$ where the v_is are neighbors of v (or 0s if $d(v) < k$). Delete v from G and repeat. The labels assigned to vertices use $(k + 1) \lg n$ bits, and adjacency testing is the same as for planar graphs.

Complete Multipartite (Turan) Graphs

It is no surprise that interval graphs have efficient local representations; this is nearly immediate from their definition. The existence of efficient local representations for trees and planar graphs are instances of the same idea that shows graphs in $\mathscr{D}_k$ have efficient local representations. Here is another example to illustrate the central idea.

Recall that a graph G is a *complete multipartite* graph if we can partition its vertex set as $V(G) = I_1 \,\dot\cup \cdots \dot\cup\, I_k$ where each I_j is an independent set, and for $i \neq j$, every vertex in I_i is adjacent to every vertex in I_j.

Here's an efficient local representation. Given the partition of $V(G)$, let $\ell(v) = j$ if $v \in I_j$. We have $v \sim w$ exactly when $\ell(v) \neq \ell(w)$. Since the number of parts in the partition of G is at most n, the number of bits used in the labels is bounded by $\lg n$, and so this is an efficient local representation.

Circular Arc Graphs

The family of *circular arc graphs* [32] are a natural extension of interval graphs. In this case, we assign to each vertex v of a graph G an arc A_v of some fixed circle. Vertices v and w are adjacent exactly when their arcs A_v and A_w intersect.

As in the case of interval graphs, there is no loss of generality in assuming that the $2n$ end points of the arcs representing G are distinct and therefore may placed at the corners of a regular $2n$-gon. Therefore, the arcs can be described using just $2 \lg n$ bits, and hence we have an efficient local representation.

String Graphs

The family of *string graphs* [16, 17] generalize the circular arc graphs. In a circular arc graph, each vertex is represented by a special type of curve: an arc on a fixed circle. In a string graph, we use arbitrary planar curves. A graph G is a *string graph* provided we can assign to each vertex v a curve S_v in the plane so that vertices v and w are adjacent if and only if $S_v \cap S_w \neq \emptyset$. As shown by Sinden [29], not all graphs are string graphs.

Clearly a string representation of a graph is local, but is it efficient? Clearly we cannot represent an arbitrary curve with just a handful of bits. Is there a trick (such as the one we used for interval and circular arc graphs) that enables us to bound the number of bits needed for each vertex? We shall see that the answer is no (Proposition 2.5).

6.2   Efficient Local Representations

6.2.1   Main Definitions

The problem can be described as follows. Let $\mathscr{P}$ be a property of graphs. We want to know if $\mathscr{P}$-graphs have efficient local representations. This means n-vertex graphs in $\mathscr{P}$ can be labeled with "short" labels—$O(\log n)$ bits—and adjacency can be tested just by comparing the labels on two vertices. The adjacency test is the same for all graphs with property $\mathscr{P}$.

Let's be more precise. By a *graph property* we mean an isomorphism-closed set of graphs: $G \in \mathscr{P} \wedge H \cong G \implies H \in \mathscr{P}$. It is sometimes more comfortable to refer to a property as a *class of graphs*.

A graph property $\mathscr{P}$ is *hereditary* if it is closed under taking induced subgraphs; that is, $G \in \mathscr{P} \wedge H \leq G \implies H \in \mathscr{P}$ (where $H \leq G$ denotes that H is an induced subgraph of G). It is natural in this context to focus on hereditary properties because if G has an efficient local representation, then $G - v$ does as well.

Local representations are associated with graph classes (and not with individual graphs).

Definition 2.1. Let $\mathscr{P}$ be a hereditary property of graphs. A *local representation* for $\mathscr{P}$ is a symmetric function $A : \mathbb{Z}^+ \times \mathbb{Z}^+ \to \{0, 1\}$ such that for every graph $G \in \mathscr{P}$ there is a labeling function $\ell : V(G) \to \mathbb{Z}^+$ such that for distinct vertices v, w of G, we have $v \sim w \iff A[\ell(v), \ell(w)] = 1$.

In other words, A is an adjacency test used by all graphs in $\mathscr{P}$. Each graph G in $\mathscr{P}$ has its own labeling function ℓ, and adjacency of two distinct vertices can be tested by applying the test A to the labels assigned to the pair.

[Note that Definition 2.1 requires that the vertex labels be positive integers, but in the examples in Section 6.1, the labels are tuples of integers. This is a minor technicality as we could perform kludges such as converting a pair of nonnegative integers (a, b) into a ternary number by writing a and b in binary and separating the values with a 2, like this: $(18, 11) \mapsto 1001021011_{\text{three}}$.]

Definition 2.1 omits any notion of efficiency and, consequently, is not interesting. Indeed, *all* hereditary properties have local representations using this definition; here's how.

Let G be any graph and assume that $V(G) = [n]$. We label vertex v by $\ell(v) = [v; a(v)]$ where $a(v)$ is an n-tuple of 0s and 1s that indicates the neighbors of v. That is, the jth entry in $a(v)$ is 1 exactly when $v \sim j$. Stated differently, $a(v)$ is the vth row of G's adjacency matrix. The test function A applied to labels $\ell(v) = [v; a(v)]$, and $\ell(w) = [w; a(w)]$ simply returns the wth entry in $a(v)$.

In other words, if we allow long labels, the neighborhood of a vertex can be trivially encoded in the label. We therefore impose a bound on the size of the vertex labels. We have previously expressed this as a bound on the number of bits that specify the label as being $O(\log n)$. These bits can be merged to form a single positive integer, and the restriction on the number of bits translates to requiring labels to lie in a set of the form $[n^k]$ where k is a given positive integer.

Definition 2.2. Let $\mathscr{P}$ be a hereditary property of graphs. An *efficient local representation* for $\mathscr{P}$ is a symmetric function $A : \mathbb{Z}^+ \times \mathbb{Z}^+ \to \{0, 1\}$ and a positive integer k such that every graph $G \in \mathscr{P}$ there is a labeling function $\ell : V(G) \to [n^k]$ (where $n = |V(G)|$) such that for distinct vertices v, w of G we have $v \sim w \iff A[\ell(v), \ell(w)] = 1$.

As before, the adjacency test A depends only on the property $\mathscr{P}$. Likewise the positive integer k is fixed for the class (does not depend on n). That labels take values in $[n^k]$ is tantamount to saying that the labels are (at most) $k \lg n$ bits in length. It is in this sense, the representation is (space) *efficient*.

6.2.2 The Problem

When we failed to restrict the number of bits in the labels, the class of all graphs $\mathscr{G}$ has a local representation. The interesting question is: What happens when we require efficiency? Is there an efficient local representation for all graphs?

This would be fantastic, but not surprisingly, the answer is no. To see why, we use a counting argument. To that end, we recall the definition of the *speed* of a hereditary property of graphs [1, 28].

Definition 2.3. Let $\mathscr{P}$ be a hereditary property of graphs and let n be a positive integer. Define $\mathscr{P}(n)$ to be the number of graphs in $\mathscr{P}$ with vertex set $[n]$. The function $\mathscr{P}(\cdot)$ is called the *speed* of the property.

For example, let $\mathscr{G}$ denote the class of all graphs. Then $\mathscr{G}(n) = 2^{\binom{n}{2}}$. Or let $\mathscr{P}$ be the property of having at most one edge. The set of graphs with vertex set $[n]$ that has at most one edge consists of the edgeless graph $\overline{K_n}$ and $\binom{n}{2}$ graphs with exactly one edge. Therefore $\mathscr{P}(n) = 1 + \binom{n}{2}$.

To show that not all hereditary properties admit an efficient local representation, we prove the following.

Proposition 2.4. *Let $\mathscr{B}$ be the class of bipartite graphs. Then $\mathscr{B}$ does not admit an efficient local representation.*

Proof. Suppose for contradiction that $\mathscr{B}$ admits an efficient local representation (A, k). Let n be a positive integer.

Every $G \in \mathscr{B}$ with $V(G) = [n]$ has a labeling function $\ell : [n] \to [n^k]$, and, necessarily, different graphs have different labeling functions.

The number of functions from $[n]$ to $[n^k]$ is n^{kn} which implies that $\mathscr{B}(n) \le n^{kn}$. However, it is easy to see that $\mathscr{B}(n) \ge 2^{n^2/4}$ which exceeds n^{kn} once n is large enough.

In a similar spirit, we show that the class of string graphs does not admit an efficient local representation.

Proposition 2.5. *Let $\mathscr{S}$ be the class of string graphs. Then $\mathscr{S}$ does not admit an efficient local representation.*

Proof. Recall that a graph G is a *split graph* [10] provided we can partition $V(G) = K \cup I$ where K is a clique and I is an independent set. We show that the class of string graphs contains all split graphs. Since the number of split graphs on vertex set $[n]$ is at least as large as the number of bipartite graphs on $[n]$, the result follows exactly as in the proof of Proposition 2.4.

Let G be a split graph with $V(G) = K \cup I$. To see that G is a string graph, assign disjoint curves to the vertices in I. For the vertices in K, we choose curves that emanate from a common point. If $v \in K$ is adjacent to $i_1, i_2, \ldots, i_d \in I$, then v's curve can be chosen to intersect the disjoint curves that represent $i_1, i_2, \ldots, i_d$ while avoiding all other I-curves as in Figure 6.2.   □

Note that the key fact we used in the proof of Proposition 2.4 is that *there are too many graphs on n vertices*. In order for a hereditary property $\mathscr{P}$ to admit an efficient local representation, the speed of $\mathscr{P}$ must be bounded by a function of the form n^{kn} for a specific integer k.

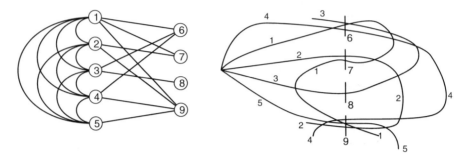

Fig. 6.2 We present a split graph and its string representation. The curves representing the vertices in the independent set $I = \{6, 7, 8, 9\}$ are the disjoint vertical line segments. The curves representing the vertices in the clique $K = \{1, 2, 3, 4, 5\}$ must intersect each other and exactly those curves for vertices in I to which they are adjacent. For example, the curve representing vertex 1 intersects vertical segments 6, 7, and 9, but not 8

Thus a hereditary property can fail to have an efficient local representation if it contains too many graphs—it violates the "speed limit" $\mathscr{P}(n) \leq n^{kn}$ for all fixed values of k.

What else could go wrong? That's a great question. Raised by [15] and [22], this is the question we offer our readers and present as our contribution to this compendium of favorite conjectures:

Conjecture 2.6 (Muller 1987; Kannan, Naor, Rudich 1992). *Let $\mathscr{P}$ be a hereditary property of graphs. Then $\mathscr{P}$ admits an efficient local representation if and only if there is an integer k such that $\mathscr{P}(n) \leq n^{kn}$.*

6.3   Challenging Examples

Conjecture 2.6 holds for a wide swath of hereditary graph properties including some we have considered (acyclic graphs, planar graphs, interval graphs, circular arc graphs) and many more we have not (such as permutation graphs [9] and threshold graphs [20]).

The following examples are—as best we know—possible counterexamples. The challenge is to find efficient local representations for these classes or show that none exist. See [2, 6], or [21] to find other hereditary classes as possible challenges.

Line Segment Intersection Graphs

The class of line segment intersection graphs lies between interval and string graphs. A graph G is a *line segment intersection graph* if we can assign to each $v \in V(G)$ a planar line segment L_v in such a way that $v \sim w$ if and only if $L_v \cap L_w \neq \emptyset$. Let $\mathscr{L}$ denote the class of line segment intersection graphs. See [3, 5, 18, 24].

Using the techniques described in [26], one can derive an upper bound for $\mathscr{L}(n)$ of the form n^{4n}, and so—if we believe Conjecture 2.6—this class should admit an efficient local representation.

Creating a local representation is easy; checking that it is efficient is difficult! Here's the representation. If $G \in \mathscr{L}$, then we know there is an assignment $v \mapsto L_v$ mapping vertices to line segments. These line segments can be represented as a 4-tuple of numbers (x, y, z, w) that specify the end points of the segment as (x, y) and (z, w). It's easy to check that there is no loss of generality in assuming that these coordinates are positive rational numbers (because $\mathbb{Q}^2$ is dense in $\mathbb{R}^2$). Therefore the labeling becomes an 8-tuple of integers. (Furthermore, by clearing denominators, we may assume all the coordinates are positive integers and a 4-tuple will suffice.)

Checking adjacency (i.e., intersection of the line segments) can be reduced to evaluating a pair of polynomial functions on the end points (details in [26]).

The only issue that remains is this: How many bits do we need to specify the end points? Do we need a high level of precision, or can this be accomplished with $O(\log n)$ bits?

Cographs

The class of *complement reducible graphs* [4], or *cographs* for short, can be described recursively as follows:

- K_1 is a cograph.
- If G is a cograph, so is its complement $\overline{G}$.
- If G and H are cographs, then so is their disjoint union. (This is the graph formed by simply taking copies of G and H on disjoint vertex sets and no additional edges.)

Let $\mathscr{C}$ denote the class of cographs. It is easy to see that $\mathscr{C}$ is a hereditary property. It is well known that cographs are exactly those graphs that do not contain the path P_4 as an induced subgraph.

With a bit of work, an upper bound of the form $\mathscr{C}(n) \leq n^{kn}$ can be derived. See sequence A000669 in [30].

Therefore, if Conjecture 2.6 holds, property $\mathscr{C}$ admits an efficient local representation.

Tolerance Graphs

The class of *tolerance graphs* [13, 14] may be considered as a generalization of interval graphs. We say that a graph G has a tolerance representation if we can assign to each vertex v of G a pair (I_v, t_v) where I_v is a closed, real interval and t_v is a positive real number so that $v \sim w$ in G if and only if the length of the intersection $I_v \cap I_w$ is at least $\min\{t_v, t_w\}$. Graphs with such a representation are called *tolerance graphs*.

An application of the methods in [26] gives a speed bound of the form n^{3n}, and therefore tolerance graphs satisfy the hypothesis of Conjecture 2.6. Do they satisfy the conclusion?

Geometric Graphs

Let $(\mathscr{X}, d)$ be a metric space. A graph G has a *geometric representation* in $(\mathscr{X}, d)$ if there is a mapping $f : V(G) \rightarrow \mathscr{X}$ such that $v \sim w$ if and only if $d[f(v), f(w)] \leq 1$. See [23].

A particularly natural example is $\mathscr{X} = \mathbb{R}^2$ together with the Euclidean metric. In this case, we can think of such graphs as having an intersection representation by unit discs.

Still with the Euclidean metric, for $\mathscr{X} = \mathbb{R}^k$, the number of geometric graphs with vertex set $[n]$ is bounded by an expression of the form n^{kn} and therefore satisfies the hypothesis of Conjecture 2.6. Do they satisfy the conclusion?

6.4   Variations

There are some natural variations on the core problem to consider.

6.4.1   Computation Concerns

We have focused on *space-efficient* representations. Each vertex holds a modest quantity of information. The combined information held by two vertices is sufficient to determine if they are adjacent. But at what cost? The authors of [15] also require that the function $A(\cdot, \cdot)$ be efficiently computable—in time polynomial in the size of the inputs. That is, the number of computational steps to check if v and w are adjacent in an n vertex graph should be bounded by an expression of the form $(\log n)^t$ for some fixed exponent t.

This is a perfectly reasonable requirement but appears to make this difficult problem only harder. Their conjecture (which we may dub the *strong efficient local representation conjecture*) is that hereditary properties that satisfy the speed limit $\mathscr{P}(n) \leq n^{kn}$ have an efficient local representation A that is polynomial-time computable. Clearly the strong version of the conjecture implies the weaker.

6.4.2   Other Label Sizes

Why do we want label sizes with $O(\log n)$ bits? This size is natural because already to name the vertices (e.g., specify vertex names in $[n]$) requires $\lg n$ bits. Thus the desire that $\ell(v)$ be represented in $O(\log n)$ bits is akin to saying that the labels are "about the same size" as the vertex names.

If, however, we are willing to modify this requirement, we can generate a host of additional problems.

Suppose we permit vertex labels to be larger—say, $O(\sqrt{n})$ bits. Then, presumably, such local representations could encompass more hereditary graph properties. If this is permitted, can we represent all hereditary properties that satisfy the speed limit $\mathscr{P}(n) \leq n^{kn}$? For the case of $O(\sqrt{n})$ bits, the counting argument shows that the properties must obey a bound of the form $n^{k\sqrt{n}}$; if $\mathscr{P}$ does obey such a bound, must it admit an $O(\sqrt{n})$ bits-per-vertex local representation?

On the other hand, we might consider using smaller labels, i.e., with $o(\log n)$ bits per vertex. As an extreme example, suppose $\mathscr{P}$ is the property of being a complete bipartite graph. Then we can label vertices with just a single bit (0 for vertices in one part of the bipartition and 1 for vertices in the other) and use the function $A(x, y) = \mathbf{1}[x \neq y]$.

Using the results in [28], we have the following result from [27].

Theorem 4.1. *Let $\mathscr{P}$ be a hereditary property of graphs and let k be a positive number with $k < \frac{1}{2}$. If, for all n sufficiently large, we have $\mathscr{P}(n) \leq n^{kn}$, then $\mathscr{P}$ admits a local representation in which the labels have $O(1)$ bits.* $\qquad\square$

In other words, the "super efficient" [$o(\log n)$ bits per vertex] local representation conjecture is true. Furthermore, if a property has a representation using $o(\log n)$ bits per vertex, then it has a representation using a constant number of bits per vertex.

Acknowledgements Many thanks to Ralucca Gera and Craig Larson for the invitation to present this problem at the 2012 SIAM Discrete Mathematics Conference in Halifax, Nova Scotia, and to contribute to this volume. Thanks also to my students Elizabeth Reiland and Yiguang Zhang, as well as to a highly dedicated referee, for many helpful comments on drafts of this chapter.

References

1. Balogha, J., Bollobás, B., Weinreich, D.: The speed of hereditary properties of graphs. J. Comb. Theory Ser. B **79**, 131–156 (2000)
2. Brandstädt, A., Le, V.B., Spinrad, J.: Graph Classes: A Survey. Society for Industrial and Applied Mathematics, Philadelphia (1999)
3. Chalopin, J., Gonçalves, D.: Every planar graph is the intersection graph of segments in the plane. In: STOC '09 Proceedings of the Forty-First Annual ACM Symposium on Theory of Computing, pp. 631–638 (2009)
4. Corneil, D.G., Lerchs, H., Burlingham, L.S.: Complement reducible graphs. Discret. Appl. Math. **3**(3), 163–174 (1981)
5. de Fraysseix, H., de Mendez, P.O., Pach, J.: Representation of planar graphs by segments. Int. Geogr. **63**, 109–117 (1991)
6. de Ridder, H.N., et al.: Information system on graph classes and their inclusions. http://www.graphclasses.org/
7. Erdős, P., Goodman, A.W., Pósa, L.: The representation of a graph by set intersections. Can. J. Math. **18**(1), 106–112 (1966)
8. Even, S.: Graph Algorithms, 2nd edn. Cambridge University Press, Cambridge (2011)
9. Even, S., Pneuli, A., Lempel, A.: Permutation graphs and transitive graphs. J. ACM **19**(3), 400–410 (1972)
10. Földes, S., Hammer, P.L.: Split graphs. Congr. Numer. **19**, 311–315 (1977)

11. Gilmore, P.C., Hoffman, A.J.: A characterization of comparability graphs and of interval graphs. Can. J. Math. **16**, 539–548 (1964)
12. Golumbic, M.C.: Algorithmic Graph Theory and Perfect Graphs, 2nd edn. North Holland, Amsterdam (2004)
13. Golumbic, M.C., Trenk, A.N.: Tolerance Graphs. Cambridge University Press, Cambridge (2004)
14. Golumbic, M.C., Monma, C.L., Trotter, W.T.: Tolerance graphs. Discret. Appl. Math. **9**(2), 157–170 (1984)
15. Kannan, S., Naor, M., Rudich, S.: Implicit representation of graphs. SIAM J. Discret. Math. **5**, 596–603 (1992)
16. Kratochvíl, J.: String graphs I. The number of critical nonstring graphs is infinite. J. Comb. Theory (B) **52**, 53–66 (1991)
17. Kratochvíl, J.: String graphs II. Recognizing string graphs is NP-hard. J. Comb. Theory (B) **52**, 67–78 (1991)
18. Kratochvíl, J., Nesetril, J.: Independent set and clique problems in intersection defined graphs. Comment. Math. Univ. Carol. **31**(1), 85–93 (1990)
19. Lekkerkerker, C.G., Boland, J.C.: Representation of a finite graph by a set of intervals on the real line. Fundam. Math. **51**, 45–64 (1962)
20. Mahadev, N.V.R., Peled, U.N.: Threshold Graphs and Related Topics. North-Holland, Amsterdam (1995)
21. McKee, T., McMorris, F.R.: Topics in Intersection Graph Theory. Society for Industrial and Applied Mathematics, Philadelphia (1999)
22. Muller, J.H.: Local structure in graph classes. Ph.D. thesis, Georgia Institute of Technology (1987)
23. Penrose, M.: Random Geometric Graphs. Oxford University Press, Oxford (2003)
24. Scheinerman, E.: Intersection classes and multiple intersection parameters of graphs. Ph.D. thesis, Princeton University (1984)
25. Scheinerman, E.: Characterizing intersection classes of graphs. Discret. Math. **55**(2), 185–193 (1985)
26. Scheinerman, E.: Geometry. In: Beineke, L., Thomas, R. (eds.) Graph Connections, pp. 141–154. Clarendon Press, Oxford (1997)
27. Scheinerman, E.: Local representations using very short labels. Discret. Math. **203**, 287–290 (1999)
28. Scheinerman, E., Zito, J.: On the size of hereditary properties of graphs. J. Comb. Theory (B) **61**, 16–39 (1994)
29. Sinden, F.: Topology of thin-film RC-circuits. Bell. Syst. Technol. J. **45**(9), 1639–1662 (1966)
30. Sloane, N.J.: The on-line encyclopedia of integer sequences. http://oeis.org/
31. Spinrad, J.: Efficient Graph Representations. The Fields Institute for Research in Mathematical Sciences. The American Mathematical Society, Providence (2003)
32. Tucker, A.C.: Matrix characterizations of circular-arc graphs. Pac. J. Math. **39**, 535–545 (1971)

Chapter 7
Some of My Favorite Coloring Problems for Graphs and Digraphs

John Gimbel

Abstract One of the central notions in graph theory is that of a coloring–a partition of the vertices where each part induces a graph with some given property. The most studied property is that of inducing an empty graph–a graph without any edges. Changing the property slightly creates interesting variations. In this paper I will discuss a few of my favorite coloring problems and variations. This discussion is not meant to be comprehensive. The field is so massive that attempts to catalog all important developments were abandoned many years ago. So I will restrict this to a very small set of problems that reflect my personal interests and perhaps nothing more.

Mathematics Subject Classification 2010: 05C15

7.1 Introduction

In this paper I will discuss a few of my favorite coloring problems. All graphs will be simple, that is, without loops, arcs, and parallel edges, but may or may not be finite. A *coloring* is simply a partition of the vertex set into parts, each of which induces a graph having a given property. For example, in the most studied form of coloring, each part induces an empty graph. I will refer to this as a *traditional coloring*. It is often the case that mathematicians are interested in the minimum number of parts possible in such a partition. In this particular example, the minimum number is referred to as the *chromatic number*, denoted by $\chi(G)$. More generally, we will refer to the parts in such a partition as *color classes* and a given partitioning as a *coloring*.

J. Gimbel (✉)

Department of Mathematic and Statistics, University of Alaska Fairbanks,
Fairbanks, AK 99775-6660, USA
e-mail: jggimbel@alaska.edu

© Springer International Publishing Switzerland 2016

R. Gera et al. (eds.), *Graph Theory*, Problem Books in Mathematics,
DOI 10.1007/978-3-319-31940-7_7

7.2   Coloring the Plane

I am told that I do not need to restrict these remarks to my own problems. So I'll
first mention a problem I heard from Paul Erdös in 1980. The first place it appears
in print is in Martin Gardner's monthly column in *Scientific American* in 1960 [26].
(All unreferenced results in this section can be found in [48].) The problem seems
to have been discussed beforehand in private communications between a number
of fine mathematicians. It is often referred to as the problem of *Coloring the Plane*
and also as the *Hadwiger-Nelson Problem*. You may also hear it referred to as "the
other four-color problem." The problem can be simply stated. Let G_p be the infinite
graph whose vertices are all points in the Cartesian plane and where two vertices
are adjacent if and only if the distance between them is exactly one unit.

Problem 1 (Coloring the Plane).   *What is the chromatic number of G_p?*

For a lively discussion of the origins of the this problem, see [48]. This problem
was probably first considered in 1950 by the eighteen-year-old Edward Nelson, later
of Princeton University, who showed that four colors are necessary to color this
graph; that is, $\chi(G_p) \geq 4$. As far as I know, he never published that result but
communicated it privately.

The bound is not difficult to demonstrate. Start with the three vertices of an
equilateral triangle, where each side has length one. If we color the vertices of this
triangle, then all three colors will appear. Now add to this triangle a vertex that has
one unit distance from two of these three vertices. If we *three*-color this "diamond,"
then the two, nonadjacent vertices must be given the same color. Now, make two
copies of this diamond and identify a vertex of degree two in each one, making them
a single vertex, but do this in such a way that the distance between the remaining
two vertices of degree two is one unit, and add an edge between them. Clearly, this
graph, known as the *Moser Spindle* [41], has chromatic number four (cf. Figure 7.1).

In 1950, John Isbell showed that $\chi(G_p) \leq 7$. Several elegant demonstrations of
this can be given [33, 50]. To see this, consider tiling the plane with a collection of
hexagons, all of which have a diameter slightly less than one. As demonstrated in
Figure 7.2, it is not too difficult to assign seven colors to these hexagons so that the
distance between hexagons with the same color is greater than one.

A number of eminent mathematicians have offered speculation on this question.
Erdös felt that G_p almost surely has chromatic number greater than four. Solid
evidence supports this. For example, nobody has found a collection of points
occupying 25% of the plane, where no distance between any two points in the
collection is one. Alexander Soifer believes [48] that $\chi(G_p)$ is either four or seven,
but conjectures it to be seven.

Further, a remarkable coloring of G_p by puzzle expert Edward Pegg, Jr. uses six
colors on the majority of the plane. The remaining seventh color occupies only about
one third of one percent of the plane.

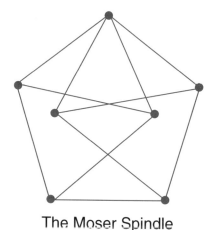

The Moser Spindle

Fig. 7.1 The moser spindle

Fig. 7.2 A seven coloring of the plane

The problem of Coloring the Plane can be "reduced" to a finite graph in the following sense. From de Bruijn [13] we know that the chromatic number of the infinite graph G_p is k if an only if G contains a finite subgraph having chromatic number k. The proof of de Bruijn's result uses the Axiom of Choice. Hadwiger [32] showed that if the plane is covered by five, congruent-closed sets, then a pair of points at distance one can be found in one of the sets. As a graduate student, Kenneth Falconer showed that if all color classes are required to be Lebesgue measurable, then at least five colors are required. If each color class consists of unions of sets, each bounded by a Jordan curve, then at least six colors are required.

The problem easily extends to higher dimensions. It is known that the graph G_p in 3-space has chromatic number somewhere between six and fifteen [8, 47].

Erdös asked, if we restrict ourselves to subgraphs of G_p with high girth, where the *girth* of a graph equals the length of a shortest cycle, can we push the chromatic number down to three? Wormald [54] showed such a girth must be at least six. In doing so, Wormald proved the existence of a subgraph of G_p with 6,448 vertices and girth five that has chromatic number four. While the existence of this graph is known, no such explicit construction of this graph has been found.

7.3   Three Coloring Planar Graphs

The next question is also not mine. In fact, it isn't even a question, but rather a meta-question. Under what conditions can a planar graph be *three*-colored? I'll call this the *Three-Color Problem*. There are a number of fascinating results about this problem. Perhaps the most famous is often referred to as Grötzsch's Theorem, which states that any planar, triangle-free graph can be three-colored. Several elegant proofs of this exist including a beautiful proof by Carsten Thomassen [53]. In addition, Thomassen [51] shows that a toroidal graph (e.g., a graph that embeds on the torus without the crossing of any edges), without cycles of lengths three or four, can be three-colored. In [16] Dvorak et al. present a linear-time algorithm which three colors planar, triangle-free graphs.

Another result in this area is mentioned, almost as an aside, in Heawood's disposition of Kempe's fallacious proof of the Four-Color Theorem [36]. To slightly reformulate the statement of this result, any planar triangulation can be three-colored if it has no vertices of odd degree. In Heawood's terms, "The proof of this is not difficult, but it appears to shed no light on the main proposition." And here, the "main proposition" is a correct proof of the Four-Color Theorem. Despite the observation of having a simple proof for three-coloring these planar triangulations, Heawood didn't offer one. Some discussion as to who first formulated a proof continues. An interesting description can be found in [49]. Dirac mentioned in an obituary for Heawood that surely Heawood had the proof but chose not to include it in any of his published work on the subject. Soifer also believes that Kempe [48] had a proof but omitted it, since it was not the main concern of his paper.

Allow me to mention a few related theorems that I find interesting. The first of these appears to be an extension of Heawood's result and is a special case of a more general remark discovered by Ore ([46], Chapter 13). Given a plane embedding of a graph G, where the number edges bounding each face is a multiple of three, then G is *three*-colorable if each vertex has even degree. Further, a planar graph is *three*-colorable if and only if it is a subgraph of a planar triangulation having no odd vertices. This result is shown independently in a variety of places, including [40].

Grünbaum showed [30] that any planar graph with at most three triangles is *three*-colorable. He then asked, if all triangles are distanced at least one apart in a planar graph, can it be *three*-colored? Ivan Havel [34] disproved this in 1969. However, Havel wondered if pushing the triangles further apart would guarantee *three*-colorability. This inspired an entire mathematical industry of related results (see [48], Chapter 19). We now know, for example, that there is a very large integer d such that if G is a planar graph in which any two triangles are at least distance d apart, then G is *three*-colorable [15]. It is conjectured by Borodin [5] that the distance $d = 4$ is sufficient to guarantee *three*-colorability. This is referred to as the "strong version of Havel's conjecture." Supporting this conjecture is the fact that if G is planar with no *five*-cycles nor any triangles within distance 3, then G is *three*-colorable. Since testing a planar graph with maximum degree four for *three*-colorability is NP-complete [27], it is probable that the problem of *three*-coloring planar graphs will be with us for some time to come.

7.4 Erdös Problems

Perhaps all research mathematicians have a first love. For me, it is cochromatic numbers. This was the topic of my doctoral dissertation. These numbers were introduced by Linda Lesniak, my dissertation advisor, along with her friend H. Joseph Straight. In their seminal paper [39], they generalized the definition of split graphs. Split graphs were originally defined [24] in a University of Waterloo technical report, by S. Foldes and P. Hammer in 1976. They later published a paper [25] with the same title, in the 1977 *Proceedings of the Eighth Southeastern Conference on Combinatorics, Graph Theory, and Computing*. In fact, two nonequivalent definitions are given in these two papers. In both definitions, the vertices of a split graph can be partitioned into two sets. In the first definition, each set must induce either an empty subgraph or a complete subgraph. In the second definition, exactly one set induces an empty graph in the original graph, and the second set induces a complete graph. Hence, using the first definition, a bipartite graph is a split graph. However, using the second definition, nontrivial bipartite graphs are not split graphs. It is this second definition that "stuck," and now over 300 papers have been written on this topic. In fact, the notion of a split graph was discussed earlier by Gyárfás and Lehel [31], but they did not name these graphs, as they were not the central idea of the paper. A popular characterization of split graphs, sometimes attributed to [24] in terms of forbidden subgraphs, can be found in [31].

It was Lesniak and Straight [39] who defined the *cochromatic number* $z(G)$ of a graph G to be the fewest number of colors needed to color the vertices of G so that each color class induces either a complete graph or an empty graph. Thus, using either definition of a split graph, a split graph will have cochromatic number at most two.

The topic of cochromatic numbers continues to be explored (see [6, 9]). In 1993 Paul Erdös and I published a paper [19] of unsolved problems in cochromatic theory.

Many of the problems have since been solved. For example, we asked if a graph has chromatic number m, what is the largest cochromatic number of all (not necessarily induced) subgraphs? Using an elegant application of probability, this was given a best answer up to a constant coefficient by Alon et al. [2]. We asked another question which remains open and is still one of my favorite problems. It concerns the random graph R_n, formed by taking n labeled vertices and with probability one half, adding an edge between each pair of distinct vertices. Relying on sophisticated results proved by others, it isn't difficult to show that as n goes to infinity, almost surely

$$\chi(R_n)/z(R_n) \to 1.$$

So our related question concerns not the quotient but the difference.

Problem 2. *Is it true that as n goes to infinity, almost surely*

$$\chi(R_n) - z(R_n) \to \infty?$$

At a random graph conference in Poznán, Paul Erdös offered $100 if the answer was yes and $1000 if the answer was no. He later told me he thought the $1000 prize was too high.

When I was young, Erdös visited Western Michigan University, where I was a student. He encouraged me to study dichromatic numbers and told me they were related to cocolorings. Initially, the connection wasn't clear, but it came into focus over time. Given D, a digraph without cycles of lengths one nor two, let the *dichromatic number* of D, denoted $d(D)$, be the smallest order of a coloring of the vertices of D, for which there are no monochromatic cycles. This notion was introduced [43] by the late, eminent Mexican mathematician, Professor Victor Neumann-Lara in 1982. Unfortunately, the term is used elsewhere in graph theory with another meaning. Independently, the dichromatic number was also developed by Meyniel [37], who referred to these as a *quasicolorings*.

Given an oriented graph D of order n, label the vertices with the integers 1 through n. Then, construct a graph $G(D)$ on the same vertex set and let vertices i and j be adjacent in $G(D)$ if $i < j$ and i is oriented toward j in D. Note that $d(D) \le z(G(D))$. With this observation, along with a probabilistic argument, I was able to show in joint work with Erdös and Dieter Kratsch [20] that if $d(D) = m$, then D has at least $c_1 m log(m)$ vertices and at least $c_2 m 2 log 2(m)$ edges, for some positive constants c_1 and c_2, and this result is best possible, except for a changes in the constants. This answered an earlier question of Erdös [18].

Erdös also asked [18] for a fixed m, if D is an oriented graph of smallest order with dichromatic number m, must D be a tournament? Clearly, if G is an undirected graph of smallest order with a fixed chromatic number, then it must be a clique. Erdös also asked if Δ is the maximum degree of the underlying graph of D, must $d(D)$ be $o(\Delta)$? He speculated that for some constant c, it might be less than $c\Delta/log(\Delta)$. Erdös also asked [18], if a graph G has large chromatic number, can we put an orientation on it with large dichromatic number? This interesting question remains unsolved.

Professors Neumann-Lara and Urrutia showed that if D is planar, then $d(D) \leq 3$ [45] but could not show that this bound is sharp. In fact, Neumann-Lara conjectured [44] the following.

Conjecture 3 (Neumann-Lara). *If D is a planar digraph then $d(D) \leq 2$.*

This question has been pondered by Ararat Harutyunyan, who tells me he believes it is more difficult than the Four-Color Problem. I've spent a bit of time with this seemingly simple question and cannot answer it. It seems to be very slippery.

7.5 Danish Problems

I have had several opportunities to work in Copenhagen with Carsten Thomassen, the eminent Danish graph theorist. We developed several coloring results. For example, we were able to find the largest chromatic number of a triangle-free graph with fixed size, asymptotically speaking. We were also able to show that if G is a graph embedded on the projective plane, such that all contractible cycles have length at least four, then G is *three*-colorable if and only if G doesn't contain a non-bipartite quadrangulation. We also showed that, for k at least four, there are only a finite number of k-critical graphs with girth at least six that embed on a given surface. In a similar vein, for k at least five, there are only a finite number of k-critical, triangle-free graphs that embed on a given surface. These last two results were independently discovered by my good friends Bojan Mohar and the late Steve Fisk [23].

These ideas seem to have generated a number of other results. Several open problems rose from our work. For instance, for each surface S, we can compute in polynomial time the chromatic number of any graph embedded on S that has girth six or greater. And for triangle-free graphs G on this surface, we can decide in polynomial time if the chromatic number of G is at most four. But we do not know the following.

Problem 4. *Does there exist a polynomial time algorithm that decides if a triangle-free graph embedded on a fixed surface is three-colorable?*

Problem 5. *Does there exist a surface S and an infinite number of four-critical graphs of girth five that embed on S?*

Thomassen subsequently showed that there are no such graphs on the torus nor on the projective plane [51]. Recent interesting results along these lines were announced [16] by Zdeněk Dvořák and Bernard Lidický.

My favorite open problem following from the work with Thomassen is the following. An S_g-polytope is an orientable surface of genus g made by gluing convex polygons together so that adjacent faces are not coplanar. The chromatic number of such a surface is the fewest number of colors needed to color the regions so that adjacent regions are given different colors. Another way to think of this is to color the dual of this surface. Thus, a vertex is placed in each region, and vertices are adjacent if their corresponding regions are adjacent. For a fixed g, let us refer to the largest chromatic number needed to color an S_g-polytope as the chromatic number of S_g and denote it $\chi(S_g)$.

We were able to show that $\chi(S_g) = o(g^{3/7})$, answering a question of Croft et al. [10]. Perhaps the fraction $3/7$ can be lowered. But we are unable to say if

Problem 6. *Is $\chi(S_g) = O(1)$?*

Some days I think that the answer to this question is yes, while on other days I am certain it isn't. David Barnett [4] showed that $\chi(S_1) \leq 6$. And Thomassen replaced this six with five [52]. This leads me to ask:

Problem 7. *Is $\chi(S_1) \leq 4$?*

In any case, I have yet to see an S_g-polytope that needs five colors. If you know of one, please let me know.

7.6 Turkish Problems

I have never been to Turkey but hope to visit there someday. I understand it is a beautiful place, and the mixture of many interesting cultures is well worth experiencing. On several occasions I've had the opportunity to meet Tinaz Ekim of Boğazici University. Together, we are interested in extending the ideas in defective coloring and combining them with the notion of cocoloring. Recall that a defective coloring is one where each color class has a bound on the maximum degree of the induced subgraph. To formalize slightly, let us say a collection of vertices is k *dependent* if it induces a subgraph with maximum degree at most k. This notion was originally formulated in [22]. A partition of the vertex set is k *defective* if each part is k dependent and the k-*defective chromatic number* of G, denoted $\chi_k(G)$, is the minimum order of all k-defective colorings. A k-*defective cocoloring* of G is a partition of $V(G)$ where each part induces a k-dependent graph in G or in the complement of G. The minimum order of all k-defective cocolorings is the k-*defective cochromatic number* and is denoted $z_k(G)$. It is not hard to show that if G has seven or fewer vertices, then $z_1(G)$ is at most two. Professor Ekim discovered the graph shown in Figure 7.3.

Here is a graph with *one*-defective cocoloring number three. We were also able to show [17] that if G has order 11, then $z_1(G) \leq 3$, but we do not know if this is best possible.

Problem 8. *Does there exist an integer $n > 11$ such that for all graphs of order at most n, $z_1(G) \leq 3$?*

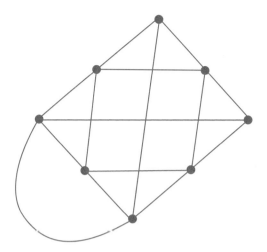

Fig. 7.3 A graph with a one-defective cocoloring number of three

A similar problem exists for z_2.

Problem 9. *Is it true that if G has order at most* 10 *then* $z_2(G) \leq 2$*?*

Given R_n, as defined in the previous section, we raise the following question, similar to Problem 2.

Problem 10. *Is it true that as n increases, almost surely*

$$z_1(R_n) - z_2(R_n) \to \infty?$$

And one can easily generalize this question. Professor Ekim and I were able to show the existence of a positive constant c with the property that for any natural numbers j and k, there exists a triangle-free graph G with $j = \chi_k(G)$, and $\lceil ck/log(k) \rceil j = \chi(G)$ [17]. With the possible change of the value of c, this is best possible. We do not have a similar result for graphs of larger girth, but certainly a similar result is true.

7.7   Alaska Problems

I would now like to comment on some results that came up in discussions with my esteemed colleagues here in Fairbanks, Alaska. These all involve tweaking the classical notion of color that is found above. In doing so, we stick to the basic underlying notion that in a coloring, the vertices are divided up in some way so that the induced subgraphs of each color class are simple, uncomplicated, and in some way or another orderly. This applies to the types of colorings already discussed. In cocolorings the color classes are uniform in the sense that they are either empty or complete graphs. In defective colorings, they are simple in the sense that the induced subgraphs do not have any vertices of large degree.

A *subcoloring* of G is a coloring of the vertices of G where each color class induces a disjoint union of cliques. (Disjoint unions of cliques are sometimes called "equivalence graphs.") We'll let $\chi_s(G)$ denote the minimum order of all subcolorings of G, and refer to it as the *subchromatic number* of G. This notion was introduced by Albertson et al. [1] in 1989, as was an edge coloring counterpart by Domke [14]. In a traditional coloring each color class consists of the union of a group of complete graphs of order one. Hence, the subchromatic number is always bounded above by the chromatic number. By similar reasoning, the subchromatic number is bounded above by the cochromatic number, the *one*-defective chromatic number, and the chromatic number of the complement.

The subchromatic number of a planar graph is at most four and, as a corollary to a result of [11], this can be proved without resorting to the Four-Color Theorem. From Grötzsch's Theorem [29] we know that the subchromatic number of any planar triangle-free graph is at most three. Professor Chris Hartman and I showed that the question, Is $\chi_s(G) \leq 2$? is NP-complete, even when restricted to planar, triangle-free graphs with a maximum degree of four. (And here "four" cannot be replaced by "three.") Since the subchromatic number of a triangle-free graph equals the *one*-defective chromatic number, we note that the question "Is $\chi_1(G) \leq 2$?" is also NP-complete, even when restricted to planar, triangle-free graphs with a maximum degree of four, answering a question of [12].

So what happens if we push the girth of such graphs even higher? If a planar graph has girth 11 then by a result of Nešetřil, et al. [42], it must contain an isolated vertex, a pendant vertex, or two adjacent vertices of degree exactly two. By induction, such graphs have *one*-defective chromatic number and subchromatic number at most two. So, for what maximum g can we say that if G is planar and has girth g, do the questions "Is $\chi_s(G) \leq 2$?" and "Is $\chi_1(G) \leq 2$?" remain NP-complete?

The complete *four*-partite graph $K_{1,2,3,4}$ clearly has a subchromatic number equal to four. Professor Hartman and I showed that any graph with fewer vertices has a subchromatic number of at most three [28]. This makes me wonder, how many vertices are needed in a graph with subchromatic number five?

A related parameter involves k-divided colorings. For a fixed number k, a graph is *k-divided* if each component has order at most k. A *k-divided coloring* is a partition of the vertex set where each part is k-divided. The *k-divided chromatic number*, $c_k(G)$ of G, is the minimum order of all k-divided colorings of G. In a traditional coloring, each color class is 1-divided. Thus, $\chi(G) = c_1(G)$ and similarly $\chi_1(G) = c_2(G)$ for all graphs.

Some interesting results exist for this parameter. For example, Alon et al. [3] showed that if G has a maximum degree of four, then $c_{57}(G) \leq 2$. This was later improved by Haxell et al. [35] to *six*-divided colorings. It is known that for every surface S and integer Δ, there is a k such that if G has maximum degree at most Δ and embeds on S, then $c_1(G) \leq 3$ [21]. On the other hand, Professor Chappell and I showed that for any fixed k, the decision problem "Is $c_k(G) \leq 2$?" is NP-complete, even for planar graphs. Also, for $k \geq 2$ the problem "Is $c_k(G) \leq 3$?" is NP-complete, even for planar, triangle-free graphs with maximum degree four. And in general, for fixed k and m, the problem "Is $\chi_k(G) \leq m$?" is NP-complete. So what sorts of other properties can be placed on graphs to turn these last three problems from NP-complete to P?

Glenn Chappell and I were also able to show [7] that for every fixed surface S, there is a k so that if G embeds on S, then $c_k(G) \leq 6$. We believe that six can be reduced, but have been unable to show this. We would like to know what is the largest girth of a planar graph G with $3 = c_2(G)$.

7.8 A Prague Problem

Let me close by mentioning a parameter developed with my close friend, Professor Jaroslav Nešetřil, the long-serving lion of Czech combinatorics, who has spent much of his distinguished career at Charles University. Suppose we are given an unbounded number of isolated vertices to begin with. With these, we allow two operations. We can take the disjoint union of graphs as well as the complement. Through multiple applications of these two operations, we can build a family of graphs known as *cographs*. There is a vast and significant body of literature on these graphs. Professor Nešetřil and I studied colorings of graphs where each color class induces a cograph. We call the minimum number of colors needed for such a coloring the c-chromatic number of G, and denote it $c(G)$. We were able to show that the problem "Is $c(G) \leq 3$?" is NP-complete for planar graphs. We would be interested in answers to the following two problems.

Problem 11. *Does there exist a planar, triangle-free graph G with $c(G) = 3$?*

Problem 12. *Is the decision problem, Is $c(G) = 2$?, NP-complete for planar, triangle-free graphs?*

7.9 Conclusion

As mentioned at the start, this paper is not meant to be comprehensive. And if you read it completely you almost surely said, "Why didn't he include one or two of my favorite problems???" To you, dear reader, I offer my apologies and yet offer my profuse thanks for your attention. A number of years ago, Professors Jensen and Toft produced a marvelous book [38] on graph colorings, and by most accounts, it was complete. It certainly cataloged all my favorite variations on the notion of coloring. For a number of years, they maintained a website, where we were able to add additional material on the topic including solutions to open problems listed in their book. But the subject became too broad even for those specialists, and the web project was eventually jettisoned. How can I be expected to cover the entire topic?

Allow me one last observation. I list a large number of papers in the following bibliography. Some of these I am referencing from other papers. I have looked at some of them, but not all. I offer this as a simple caution. In gathering this material together, I was often surprised by how many references I checked that proved to be erroneous. Indeed, it seems that many of us are quoting from papers that quote from others that quote from others, and a variety of errors in referencing get introduced in this form of academic gossip. I don't mean to suggest that this is unique to coloring theory, nor more broadly to graph theory. So let me ask your forgiveness in advance for errors in location and attribution. This is especially true if I failed to offer you proper credit for some superb result you are very proud of and for which you richly deserve praise.

References

1. Albertson, M., Jameson, R., Hedetniemi, S.: The subchromatic number of a graph. Graph colouring and variations. Discret. Math. **74**(1–2), 33–49 (1989)
2. Alon, N., Krivelevich, M., Sudakov, B.: Subgraphs with a large cochromatic number. J. Graph Theory **25**(4), 295–297 (1997)
3. Alon, N., Ding, G., Oporowski, B., Vertigan, D.: Partitioning into graphs with only small components. J. Comb. Theory Ser. B **87**(2), 231–243 (2003)
4. Barnette, D.: Coloring polyhedra manifolds. In: Discrete Geometry and Convexity (New York, 1982). Annals of the New York Academy of Sciences, vol. 440, pp. 192–195. New York Academy of Sciences, New York (1985)
5. Borodin, O., Glebov, A., Jensen, T.: A step towards the strong version of Havel's three color conjecture. J. Comb. Theory Ser. B **102**(6), 1295–1320 (2012)
6. Campos, V., Klein, S., Sampaio, R., Silva, A.: Fixed-parameter algorithms for the cocoloring problem. Discret. Appl. Math. **167**, 52–60 (2014)
7. Chappell, G., Gimbel, J.: On subgraphs without large components. Preprint
8. Coulson, D.: A 15-colouring of 3-space omitting distance one. Discret. Math. **256**, 83–90 (2002)
9. Chudnovsky, M., Seymour, P.: Extending the Gyárfás-Sumner conjecture. J. Comb. Theory Ser. B **105**, 11–16 (2014)
10. Croft, H., Falconer, K., Guy, R.: Unsolved problems in geometry. In: Unsolved Problems in Intuitive Mathematics, II. Springer, New York (1991)
11. Cowen, L., Cowen, R., Woodall, D.: Defective colorings of graphs in surfaces: partitions into subgraphs of bounded valency. J. Graph Theory **10**(2), 187–195 (1986)
12. Cowen, L., Goddard, W., Jerusem, C.: Defective coloring revisited. J. Graph Theory **24**(3), 205–219 (1997)
13. de Bruijn, N.G., Erdös, P.: A colour problem for infinite graphs and a problem in the theory of relations. Ned. Akad. Wet. Proc. A **54**(13), 369–373 (1951)
14. Domke, G., Laskar, R., Hedetniemi, S.: The edge subchromatic number of a graph. 250th Anniversary Conference on Graph Theory (Fort Wayne, IN, 1986). Congressus Numerantium, vol. 64, pp. 95–104. Utilitas Mathematica Publishing, Winnipeg (1988)
15. Dvořák, Z., Král, D., Thomas, R.: Coloring Planar Graphs with Triangles Far Apart. Citeseer (2009)
16. Dvořak, Z., Kawarabayashi, K.-I., Thomas, R.: Three-coloring triangle-free planar graphs in linear time. ACM Trans. Algoritm. (TALG) **7**(4), 41 (2011)
17. Ekim, T., Gimbel, J.: Some defective parameters in graphs. Graphs Comb. **29**(2), 213–224 (2013)

18. Erdös, P.: Problems and results in number theory and graph theory. Proceedings of the Ninth Manitoba Conference on Numerical Mathematics and Computing (University of Manitoba, Winnipeg, 1979). Congressus Numerantium, vol. XXVII, pp. 3–21. Utilitas Mathematica, Winnipeg (1980)

19. Erdös, P., Gimbel, J.: Some problems and results in cochromatic theory. In: Quo Vadis, Graph Theory? Annals of Discrete Mathematics, vol. 55, pp. 261–264. North-Holland, Amsterdam (1993)

20. Erdös, P., Gimbel, J., Kratsch, D.: Some extremal results in cochromatic and dichromatic theory. J. Graph Theory **15**(6), 579–585 (1991)

21. Esperet, L., Joret, G.: Colouring planar graphs with three colours and no large monochromatic components. Comb. Probab. Comput. **23**(4), 551–570 (2014)

22. Fink, J., Jacobson, M.: On n-domination and n-dependence and forbidden subgraphs. In: Alavi, Y., Schwenk, A. (eds.) Graph Theory with Applications to Algorithms and Computer Science (Kalamazoo, 1984), pp. 301–311. Wiley, New York (1985)

23. Fisk, S., Mohar, B.: Coloring graphs without short non-bounding cycles. J. Comb. Theory Ser. B **60**, 268–276 (1994)

24. Foldes, S., Hammer, P.: Split graphs. Technical Report. Department of Combinatorics and Optimization, University of Waterloo. CORR 76–3 (1975)

25. Foldes, S., Hammer, P.: Split graphs. In: Proceedings of the Eighth Southeastern Conference on Combinatorics, Graph Theory and Computing (Louisiana State University, Baton Rouge, 1977), Congressus Numerantium, vol. XIX, pp. 311–315. Utilitas Mathematica, Winnipeg (1977)

26. Gardner, M.: Mathematical games. Sci. Am. **203**, 172–180 (1960)

27. Garey, M., Johnson, D., Stockmeyer, L.: Some simplified NP-complete problems. Theor. Comp. Sci. **1**(3), 237–267 (1976)

28. Gimbel, J., Hartman, C.: Subcolorings and the subchromatic number of a graph. Discret. Math. **272**(2–3), 139–154 (2003)

29. Grötzsch, H.: Zur Theorie der diskreten Gebilde, VII: Ein Dreifarbensatz für dreikreisfreie Netze auf der Kugel, Wiss. Z. Martin-Luther-U., Halle-Wittenberg, Math.- Nat. Reihe **8**, 109–120 (1959)

30. Grünbaum, B.: Grötzsch's theorem on 3-colorings. Mich. Math. J. **10**, 303–310 (1963)

31. Gyárfás, A., Lehel, J.: A Helly-type problem in trees. In: Combinatorial Theory and Its Applications, II (Proceedings of the Colloquium, Balatonfüred, 1969), pp. 571–584. North-Holland, Amsterdam (1970)

32. Hadwiger, H.: Überdeckung des euklidischen Raumes durch kongruente Mengen. Port. Math. **4**, 238–242 (1945)

33. Hadwiger, H.: Ungelöste probleme, Nr 11. Elemente der Mathematik **16**, 103–106 (1961)

34. Havel, I.: The coloring of planar graphs by three colors. In: (Czech) Mathematics (Geometry and Graph Theory), pp. 89–91. University Karlova, Prague (1970)

35. Haxell, P., Szabo, T., Tardos, G.: Bounded size components, partitions and transversals. J. Comb. Theory Ser. B **88**(2), 281–297 (2003)

36. Heawood, P.: Map-colour theorem. Q. J. Pure Appl. Math. **24**, 332–338 (1890)

37. Jacob, H., Meyniel, H.: Extension of Turán's and Brooks' theorems and new notions of stability and coloring in digraphs. In: Combinatorial Mathematics (Marseille-Luminy, 1981), vol. 75, pp. 365–370. North-Holland, Amsterdam (1983)

38. Jensen, T., Toft, B.: Graph Coloring Problems. Wiley-Interscience Series in Discrete Mathematics and Optimization, xxii+295 pp. Wiley, New York (1995). ISBN:0-471-02865-7

39. Lesniak, L., Straight, H.J.: The cochromatic number of a graph. Ars Comb. **3**, 39–45 (1977)

40. Martinov, N.: 3-colorable planar graphs. Serdica **3**(1), 11–16 (1977)

41. Moser, L., Moser, W.: Solution to problem 10. Can. Math. Bull. **4**, 187–189 (1961)

42. Nešetřil, J., Raspaud, A., Sopena, E.: Colorings and girth of oriented planar graphs. Discret. Math. **165/166**, 519–530 (1997)

43. Neumann-Lara, V.: The dichromatic number of a digraph. J. Comb. Theory Ser. B **33**(3), 265–270 (1982)

44. Neumann-Lara, V.: Vertex colorings in digraphs. Technical Report, University of Waterloo (1985)
45. Neumann-Lara, V., Urrutia, J.: Vertex critical r-dichromatic tournaments. Discret. Math. **49**(1), 83–87 (1984)
46. Ore, O.: The Four-Color Problem. Pure and Applied Mathematics, vol. 27. Academic, New York (1967)
47. Radoičić, R., Tóth, G.: Note on the chromatic number of space. In: Discrete and Computational Geometry: the Goodman-Pollack Festschrift, vol. 25, pp. 695–698. Springer, New York (2002)
48. Soifer, A.: The Mathematical Coloring Book. Mathematics of Coloring and the Colorful Life of Its Creators. Springer, New York (2009)
49. Steinberg, R.: The state of the three color problem. In: Quo Vadis, Graph Theory? Annals of Discrete Mathematics, vol. 55. pp. 211–248. North-Holland, Amsterdam (1993)
50. Székely, L., Wormald, N.: Bounds on the measurable chromatic number of $\mathbb{R}^n$. Graph theory and combinatorics. Discret. Math. **75**(1–3), 343–372 (1989)
51. Thomassen, C.: Grötzsch's 3-color theorem and its counterparts for the torus and the projective plane. J. Comb. Theory Ser. B **62**(2), 268–279 (1994)
52. Thomassen, C.: Color-critical graphs on a fixed surface. J. Comb. Theory Ser. B **70**(1), 67–100 (1997)
53. Thomassen, C.: A short list color proof of Grötzsch's theorem. J. Comb. Theory Ser. B **88**, 189–192 (2003)
54. Wormald, N.: A 4-chromatic graph with a special plane drawing. J. Aust. Math. Soc. Ser A **28**(1), 1–8 (1979)

Chapter 8
My Top 10 Graph Theory Conjectures and Open Problems

Stephen T. Hedetniemi

Abstract This paper presents brief discussions of ten of my favorite, well-known, and not so well-known conjectures and open problems in graph theory, including (1) the 1963 Vizing's Conjecture about the domination number of the Cartesian product of two graphs [47], (2) the 1966 Hedetniemi Conjecture about the chromatic number of the categorical product of two graphs [28], (3) the 1976 Tree Packing Conjecture of Gyárfás and Lehel [23], (4) the 1981 Path Partition Conjecture of Lovász and Mihók [8], (5) the 1991 Inverse Domination Conjecture of Kulli and Sigarkanti [34], (6) the 1995 Queens Domination Conjecture [15], (7) the 1995 Nearly Perfect Bipartition Problem [9], (8) the 1998 Achromatic-Pseudoachromatic Tree Conjecture [10], (9) the 2004 Iterated Coloring Problems and the Four-Color Theorem [30], and (10) the 2011 γ-graph Sequence Problem [16].

Mathematics Subject Classification 2010: 05C05, 05C07, 05C10, 05C12, 05C15, 05C38, 05C69, 05C78

Introduction

In the spring semester of 1961 at the University of Michigan, Prof. Frank Harary taught his first course in graph theory. I was fortunate enough to be one of about 18 graduate students taking this course. Before every class, Frank would appoint one of the students as the official notetaker. You were to write up your notes on the material that he presented in that class, take them to Frank sometime before the next class, have him read them over and make corrections if necessary, type them up on mimeograph paper, run off copies, and distribute them at the time of the next class. When the course was finished, we all had (and I still have) a complete set of

S.T. Hedetniemi (✉)
School of Computing, Clemson University, Clemson, SC 29634, USA
e-mail: hedet@clemson.edu

© Springer International Publishing Switzerland 2016
R. Gera et al. (eds.), *Graph Theory*, Problem Books in Mathematics,
DOI 10.1007/978-3-319-31940-7_8

notes for the course. It was this set of notes, and the notes produced from the second course that Frank taught in the fall of 1964, that became the basis of his book, *Graph Theory* [25].

An amusing note about this first course was that on page 1 of the first set of notes, Frank provided this joke: "Now that the University of Michigan Mathematics Department has a complete curriculum..." Even though graph theory, as a field of study, was in its infancy back in 1961, it didn't take Frank long to start presenting to us interesting conjectures and open problems. In his very first lecture, he presented to us the problem of deciding if two graphs are isomorphic and extending this to the unsolved problem of finding a complete set of invariants for graphs. Three lectures later he presented the problem of deciding if a partition of an even positive integer is graphical, that is, if such a partition is the degree sequence of some graph, and if it is graphical, to determine the number of non-isomorphic graphs having this degree sequence. He also stated as an unsolved problem that of deciding if a graphical degree sequence belongs to only one graph. He later stated as an unsolved problem that of determining which permutation groups are the automorphism groups of some graph, saying that this was known only for trees. Subsequently, Frank presented the famous Kelly-Ulam Graph Reconstruction Conjecture and the problem of characterizing Hamiltonian digraphs.

Thus, our introduction to the field of graph theory was generously sprinkled with many fascinating conjectures and open problems. In the pages that follow, I will present a collection of some of my favorite graph theory conjectures, using the idea of *just-in-time* definitions, rather than giving here, in this introduction, all of the definitions that will be needed later. I will present these conjectures in chronological order of their first appearance in the literature.

8.1   Vizing's Conjecture - 1963

Having spent most of my last 40 years researching the many aspects of the concept of domination in graphs, it is only natural that I at least mention this well-known conjecture, even though I haven't done any research on it myself. Since it is arguably the most famous conjecture in domination theory, about which more than 50 papers have been written, I feel obligated to at least present it. But in order to do so, we will need some definitions, not only for this conjecture but for those to follow.

Let us first assume that all graphs are connected. The *open neighborhood* of a vertex $v \in V$ is the set $N(v) = \{u \mid uv \in V\}$ of vertices adjacent to v. Each vertex u in $N(v)$ is called a *neighbor* of v. The *degree* of a vertex v is $deg(v) = |N(v)|$. The minimum and maximum degrees of a vertex in a graph G are denoted $\delta(G)$ and $\Delta(G)$, respectively. A vertex $v \in V$ is called an *isolated vertex* if it has no neighbors, that is, $deg(v) = 0$. The *closed neighborhood* of a vertex $v \in V$ is the set $N[v] = N(v) \cup \{v\}$.

The *open neighborhood of a set* $S \subseteq V$ of vertices is $N(S) = \bigcup_{v \in S} N(v)$, while the *closed neighborhood of a set* S is the set $N[S] = \bigcup_{v \in S} N[v]$.

A set S is a *dominating set* of a graph G if $N[S] = V$, that is, every vertex $v \in V$ is either in S or is adjacent to a vertex in S. The minimum cardinality of a dominating set in a graph G is called the *domination number* and is denoted $\gamma(G)$. A dominating set of minimum cardinality is called a γ-*set*.

A set S is *independent* if no two vertices in S are adjacent. The *independent domination number* $i(G)$ equals the minimum cardinality of a set that is both independent and dominating.

The *Cartesian product* $G \square H$ of two graphs G and H is the graph $G \square H$ having as vertex set the Cartesian product $V(G) \times V(H)$, where two vertices (u, v) and (u', v') in $V(G \square H)$ are adjacent in $G \square H$ if and only if either (i) $u = u'$ and $vv' \in E(H)$ or (ii) $uu' \in E(G)$ and $v = v'$.

Vizing's Conjecture, 1963. For any two graphs G and H, $\gamma(G \square H) \geq \gamma(G)\gamma(H)$.

Stated in words, Vizing's Conjecture asserts that the domination number of the Cartesian product of any two graphs G and H is greater than or equal to the product of the domination number of G and the domination number of H.

It is easy to construct examples where $\gamma(G \square H) > \gamma(G)\gamma(H)$, for example, $\gamma(K_3 \square P_2) = 2 > \gamma(K_3)\gamma(P_2) = 1 \times 1 = 1$, where K_3 denotes a triangle, and P_2 denotes a path of length one.

It is equally easy to construct examples where $\gamma(G \square H) = \gamma(G)\gamma(H)$, for example, the 4×4 graph, $\gamma(P_4 \square P_4) = 4 = \gamma(P_4) \times \gamma(P_4) = 2 \times 2 = 4$. If you have never seen how to dominate the 4-by-4 grid graph, $P_4 \square P_4$, with only four vertices, you might try this on a sheet of paper. The solution is unique, up to isomorphism, and is quite nice!

Vizing's Conjecture has been extensively studied by many people, and there is little that I can add to their combined efforts. For the interested reader, I recommend two excellent and comprehensive survey papers, one by Hartnell and Rall in 1998 [27] and the other by Brešar, Dorbec, Goddard, Hartnell, Henning, Klavžar, and Rall in 2012 [2]. I asked Prof. Rall if he could suggest some interesting problems related to Vizing's Conjecture for this section. He was kind enough to provide the following, which we present with his permission:

Steve, there are a number of interesting questions that are either directly related to trying to prove Vizing's conjecture or that arose in a natural way while we worked on the conjecture. The following list is certainly not complete, but I have tried to include those that I find the most interesting or that will have a direct impact on what is known about the conjecture.

We say that a graph G satisfies Vizing's Conjecture provided $\gamma(G \square H) \geq \gamma(G)\gamma(H)$ for every graph H.

1. *There is an interesting graphical invariant in our survey that has not, so far as I know, received very much attention. For a given independent set I in a graph G consider all the subsets $D \subseteq V(G)$ such that $I \subseteq N[D]$, that is, D dominates I. Obviously I is such a set D [since a set always dominates itself], but if some pairs of vertices in I are distance two apart, then we could choose a set of cardinality smaller than $|I|$ that would dominate I. On the other hand, if I is actually a 2-packing [a set of vertices no two of which are distance-two*

apart], then no subset of smaller cardinality than |I| will dominate I. Here is the invariant. We define $\gamma^i(G)$ to be the maximum, over all independent sets I in G, of the smallest cardinality of a set D such that $I \subseteq N[D]$.

Since any dominating set of G also dominates any independent set in G, it follows immediately that $\gamma^i(G) \leq \gamma(G)$. Aharoni, Berger, and Ziv showed that for any chordal graph G, $\gamma^i(G) = \gamma(G)$, and Aharoni and Szabó proved that for any pair of graphs G and H,

$$\gamma(G \Box H) \geq \gamma^i(G)\gamma(H) .$$

Together these prove that any chordal graph satisfies Vizing's Conjecture.

Here are several interesting problems in their own right. The first also has a direct application to the conjecture and is given as Question 4.8 in [2].

Problem 1. *Find other interesting classes of graphs such that γ^i and γ assume the same value for every graph in the class. Note that this will be true for all graphs for which the 2-packing and domination numbers are equal.*

Problem 2. *Are there classes of graphs where γ^i can be computed efficiently?*

2. *This question arose in attempts to find a counterexample to the conjecture. In Section 8.6 we made a list of a number of properties that any minimal counterexample to the conjecture must hold. By minimal counterexample we mean a graph G of smallest order such that for some graph H, $\gamma(G \Box H) < \gamma(G)\gamma(H)$.*

One of these properties is what Burton and Sumner call totally dot-critical. A graph G is totally dot-critical if for every pair of vertices u and v in G, $\gamma(G_{uv}) < \gamma(G)$, where G_{uv} is the graph obtained from G by identifying u and v and then removing any parallel edges. A graph G is domination edge-critical if $\gamma(G + xy) < \gamma(G)$ whenever x and y are nonadjacent vertices in G. It turns out that any minimal counterexample to Vizing's conjecture is totally dot-critical, and you can assume it is domination edge-critical.

Problem 3. *Give a structural characterization of graphs that are both domination edge-critical and totally dot-critical. It would be interesting just to know the structure of graphs having domination number 4 that are both domination edge-critical and totally dot-critical.*

3. *Here is an interesting question that is probably easier to settle than Vizing's conjecture. Of course, its answer is "yes" if Vizing's conjecture is true, but one might be able to prove it without first proving Vizing's conjecture.*

Question. *Is it the case that for every pair of graphs G and H, $i(G \Box H) \geq \gamma(G)\gamma(H)$?*

4. *In 2000 Clark and Suen proved that for every pair of graphs G and H,*

$$\gamma(G \Box H) \geq \frac{1}{2}\gamma(G)\gamma(H) .$$

This was later slightly improved by Suen and Tarr to

$$\gamma(G \Box H) \geq \frac{1}{2}\gamma(G)\gamma(H) + \frac{1}{2}\min\{\gamma(G), \gamma(H)\}$$

but no one has been able to prove a similar inequality to the one of Clark and Suen with $\frac{1}{2}$ replaced with a larger constant. To that end here is a problem that, in my opinion, would be real progress toward settling Vizing's conjecture.

Problem 4. *Find a constant $c > \frac{1}{2}$ such that for every pair of graphs G and H,*

$$\gamma(G \Box H) \geq c\,\gamma(G)\gamma(H)\,.$$

Let me thank Doug for taking the time to give us these related problems and add just a few thoughts of my own. It is interesting to compare Vizing's Conjecture with a somewhat similar problem that remained unsolved for many years: what is the domination number of the Cartesian product of two paths, that is, can the value of $\gamma(P_m \Box P_n)$ be determined? These are commonly called *grid graphs*. I once had a conversation with David Johnson about this problem, several years after he and Michael Garey had published their seminal NP-completeness book [18]. I figured that he knew as much about the complexity of combinatorial problems as anyone, but at that time it was not known if the grid domination problem was NP-complete. He told me that surely this problem must be polynomially solvable, indeed there must be a formula for this number, since we know everything there is to know about these graphs. Years later he was essentially proven correct, when in 2011, Goncalves, Pinlou, Rao, and Thomasse published such a formula [20]. One can't help but wonder what, if anything, this says about the complexity of Vizing's Conjecture. On the other hand, you might say that grid graphs are just about the simplest of all possible Cartesian products.

8.2   The Hedetniemi Conjecture - 1966

I started my PhD studies in the newly established Communication Sciences graduate program at the University of Michigan, in the fall of 1961; this was a forerunner of today's PhD programs in Computer Science and Computer Engineering. At that time I studied a decomposition theory for finite-state machines, developed by two young researchers named Kenneth Krohn and John Rhodes (see, e.g., [33]). In their decomposition theory, they showed that a finite-state machine could be decomposed into a series connection of two smaller machines, if one could define a homomorphism on the given machine. Since a finite-state machine can be viewed as a finite-directed graph, my advisor at the time, Prof. John Holland (who pioneered the study of genetic algorithms), recommended that I go see Prof. Frank Harary and ask him what was known about homomorphisms of graphs. Since I had taken Harary's first graph theory course, I knew him and he knew me. When I asked Prof. Harary what was known about homomorphisms of graphs, he told me that precisely one theorem was known, and it was by one of his former PhD students, named Geert Prins (who was at that time a professor of mathematics at Wayne State University). This theorem was later called the Homomorphism Interpolation Theorem, and it later appeared in a paper by Harary, Hedetniemi, and Prins [26]. At that time Harary gave me a copy of a typed letter he had recently received from Prins, detailing a several page proof of this result. It was difficult for me to read and understand.

While sitting in the back row of a class in coding theory, being taught by the well-known Prof. W. W. Peterson, a much simpler proof of this theorem dawned on me. After class I went back to Harary's office and showed him what was essentially a three-line proof of this simple, but interesting, result. Harary was so pleased to see this proof, that, on the spot, he invited me to do a PhD thesis with him on this subject. This was my first theorem! Subsequently I went back to my advisor, Prof. Holland, who encouraged me to pursue this line of research with Prof. Harary. When I finally defended my dissertation, both Holland and Harary were listed as my co-advisors.

I soon discovered that virtually nothing was known about homomorphisms of graphs, even though essentially homomorphisms were nothing other than colorings of graphs, about which a fair amount was known, most especially colorings of planar graphs and the famous Four-Color Conjecture. It was not long thereafter that I was led to consider homomorphisms of various products of graphs and in particular the categorical product of graphs.

The *categorical product* $G \times H$ of two graphs G and H is the graph with vertices $V(G) \times V(H)$ and edges $(u, v)(u', v') \in E(G \times H)$ if and only if $uu' \in E(G)$ and $vv' \in E(H)$; this is also called the *direct product* and the *tensor product*. A *homomorphism* from a graph G onto a graph H is a function $\phi : V(G) \to V(H)$ having the property that $uv \in E(G)$ implies $\phi(u)\phi(v) \in E(H)$. Given a homomorphism $\phi : G \to H$, we define the homomorphic image of G to be the graph $G\phi = (V\phi, E\phi)$, where $V\phi = \{\phi(v), v \in V(G)\}$ and $E\phi = \{\phi(u)\phi(v), uv \in E(G)\}$.

A *proper coloring* of a graph G is a vertex partition $V = \{V_1, V_2, \ldots, V_k\}$, into independent sets. The *chromatic number* $\chi(G)$ of a graph G equals the smallest order of a proper coloring of G.

Given these definitions it is easy to see that if $\phi : G \to H$ is a homomorphism, then $\chi(G) \leq \chi(G\phi)$; this inequality was perhaps first observed by Ringel in 1959 [41] and later by Hajós in 1961 [24]. One can also observe that the natural projection of the vertices of a categorical product $G \times H$ onto either their first or second components defines natural homomorphisms $\phi_1 : G \times H \to G$ and $\phi_2 : G \times H \to H$. Thus it follows that for any two graphs G and H, $\chi(G \times H) \leq \chi(G)$ and $\chi(G \times H) \leq \chi(H)$. This observation is what led me, without thinking much about it, to make the following conjecture [28]:

Hedetniemi Conjecture, 1966. For any graphs G and H, $\chi(G \times H) = min\{\chi(G), \chi(H)\}$.

It is easy to see that this conjecture is true if the smaller chromatic number of G and H is either 1, 2, or 3. Almost 20 years later, and with some significant effort, El-Zahar and Sauer [11] were able to prove the following result:

Theorem 1. *The chromatic number of the categorical product of two 4-chromatic graphs in 4.*

To this date, this theorem remains as perhaps the best result on this conjecture. Little did I realize at the time just how difficult this conjecture would prove to be. It is amusing to note that at the following website, this conjecture is not recommended as suitable for undergraduates!

```
http://www.openproblemgarden.org/?q=op/hedetniemis_
  conjecture
```

Forty-nine years after this conjecture first appeared, much research has been done on the deep and rich subject of colorings of products of graphs. The interested reader is referred to three comprehensive surveys that have been published on this conjecture, by Zhu in 1998 [49], Sauer in 2001 [43], and Tardif in 2008 [46]; the surveys by Zhu and by Tardif are freely available on the web (just do a Google search on "Hedetniemi conjecture").

8.3 Tree Packing Conjecture (TPC) - 1976

A sequence of graphs $G_1, G_2, \ldots, G_k$ can be *packed* into a graph H if H contains a sequence $H_1, H_2, \ldots, H_k$ of pairwise, edge-disjoint subgraphs such that for all i, $1 \leq i \leq k$, $G_i \simeq H_i$, that is, G_i is isomorphic to H_i.

Suppose you are given a collection of $n - 1$ trees, $T_2, T_3, \ldots, T_n$, where tree T_i has order i. Such a collection of $n - 1$ trees has $\sum_{i=2}^{n}(i - 1) = n(n - 1)/2$ edges. Similarly, the complete graph K_n also has $n(n - 1)/2$ edges. Thus, in theory, such a collection of trees could pack perfectly into K_n without any overlapping edges. This leads to the following conjecture, due to Gyárfás and Lehel [23]:

Tree Packing Conjecture, 1976. Any collection of $n - 1$ trees, $T_2, T_3, \ldots, T_n$, where tree T_i has order i, can be packed into K_n.

Although it appears unlikely that this nearly 40-year-old conjecture will be solved any time soon, it does seem likely that a number of partial results can be obtained concerning the packing of various kinds and numbers of trees into complete graphs. For example, it should be easy to see that if all of the trees are stars (trees of the form $K_{1,m}$), or if all of the trees are paths, then such a set of trees can be packed into K_n. There are several other results of this type.

Theorem 2 (Gyárfás and Lehel [23]). *The TPC holds with the assumption that each tree is either a path or a star.*

Theorem 3 (Roditty [42]). *The TPC holds with the assumption that all but three of the trees are stars.*

Theorem 4 (Hedetniemi, Hedetniemi, Slater [29]). *Any two trees of order n, neither of which is a star, can be packed into* K_n. .

Theorem 5 (Slater, Teo, Yap [45]). *If* G_1 *is a tree of order n and* G_2 *is any graph of order n and size at most* $n - 1$, *where neither* G_1 *nor* G_2 *is a star and* $n \geq 5$, *then* G_1 *and* G_2 *can be packed into* K_n.

Theorem 6 (Hobbs, Bourgeois, Kasiraj [31]). *Any three trees* T_n, T_{n-1}, T_{n-2}, *where tree* T_i *has order i, can be packed into* K_n.

Theorem 7 (Hobbs, Bourgeois, Kasiraj [31]). *Any set of* $n-1$ *trees* $T_2, T_3, \ldots, T_n$, *where tree* T_i *has order i can be packed into* K_n *if at most one of the trees* T_i *has diameter more than three.*

It seems that one should be able to generate several more results along these lines.

8.4   Path Partition Conjecture (PPC) - 1981

A longest path in a graph $G = (V, E)$ is called a *detour*. The number of vertices in a detour is the *detour number* of G and is denoted by $\tau(G)$. If $S \subseteq V$, then the *subgraph induced by* S is the graph $G[S] = (S, E \cap (S \times S))$.

A vertex bipartition $V = \{V_1, V_2\}$ is called an (a, b)-*partition* of G if $\tau(G[V_1]) \leq a$ and $\tau(G[V_2]) \leq b$. A graph G is said to be τ-*partitionable* if for every pair of positive integers a, b such that $a + b = \tau(G)$, G has an (a, b)-partition.

Path Partition Conjecture, 1981. Every graph is τ-partitionable.

An equivalent statement of this conjecture is given by Frick in [14]:

If $G = (V, E)$ is any graph and (τ_1, τ_2) is any pair of positive integers, such that G has no path with more than $\tau_1 + \tau_2$ vertices, then there exists a bipartition $V = \{V_1, V_2\}$ of the vertex set of G such that $G[V_i]$ has no path with more than τ_i vertices, for $i = 1, 2$.

This conjecture was first mentioned by Lovász and Mihók in 1981 and appeared in a paper by Laborde, Payan, and Xuong in 1983 [35]. In [9], Dunbar and Frick present 14 cases in which the PPC is true, and in [14] Frick presents several more cases in which the PPC is true. We present just a few of these here.

The Path Partition Conjecture is true when:

- Every cyclic block of G is Hamiltonian; obviously, this includes all Hamiltonian graphs.
- $G = G_1 + G_2$ is the join of two graphs G_1 and G_2.
- G is 2-degenerate, i.e., every induced subgraph of G has a vertex of degree at most 2.
- $\tau(G) \leq 13$.
- $\Delta(G) \leq 3$.
- $\Delta(G) \geq |V(G)| - 8$.

- G is weakly pancyclic, i.e., G contains a cycle of every length between the *girth* $g(G)$ and the *circumference* $c(G)$ of G, where the *girth* $g(G)$ equals the minimum length of a cycle in G, and the *circumference* $c(G)$ equals the maximum length of a cycle in G.
- G is claw-free, that is, G contains no induced subgraph isomorphic to the graph $K_{1,3}$.
- G is planar and has girth 5, 8 9, or 16.

Dunbar and Frick [9] also prove that if the PPC is true for all 2-connected graphs, then it is true for all graphs. They suggest as an interesting problem that of deciding if the PPC is true for all planar graphs.

The reader is also referred to a recent survey of the PPC by Frick [14], in which a related, but weaker, conjecture is given, one in which Gary Chartrand, Dennis Geller, and I had a role in creating. In 1968, we published a paper [4] in which we introduced the idea of partitioning the vertices of a graph $V = \{V_1, V_2, \ldots, V_k\}$ into a minimum number of subsets such that no induced subgraph $G[V_i]$ contains a path of length greater than n, for some nonnegative integer $n \leq |V(G)|$. We denoted this minimum number $\chi_n(G)$. In [9] this is reworded as follows: an *n-detour coloring* of a graph G is a coloring of the vertices $V(G)$ such that no path in G of length greater than n is monocolored. The *nth detour chromatic number* $\chi_n(G)$ equals the minimum number of colors that can be used to produce an *n*-detour coloring of G. In [9], the authors state that "Our initial interest in the PPC was based on the fact that, if the PPC is true, then the following conjecture is also true."

Conjecture. For every graph G and every positive integer $n \leq |V(G)|$, $\chi_n(G) \leq \lceil \tau(G)/n \rceil$.

In [14] Frick gives some interesting evidence that the PPC might be true. Let $\mathcal{I}$ denote the class of all finite, simple graphs. A property $\mathcal{P}$ of graphs is called *hereditary* if whenever a graph G has property $\mathcal{P}$, so does every subgraph of G. For example, if a graph G is planar, so is every subgraph of G.

A property $\mathcal{P}$ is called *additive* if whenever two graphs G and H have property $\mathcal{P}$, so does the disjoint union $G \cup H$. For example, if graphs G and H are both acyclic, then the disjoint union $G \cup H$ is also acyclic.

Frick presents the following examples of additive, hereditary families of graphs (we take the liberty of changing the notation slightly here):

$\mathcal{O} = \{G \in \mathcal{I} : E(G) = \emptyset\}$.
$\mathcal{C}_k = \{G \in \mathcal{I} : \text{for every connected component } C \text{ of } G, |V(C)| \leq k + 1\}$.
$\Delta_k = \{G \in \mathcal{I} : \Delta(G) \leq k\}$.
$\mathcal{D}_k = \{G \in \mathcal{I} : G \text{ is } k\text{-degenerate, every subgraph } G' \subset G \text{ has } \delta(G') \leq k\}$.
$\mathcal{P}_k = \{G \in \mathcal{I} : \tau(G) \leq k + 1, \text{ every path in } G \text{ has order at most } k + 1\}$.

Notice that the following families of graphs are equal:

$$\mathcal{O} = \mathcal{C}_0 = \Delta_0 = \mathcal{D}_0 = \mathcal{P}_0.$$

In the following let $\mathcal{P} \circ \mathcal{Q}$ denote the family of graphs G whose vertices can be partitioned into two sets $V = \{V_1, V_2\}$ such that the induced subgraph $G[V_1]$ has property $\mathcal{P}$ and the induced subgraph $G[V_2]$ has property $\mathcal{Q}$. Given this, the PPC can be formulated as follows:

Path Partition Conjecture, 1981. For all integers p and q, $\mathcal{P}_{p+q+1} \subseteq \mathcal{P}_p \circ \mathcal{Q}_q$.

It is interesting that several theorems of this form have been proved, also for additive, hereditary properties.

Theorem 8 (Lovasz [37], 1966). *For all positive integers p and q, $\Delta_{p+q+1} \subseteq \Delta_p \circ \Delta_q$.*

Theorem 9 (Borodin [1], 1976). *For all integers p and q, $\mathcal{D}_{p+q+1} \subseteq \mathcal{D}_p \circ \mathcal{D}_q$.*

Theorem 10 (Jensen-Toft [32], 1995). *For all integers p and q, $\mathcal{C}_{p+q+1} \subseteq \mathcal{C}_p \circ \mathcal{C}_q$.*

8.5 Inverse Domination Conjecture - 1991

The minimum cardinality of a maximal independent set is called the *independent domination number* and is denoted $i(G)$, while the maximum cardinality of an independent set is called the *vertex independence number* and is denoted $\beta_0(G)$. Independent sets of cardinality $i(G)$ and $\beta_0(G)$ are called *i*-sets and β_0-sets, respectively. The maximum cardinality of a minimal dominating set is called the *upper domination number* and is denoted $\Gamma(G)$.

In 1962 Ore [39] proved the following theorem, from which a number of corollaries can be derived.

Theorem 11 (Ore). *In any graph $G = (V, E)$ having no isolated vertices, the complement $V - D$ of any minimal dominating set D is a dominating set.*

Corollary 1. *In any graph $G = (V, E)$ having no isolated vertices, the vertices V can be partitioned $V = \{D, V - D\}$ into two sets, each of which is a dominating set; furthermore, the first set D can be chosen to be a minimal dominating set, a γ-set, a Γ-set, a maximal independent set, an i-set, or a β_0-set, as desired.*

Thus, in particular, in any graph $G = (V, E)$ without isolated vertices, the complement $V-D$ of every γ-set D is a dominating set, and therefore, $V-D$ contains a minimal dominating set. In 1991, Kulli and Sigarkanti [34] defined the *inverse domination number* $\gamma^{-1}(G)$ to equal the minimum cardinality of a dominating set in the complement $V - D$ of a γ-set D; such a dominating set is called an *inverse dominating set* of G.

In this paper they state the following as a theorem and give the following "proof."

Theorem 12. *For any graph G having no isolated vertices, $\gamma^{-1}(G) \leq \beta_0(G)$.*

Proof. Let $D \subset V$ be a γ-set of G, and let $S \subset V - D$ be a maximal independent set in the subgraph $G[V - D]$ induced by the complement $V - D$ of D. Consider the following two cases:

Case 1. $S = V - D$, that is, $V - D - S = \emptyset$. In this case S is an independent
   dominating set of G, since every vertex in D must have at least one neighbor in
   $V - D$, because, by assumption, G has no isolated vertices. Thus, S is an inverse
   dominating set and being independent $\gamma^{-1}(G) \leq |S| \leq \beta_0(G)$, as required.
Case 2. $(V - D) - S \neq \emptyset$. In this case, since S is a maximal independent set in the
   induced subgraph $G[V - D]$, then every vertex in $(V - D) - S$ must be adjacent to
   at least one vertex in S. Consider therefore how many vertices in D are dominated
   by vertices in S.

   Case 2.a. $D \subset N(S)$, that is, every vertex in D is dominated by a vertex in S.
      In this case S is an independent dominating set in G and, therefore, an inverse
      dominating set in G. Thus, $\gamma^{-1}(G) \leq |S| \leq \beta_0(G)$, as required.
   Case 2.b. S does not dominate every vertex in D. Let $D' = D - (N(S) \cap D) \neq \emptyset$
      be the set of vertices in D not dominated by S. Since D is a minimum
      dominating set in a graph having no isolated vertices, every vertex in D must
      have a neighbor in $V - D$, and in particular every vertex in D' must have a
      neighbor in $(V - D) - S$, since no vertex in D' is adjacent to any vertex in S.
      Let $S' \subset (V - D) - S$ be a minimum cardinality set of vertices in $(V - D) - S$
      that dominates every vertex in D'.

It follows that $S \cup S'$ is an inverse dominating set of G, that is, a dominating set
in the complement $V - D$ of a γ-set D of G.

At this point, all that remains is to show that $|S \cup S'| \leq \beta_0(G)$. But here is where
the proof by Kulli and Sigarkanti seems to have a problem. The authors correctly
state that $|S \cup S'| \leq |S \cup D'|$.

The authors then state that $|S \cup D'| \leq \beta_0(G)$, but provide no reason why this is
true. Indeed, if $S \cup D'$ is an independent set, then this inequality is true. But there
is no guarantee that $S \cup D'$ is an independent set, since the set D' might not be
independent. Furthermore, there is no reason we can see why this inequality should
be true.

Thus, the Kulli-Sigarkanti "proof" is not a proof, and we are left with the
following conjecture:

Inverse Domination Conjecture, 1991. For any graph G having no isolated
vertices, $\gamma^{-1}(G) \leq \beta_0(G)$.

As of this writing, we know of no proof that $\gamma^{-1}(G) \leq \beta_0(G)$, and we are not
aware that any counterexample has been constructed. But there is a fair amount of
evidence that suggests that this conjecture might be true.

1. One can easily show that whenever $\gamma(G) = i(G)$, then $\gamma^{-1}(G) \leq \beta_0(G)$. In this
   case, the set $S \cup D'$ can always be made to be an independent set, by choosing
   the γ-set D to be an independent dominating set. It has been shown that if G is a
   claw-free graph, i.e., $K_{1,3}$-free, then $\gamma(G) = i(G)$. Thus, the Inverse Domination
   Conjecture holds for claw-free graphs; this has also been shown by Frendrup
   et al. [13].
2. One can easily show that whenever $\beta_0(G) = \Gamma(G)$, then $\gamma^{-1}(G) \leq \beta_0(G)$. In this
   case, any minimal dominating set, say S, in the complement $V - D$ of a γ-set D

satisfies $|S| \leq \Gamma(G) = \beta_0(G)$. Thus, any such set S is an inverse dominating set of cardinality at most $\beta_0(G)$. It has been shown that quite a few classes of graphs satisfy $\beta_0(G) = \Gamma(G)$ and therefore satisfy the Inverse Domination Conjecture. These include the following classes (not defined here):

   (i) Trees,
   (ii) Bipartite graphs, including all grid graphs,
  (iii) Chordal graphs,
  (iv) Circular arc graphs,
   (v) C_4-free graphs,
  (vi) Net-free graphs,
 (vii) Upper bound graphs,
(viii) Trestled graphs,
  (ix) Strongly perfect graphs,
   (x) Cographs,
  (xi) Permutation graphs, .
 (xii) Comparability graphs,
(xiii) Co-chordal graphs,
(xiv) Peripheral graphs,
 (xv) Parity graphs,
(xvi) Gallai graphs,
(xvii) Perfectly orderable graphs,
(xviii) Middle graphs.

In [13], Frendrup, Henning, Randerath, and Vestergaard show that the Inverse Domination Conjecture holds, not only for bipartite graphs, chordal graphs, and claw-free graphs but also for the following classes of graphs:

(xix) Split graphs,
 (xx) Very well-covered graphs,
(xxi) Cactus graphs.

3. It is trivial to show that any graph having two disjoint γ-sets satisfies the Inverse Domination Conjecture; in this case $\gamma^{-1}(G) = \gamma(G) \leq i(G) \leq \beta_0(G)$. If a graph does not have two disjoint γ-sets, but does have a disjoint γ-set and i-set, or a disjoint γ-set and β_0-set, then the Inverse Domination Conjecture is also true.

4. From the attempted proof of this conjecture by Kulli-Sigarkanti above, any graph having an independent set S which is maximal independent in the complement $G[V - D]$ of a γ-set D, that dominates every vertex in D except those in D', and D' is an independent set, which satisfies the Inverse Domination Conjecture.

5. A *vertex cover* is a set S of vertices having the property that for every edge $uv \in E$, either $u \in S$ or $v \in S$. The *vertex covering number* $\alpha_0(G)$ equals the minimum cardinality of a vertex cover in G. It is well known that for any graph G, the complement $V - S$ of a vertex cover S is an independent set, and conversely, the complement of any independent set is a vertex cover. It follows, therefore, that for any graph G of order n, $\alpha_0(G) + \beta_0(G) = n$; this is a well-known theorem of Gallai [17]. It is also easy to see that for any graph G without

isolated vertices, $\gamma(G) \le \alpha_0(G)$. From this it follows that if a graph G has an α_0-set S that contains as a subset a γ-set $S' \subseteq S$, then the Inverse Domination Conjecture holds, since the complement $V - S$ of S is a β_0-set, and therefore the complement of S' contains a β_0-set; thus, $\gamma^{-1}(G) \le \beta_0(G)$. Because of this, it also follows that if $\gamma(G) = \alpha_0(G)$, then $\gamma^{-1}(G) \le \beta_0(G)$, and the conjecture is true.

8.6   Queens Domination Conjecture - 1994

The *Queen's graph* Q_n is the graph obtained from an n-by-n chessboard from the moves of a queen, namely, the vertices of Q_n correspond one to one with the n^2 squares of the n-by-n chessboard, and two vertices are adjacent in Q_n if and only if the corresponding squares lie on a common row, column, or diagonal. Thus, the vertices in any row, column, or diagonal form a clique (or complete subgraph) of Q_n. The *Queens domination number* $\gamma(Q_n)$ equals the domination number of the graph Q_n. Equivalently, it equals the minimum number of queens necessary to cover, or dominate, all squares of the chessboard not containing a queen.

The problem of determining the Queens domination number seems to be extremely difficult, as exact values of $\gamma(Q_n)$ have been determined for relatively few values of n. An excellent discussion of the computational complexity of computing the values of $\gamma(Q_n)$ is given by Fernau in [12].

A closely related parameter is the *independent Queens domination number* $i(Q_n)$, in which you seek a minimum number of queens on an n-by-n board that dominate all squares, but no two of these queens can appear in the same row, column, or diagonal, i.e., they cannot attack each other.

Figures 8.1 and 8.2 show solutions for $n = 8$ and $n = 11$. In fact, it is known that $\gamma(Q_8) = \gamma(Q_9) = \gamma(Q_{10}) = \gamma(Q_{11}) = 5$. I have always thought it is remarkable that five Queens suffice to dominate the 11-by-11 chessboard. It is known that *six* Queens are needed to dominate the 12-by-12 board, by the way.

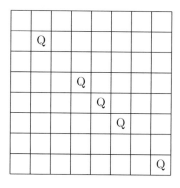

Fig. 8.1 8×8 Queens domination

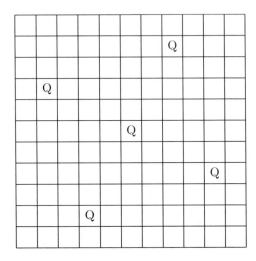

Fig. 8.2 11 × 11 Queens domination

There is an interesting story to tell about the particular solution shown above in Figure 8.1 that $\gamma(Q_8) = 5$. Since the 1850s many people have studied what is generally known as "The Eight Queens Problem" (or in general, the "N Queens Problem), in which one is asked to place eight queens on the 8-by-8 chessboard in such a way that no queen can attack another queen. By contrast, in the "The Five Queens Problem," one is asked to place only five queens on the chessboard in such a way that all unoccupied squares can be attacked by at least one queen, but it doesn't matter if two of the five queens can attack each other.

In January of 1977 I became Professor and Head of the Department of Computer and Information Science at the University of Oregon. About a year later I was invited to give a colloquium in the University of Oregon's Department of Mathematics. At that time I was working on a variety of domination problems, and this was the topic of the colloquium talk that I gave. Unknown to me at the time, sitting somewhere in the audience was a Professor Robin Dawes, who was in University of Oregon's Department of Psychology. In the colloquium talk, I happened to discuss the application of domination in graphs to chessboards, and I mentioned the Queens domination problem and showed several solutions for boards of different sizes.

A day later, I received through campus mail a short note from Professor Dawes, in which he stated that he enjoyed the talk I had given and then showed me the solution you see above in Figure 8.1, in which all five queens lie on the main diagonal. I had never seen this solution before. Prof. Dawes simply asked, "Is this interesting?"

It was indeed quite interesting, for it took me only a few seconds to realize that this solution had the property that not only were all unoccupied squares dominated by these five queens but that all five queens were themselves dominated by another queen! I could see that this was a type of domination that had not been studied before.

The *domination number* $\gamma(G)$ of a graph $G = (V, E)$ can be defined as the minimum cardinality of a set $S \subseteq V$ such that $N[S] = V$, where this equality involves the closed neighborhood $N[S]$ of set S. In this definition there is no requirement that every vertex $v \in S$ be dominated by, or adjacent to, another vertex in S. But Prof. Dawes' solution, in Figure 8.1, suggested to me the following definition and name: the *total domination number* $\gamma_t(G)$ of a graph G equals the minimum cardinality of a set $S \subseteq V$ such that $N(S) = V$, where this equality involves the open neighborhood $N(S)$ of set S.

At that time I had been doing a lot of research with Prof. E. J. Cockayne at the University of Victoria, and so I mailed Ernie a letter (we did not have email in those days!) telling him about this new definition and suggesting that we might work up a paper introducing the idea of "total domination." He quickly agreed to do this, and we knew that we simply had to ask Prof. Dawes to be a co-author on this paper, to which he kindly agreed. This paper [6] was the first to introduce the concept of total domination, a concept on which more than 400 papers have now been published.

Over the next dozen years or so, I continued to explore various graph theory parameters for graphs defined by the movements of various chess pieces, for example, the bishops graph, the rooks graph, and the knights graph. At the 1994 Western Michigan University Graph Theory Conference, I gave a talk [15] in which I made the following "obvious" conjecture:

Queens Domination Conjecture, 1994. For any $n \geq 1$, $\gamma(Q_n) \leq \gamma(Q_{n+1})$.

Surely this must be true! How hard can it be to prove this? I have always felt that some simple proof of this must exist. Yet, surprisingly, no such proof has appeared. Why is this?

One possible answer is that a queen placed in the $(n + 1)^{st}$ row or column of an $(n + 1)$-by-$(n + 1)$ Queens graph attacks, or dominates, a set of squares within the n-by-n Queens graph that cannot be attacked by any one queen inside the n-by-n Queens graph. Thus, it is theoretically possible that fewer queens might be needed to dominate Q_{n+1} than are required to cover Q_n. So far, no counterexample to this simple conjecture has been found.

Without a doubt, the definitive paper on Queens domination is the 2001 paper by Östergard and Weakley [40]. I need not repeat here all of the many results in this comprehensive paper, but perhaps a summary of known results would be helpful to the reader. Can you see any pattern(s) in the following numbers?

(i) For all $n \leq 120$, the value of $\gamma(Q_n)$ is either known or known to be one of two consecutive values.
(ii) For all n, $\gamma(Q_n) \leq 69n/133 + O(1)$.
(iii) For all n, $(n - 1)/2 \leq \gamma(Q_n) \leq i(Q_n)$.
(iv) $\gamma(Q_n) = (n - 1)/2$, for $n = 3, 11$.
(v) $\gamma(Q_n) = \lceil n/2 \rceil$, for $n = 1, 2, 4 - 7, 9, 10, 12, 13, 17 - 19, 21, 23, 25, 27, 29 - 31, 33, 37, 39, 41, 45, 49, 53, 57, 61, 65, 69, 71, 73, 117, 121, 125, 129 - 131$.
(vi) $\gamma(Q_n) = \lceil n/2 \rceil + 1$, for $n = 8, 14, 15, 16$.

(vii) $\gamma(Q_n) \in \{\lceil n/2 \rceil, \lceil n/2 \rceil + 1\}$ for $n = 20, 22, 24 - 26, 28, 32, 34, 35, 36, 38,$
40, 42, 43, 44, 46, 47, 48, 50, 51, 52, 54, 55, 56, 58, 59, 60, 62, 63, 64, 66, 67,
68, 70, 72, 74, 75, 76, 78, 79, 80, 82, 83, 84, 86, 87, 88, 90, 92, 94 − 96, 98 −
100, 102 − 104, 106 − 108, 110 − 112, 114, 116, 118 − 120, 122, 126, 132.

Thus, the smallest value of n for which the value $\gamma(Q_n)$ is not known is $n = 20$; either $\gamma(Q_{20}) = 10$ or $\gamma(Q_{20}) = 11$.

This brings to mind the following story. Back around 1985, Prof. Alice McRae, now at Appalachian State University, was a PhD student of mine. She had been taking a graduate course in which they were learning about genetic algorithms. In my office one day, I asked her if she could write a genetic algorithm for approximating the value of $\gamma(Q_{18})$. At that time it was known that $\gamma(Q_{18})$ was either 9 or 10. So off she went. A day later she returned to my office and said that $\gamma(Q_{18}) = 9$! She had written her genetic algorithm and executed it for $n = 18$, and within just a few seconds, her program had found a placement of *nine* queens that dominated the 18-by-18 chessboard. Needless to say I was quite pleased and surprised to hear this from Alice. Over the next weekend I continued to think about this result. Early the next week when I saw Alice again, I said, "Alice, could you run that program again, for $n = 18$?" I couldn't believe that her program found an exact solution that fast. The next day, Alice came back into my office and told me that her program once again found the same placement of *nine* queens, but it took her program several hours of machine time to find it!

In their paper [40], Östergard and Weakley make the following conjecture:

Conjecture (Östergard and Weakley). For any $n \geq 1$, $i(Q_n) \leq \lceil n/2 \rceil + 1$.

They also wonder if there is any value of n, other than 3 and 11, for which $\gamma(Q_n) = (n-1)/2$.

Here is one more thought about the Queens Domination Conjecture. In 2008, Sinko and Slater [44] introduced the concept of Queens domination using only border squares $bor(Q_n)$. Surprisingly, they showed that $bor(Q_{13}) = 9 < bor(Q_{12}) = 10$, and thus this type of Queens domination is not monotone!

8.7 Nearly Perfect Bipartition Problem - 1995

In 1969, R. L. Graham [21] defined a cutset of edges to be *simple* if no two edges in the cutset have a vertex in common, that is, the set of edges in the cutset is a disconnecting matching. Graham defined a graph to be *primitive* if G has no simple cutset, but every proper subgraph of G has a simple cutset. He then asked: what are the primitive graphs? The problem in this section is inspired by this question.

A set $S \subset V$ of vertices in a graph $G = (V, E)$ is *nearly perfect* if every vertex in $V - S$ is adjacent to at most one vertex in S. Nearly perfect sets in graphs were first defined and studied by Dunbar et al. in 1995 [9]. At that time S. T. Hedetniemi and

McRae spent a fair amount of time considering, unsuccessfully, the complexity of the following decision problem:

NEARLY PERFECT BIPARTITION
INSTANCE: Connected graph $G = (V, E)$.
QUESTION: Can the vertices of G be partitioned into two sets $V = \{S, V - S\}$ such that both S and $V - S$ are nearly perfect sets?

For the next 20 years, we knew of no progress that had been made on settling the complexity of this problem.

Let us assume that the graph G in question is connected. If it is not connected, then such a bipartition trivially exists in which there are no edges between V_1 and V_2, as long as either V_1 or V_2 is a disjoint component of G.

If such a bipartition exists, then consider all edges between V_1 and V_2. This collection of edges must define a matching whose removal disconnects G. Thus, we seek a *disconnecting matching*, that is, a matching M, the removal of the edges in which it disconnects G. This in turn gives rise to the following, equivalent, decision problem:

DISCONNECTING MATCHING
INSTANCE: Connected graph $G = (V, E)$
QUESTION: Does G contain a disconnecting matching?

It is immediately obvious that every edge in a tree forms a disconnecting matching. Any graph with a leaf has a single edge, disconnecting matching. All grid graphs have many disconnecting matchings. The well-known Petersen graph has a disconnecting matching. On the other hand, no complete graph of order 3 or more has a disconnecting matching.

Nearly Perfect Bipartition Problem, 1995. What is the complexity of Nearly Perfect Bipartition or Disconnecting Matching?

I decided to include this problem in my top 10 largely because of Alice McRae. In her 1988 PhD thesis [38], Alice had more than 80 original NP-completeness theorems! Early in her graduate studies, she showed a knack for developing creative solutions to all sorts of NP-completeness questions. In the many years since then, whenever we are confronted with an NP-completeness question, we "Just ask Alice." Usually within 24 hours, and often overnight, Alice will send back an NP-completeness proof! If Alice has trouble showing that some problem is NP-complete, then it is probably a tough problem. Alice tried very hard to settle Nearly Perfect Bipartition, but its complexity eluded us.

Addendum: After this paper had been completed, I heard from Alice that one of her undergraduate students at Appalachian State University, named Neil Butcher, had just solved this 20-year-old problem! Neil uses a transformation from the well-known 1-in-3 SAT problem to show that Nearly Perfect Bipartition is NP-complete. I have seen his proof and it is pretty impressive. So congratulations to Neil!

8.8   Achromatic-Pseudoachromatic Tree Conjecture - 2004

As defined in Section 8.2, a proper k-coloring of a graph $G = (V, E)$ is a partition $V = \{V_1, V_2, \ldots, V_k\}$ such that each set V_i is an independent set. A proper k-coloring is called a *complete coloring* if for all i, j, $1 \le i < j \le k$, there is at least one vertex in V_i that is adjacent to a vertex in V_j.

The *achromatic number* of a graph G, denoted $\psi(G)$, equals the maximum order k of a complete proper coloring of G. The achromatic number was first identified and studied as a parameter by Harary et al. in 1967 [26] and named and further studied by Harary and Hedetniemi in 1970 [25].

The *pseudoachromatic number* of a graph G, denoted $\psi_s(G)$, equals the maximum order k of a complete coloring of G. The only difference between $\psi(G)$ and $\psi_s(G)$ is that in a pseudoachromatic partition of order $k = \psi_s(G)$, the coloring need not be proper, that is, the sets V_i need not be independent sets. The pseudoachromatic number was first studied by Gupta in 1968 [22].

Around 1998 I attempted to determine the difference between $\psi(T)$ and $\psi_s(T)$ for trees T and could not find a tree for which these values were different. Thus, on my web pages, I conjectured that these two parameters were equal for trees. Seeing this conjecture, Edwards [10] then published a paper in which he constructed a tree T of order $n = 408$, for which $\psi_s(T) - \psi(T) = 1$. In personal correspondence, I then asked Edwards if he could construct a tree for which this difference was greater than 1, but Edwards said that he could not. This then leads to the following revised conjecture:

Achromatic-Pseudoachromatic Tree Conjecture, 2004. For any tree T, $\psi(T) \le \psi_s(T) \le \psi(T) + 1$.

In [10], Edwards notes that for almost all trees, $\psi(T) = \psi_s(T)$, and says that Cairns [3] calculated the values of $\psi(T)$ and $\psi_s(T)$ for all trees of order $n \le 15$ and showed that for all of these trees, $\psi(T) = \psi_s(T)$.

At the end of Edwards' paper he says: "We have no doubt that smaller examples could be found with sufficient effort." Subsequently, my wife Sandee Hedetniemi, our son Jason Hedetniemi, and I constructed the tree of order $n = 23$ in Figure 8.3, for which $\psi_s(T) = \psi(T) + 1$. Notice that this tree has $m = 22$ edges, while the complete graph of order 7, K_7, has 21 edges. The coloring of the vertices of this tree shows that $\psi_s(T) = 7$; clearly there cannot be a pseudoachromatic 8-coloring of this tree, since this would require at least 28 edges. Notice that except for the $1 - 1$ edge between vertices x and y, this is a proper coloring.

It only remains to show that $\psi(T) = 6 < 7$. Consider the three vertices labeled x, y, z in the middle of the tree. In any proper, complete coloring of T with *seven* colors, 21 of the 22 edges must have the 21 required distinct color pairs. Thus, only one edge can have a pair of colors assigned to it that is also assigned to one other edge; call this the *duplicate edge*.

Notice that the two vertices marked x and z cannot receive the same color, else there will be 12 edges having a common color. Assume, without loss of generality, that x is colored 1 and z is colored 2. Since this must be a proper coloring, it follows that vertex y cannot be colored either 1 or 2. Suppose, without loss of generality, that vertex y is colored 3.

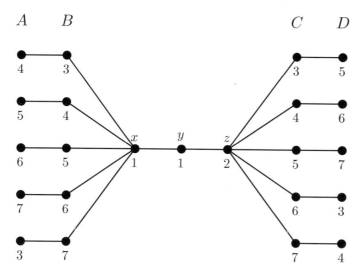

Fig. 8.3 A pseudoachromatic 7-coloring

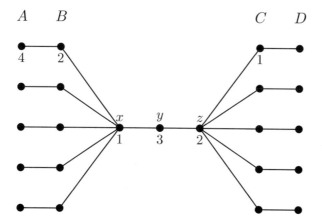

Fig. 8.4 No achromatic 7-coloring

There must be an edge whose two vertices are colored 1 and 2.

Case 1. A vertex in column B is colored 2. In this case there will be *eight* edges, one of whose vertices is colored 2; this will result in at least two duplicate edges. See Figure 8.4.

Case 2. A vertex in column C is colored 1. In this case there will be *eight* edges, one of whose vertices is colored 1; this will result in at least two duplicate edges.

We now know that a smallest tree for which $\psi_s(T) = \psi(T) + 1$ has order n, where $16 \leq n \leq 23$. The smallest order of a tree for which $\psi_s(T) = \psi(T) + 1$ remains unknown.

8.9 Iterated Coloring Problems and the Four-Color Theorem - 2004

In [30] Hedetniemi et al. defined and studied iterated colorings of graphs. These colorings, or vertex partitions, are based on the following greedy algorithm, which repeatedly removes from a graph G sets of vertices having some specified property $\mathcal{P}$, until no vertices are left.

Iterated Coloring Algorithm (ICA)
Input: graph $G = (V, E)$, property $\mathcal{P}$
Output: $\mathcal{P}$-coloring $V = \{V_1, V_2, \ldots, V_k\}$
$i = 0$:
while (V is not empty) {
  find an arbitrary $\mathcal{P}$-set S in $G[V]$;
  $i{+}{+}$;
  $V_i = S$;
  $V = V - S$;
  }
$k = i$;

It can be seen that if $\mathcal{P}$ is the property of being a maximal independent set, then the partition $V = \{V_1, V_2, \ldots, V_k\}$ produced by Algorithm ICA is a proper coloring. Define $i^*(G)$ and β_0^* to equal the minimum and maximum orders of a vertex partition produced by executing Algorithm ICA on a graph G. The following is easy to see, where $\Gamma r(G)$, the *Grundy number* of G, equals the maximum order of a partition $V = \{V_1, V_2, \ldots, V_k\}$, such that for every $1 \le i < j \le k$, every vertex in V_j is adjacent to at least one vertex in V_i. The Grundy number was first defined and studied by Christen and Selkow in 1982 [5].

Proposition 1. *For any graph G, $i^*(G) = \chi(G) \le \beta_0^*(G) = \Gamma r(G)$.*

Similarly, define $\gamma^*(G)$ and $\Gamma^*(G)$, the *iterated domination numbers*, to equal the minimum and maximum orders of a partition produced by Algorithm ICA for the property $\mathcal{P}$ of being a minimal dominating set. Thus, the following inequality chain exists, since every maximal independent set is a minimal dominating set.

Proposition 2. *For any graph G, $\gamma^*(G) \le i^*(G) = \chi(G) \le \Gamma r(G) \le \Gamma^*(G)$.*

Thus, the iterated domination number is a lower bound for the chromatic number, and the *upper iterated domination number* is an upper bound for the Grundy number. These two iterated numbers $\gamma^*(G)$ and $\Gamma^*(G)$ have not been studied very much.

We can augment this inequality chain once more, as follows: A set $S \subseteq V$ is called *irredundant* if for every vertex $u \in S$, $N[u] - N[S - \{u\}] \ne \emptyset$, that is, the closed neighborhood $N[u]$ of u contains a vertex that is not contained in the closed neighborhood $N[S - \{u\}]$ of the set S minus the vertex u. Define $ir^*(G)$, the *iterated irredundance number*, to equal the minimum order of the vertex partition produced by Algorithm ICA for the $\mathcal{P}$ property of being a maximal irredundant set.

Proposition 3. *For any graph G, $ir^*(G) \leq \gamma^*(G) \leq i^*(G) = \chi(G)$.*

The famous Four-Color Theorem asserts that for any planar graph G, $\chi(G) \leq 4$. Thus, $\gamma^*(G) \leq \chi(G) \leq 4$. The problem is this: can you prove that for any planar graph G, $\gamma^*(G) \leq 4$, *without appealing to the Four-Color Theorem?*

Similarly, since $ir^*(G) \leq \gamma^*(G) \leq \chi(G)$, it should be even easier to prove that for any planar graph G, $ir^*(G) \leq 4$, *without appealing to the Four-Color Theorem.*

We can take this progression from $\chi(G)$ to $\gamma^*(G)$ to $ir^*(G)$ one more step. Define the *irratic number irr(G)* to equal the minimum order of a vertex partition $V = \{V_1, V_2, \ldots, V_k\}$ produced by Algorithm ICA for the property of being an irredundant set (but not necessarily a maximal irredundant set). It follows by definition that

$$irr(G) \leq ir^*(G) \leq \gamma^*(G) \leq \chi(G).$$

Iterated Coloring Problems. Can you prove that for any planar graph G, $\gamma^*(G) \leq 4$, $ir^*(G) \leq 4$, or $irr(G) \leq 4$, *without appealing to the Four-Color Theorem?*

8.10 γ-graph Sequence Problem - 2011

As introduced by Fricke et al. in [16], the γ-graph of a graph $G = (V, E)$ is the graph $G(\gamma) = (V(\gamma), E(\gamma))$, whose vertices correspond one to one with the γ-sets of G, and two γ-sets, say $S_1, S_2 \subseteq V$, form an edge in $E(\gamma)$ if there exists a vertex $v_1 \in S_1$ and a vertex $v_2 \in S_2$ such that (i) v_1 is adjacent to v_2, and (ii) $S_1 = S_2 - \{v_2\} \cup \{v_1\}$ and $S_2 = S_1 - \{v_1\} \cup \{v_2\}$. Stated in other words, imagine placing a token on each of the vertices in a γ-set S_1. If you can move the token on a vertex $v_1 \in S_1$ along an edge to an adjacent vertex v_2 and the resulting set $S_2 = S_1 - \{v_1\} \cup \{v_2\}$ is another γ-set, then there is an edge between S_1 and S_2 in the γ-graph $G(\gamma)$.

Figures 8.5 and 8.6 provide illustrations of the γ-graph of the path P_{10} and the γ-graph of the cycle C_{10}. We assume that the vertices of the path P_{10} are labeled in order $1, 2, 3, \ldots, 10$. Each of the 13 γ-sets of P_{10} is indicated in brackets beside each vertex.

In [7], Connelly, Hutson, and Hedetniemi proved that for every graph H, there is a graph G such that $G(\gamma) \simeq H$, that is, every graph H is the γ-graph of some graph G and, in fact, is the γ-graph of a graph G having at most five more vertices than H.

Consider the process of repeatedly applying the γ-graph construction starting from a given graph G, that is, $G \xrightarrow{\gamma} G(\gamma) \xrightarrow{\gamma} G(\gamma)(\gamma)$, etc. We do not know much about the nature of these sequences, but we have noticed that often the sequence ends with K_1. The following examples of this process are given in [16]:

1. $K_{1,n} \xrightarrow{\gamma} K_1$.
2. $C_{3k} \xrightarrow{\gamma} \overline{K}_3 \xrightarrow{\gamma} K_1$.

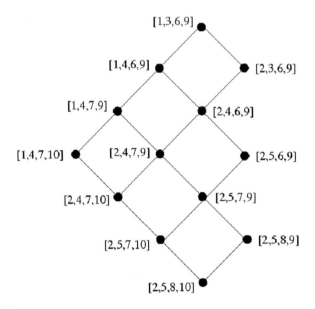

Fig. 8.5 $P_{10}(\gamma)$

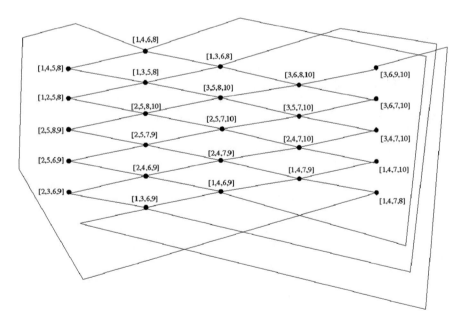

Fig. 8.6 $C_{10}(\gamma)$

3. $K_n \xrightarrow{\gamma} K_n$.
4. $C_{3k+2} \xrightarrow{\gamma} C_{3k+2}$.
5. $P_4 \xrightarrow{\gamma} C_4 \simeq P_2 \Box P_2 \xrightarrow{\gamma} K_{2,4} \xrightarrow{\gamma} K_{1,8} \xrightarrow{\gamma} K_1$.
6. $P_2 \Box P_3 \xrightarrow{\gamma} \overline{K_3} \xrightarrow{\gamma} K_1$.
7. $P_3 \Box P_3 \xrightarrow{\gamma} C_8 \cup 2K_1 \xrightarrow{\gamma} C_8 \xrightarrow{\gamma} C_8$.
8. $P_2 \Box P_{2k+1} \xrightarrow{\gamma} \overline{K_2} \xrightarrow{\gamma} K_1$.
9. $P_2 \Box P_6 \xrightarrow{\gamma} 4P_3 \cup 5K_1 \xrightarrow{\gamma} K_1$.
10. $P_{3k+2} \xrightarrow{\gamma} P_{3k} \xrightarrow{\gamma} K_1$.

Although all of the γ-graph sequences we have seen, so far, have terminated after a small number of steps, for some graphs this sequence does not terminate. Here is an example given in [16]:

$$C_3 \Box P_2 \xrightarrow{\gamma} C_3 \Box C_3 \xrightarrow{\gamma} C_3 \Box C_3 \Box C_3 \xrightarrow{\gamma} \dots$$

This leads us to ask the following question:

γ-graph Sequence Problem. Does there exist a graph G whose γ-graph sequence is a cycle of length greater than one?

Here is a simpler version of this problem.

γ-graph cycle of length two. Do there exist two distinct graphs G and H such that the γ-graph of G is H, i.e., $G(\gamma) \simeq H$, and the γ-graph of H is G, i.e., $H(\gamma) \simeq G$?

8.11   Why Can't We Solve Problems Like These?

Conjectures and problems like those above, and indeed the hundreds more in this volume, lead us to wonder why so many of them survive our attempts to settle them, some for more than 50 years. Having been involved in research for more than 50 years permits me to offer some thoughts:

1. The inherent difficulty of combinatorial or discrete mathematics. In 1956 the famous mathematician John von Neumann published an article entitled *The General and Logical Theory of Automata* [48]. In discussing "The Limitation Which Is Due to the Lack of a Logical Theory of Automata, he said: "*We are very far from possessing a theory of automata which deserves that name, that is, a properly mathematical-logical theory. There exists today a very elaborate system of formal logic, and, specifically, of logic as applied to mathematics. This is a discipline with many good sides, but also with certain serious weaknesses. This is not the occasion to enlarge upon the good sides, which I have certainly no intention to belittle. About the inadequacies, however, this may be said: Everybody who has worked in formal logic will confirm that it is one of the technically most refractory parts of mathematics. The reason for this is that it deals with rigid, all-or-none concepts, and has very little contact with the con-*

tinuous concept of the real or of the complex number, that is, with mathematical analysis. Yet analysis is the technically most successful and best-elaborated part of mathematics. Thus formal logic is, by the nature of its approach, cut off from the best cultivated portions of mathematics, and forced onto the most difficult part of the mathematical terrain, into combinatorics."

2. Important connections to other branches of knowledge have not been made that would greatly assist in developing some proofs. It has been widely recognized that researchers come to the field of graph theory from many different branches of mathematics, engineering, and science. Each of them sees problems in graph theory from different problem modeling and problem-solving perspectives. Such collaborations should be encouraged if we are to solve more of graph theory's conjectures and open problems.

3. Gödel's famous incompleteness theorem suggests that certain theorems exist that cannot be proved. Perhaps, therefore, we will have to accept the fact that some of these conjectures and open problems cannot be solved.

4. Many conjectures are of the form that some condition holds for all values of n greater than or equal to some starting value. But what if counterexamples exist for some very large values of n? How do we find these counterexamples? In one case that I recall, a conjecture was made that for all graphs, one graph parameter was always less than or equal to a second parameter. Several years later, a researcher found a counterexample consisting of a graph having some 2,500,000 vertices! But one of the referees of this paper took a look at this construction and was able to reduce the size of a counterexample to around 1,800,000 vertices! But smallest counterexamples this large pale in comparison to other examples of large, smallest counterexamples; just do a Google search on "Large counterexamples."

5. Graph theorists almost always construct relatively short proofs. However, some results can only be obtained by long proofs, say 100 pages long or longer. Very, very few mathematicians are able to construct such proofs. Ron Graham once told me that he had in his desk four theorems, each of which required more than 100 pages for him to prove.

6. As a general rule, mathematicians and their collaborators lack the available time, the necessary persistence, the extended concentration, and the maintained focus necessary to solve truly difficult problems.

7. It seems to require world-class expertise in order to solve some of the most difficult problems. Not that many mathematicians who attempt to solve these problems have achieved this level of expertise. In the recent book entitled *Outliers, The Story of Success*, the author, Malcolm Gladwell [19], discusses Chapter 2, The 10,000 Hour Rule. By this he meant the amount of time it takes to practice, in order to become world class in a given profession. Gladwell presents the following quote from an article by neurologist Daniel Levitin, *"the emerging picture from such studies is that ten thousand hours of practice is required to achieve the level of mastery associated with being a world-class expert - in anything. . . . no one has yet found a case in which true world-class expertise was accomplished in less time. It seems that it takes the brain this long to assimilate all that it needs to know to achieve true mastery"* [36].

Given this, consider the case of a young PhD student in mathematics who becomes an assistant professor. If such a person spends 10 hours every week of the year for about 20 years dutifully practicing mathematics, they will finally have met the 10,000-hour rule. In my experience, relatively few mathematicians reach this 10,000-hour plateau.

References

1. Borodin, O.V.: On decomposition into degenerate subgraphs. Diskret. Analiz. **28**, 3–11 (1976)
2. Brešar, B., Dorbec, P., Goddard, W., Hartnell, B.L., Henning, M.A., Klavžar, S., Rall, D.F.: Vizing's conjecture: a survey and recent results. J. Graph Theory **69**(1), 46–76 (2012)
3. Cairns, M.: some graph theory problems solved with C++. M.Sc. Dissertation, University of Dundee (1998)
4. Chartrand, G., Geller, D.P., Hedetniemi, S.: A generalization of the chromatic number. Proc. Camb. Philos. Soc. **64**, 265–271 (1968)
5. Christen, C.A., Selkow, S.M.: Some perfect coloring properties of graphs. J. Comb. Theory Ser. B **27**(1), 49–59 (1979)
6. Cockayne, E.J., Dawes, R.M., Hedetniemi, S.T.: Total domination in graphs. Networks **10**, 211–219 (1980)
7. Connelly, E., Hutson, K., Hedetniemi, S.T.: A note on γ-graphs. AKCE Int. J. Graphs Comb. **8**(1), 23–31 (2011)
8. Dunbar, J.E., Frick, M.: The Path Partition Conjecture is true for claw-free graphs. Discret. Math. **307**(11–12), 1285–1290 (2007)
9. Dunbar, J.E., Harris, F.C. Jr., Hedetniemi, S.M., Hedetniemi, S.T., McRae, A.A., Laskar, R.C.: Nearly perfect sets in graphs. Discret. Math. **138**, 229–246 (1995)
10. Edwards, K.J.: Achromatic number versus pseudoachromatic number: a counterexample to a conjecture of Hedetniemi. Discret. Math. **219**(1–3), 271–274 (2000)
11. El-Zahar, M., Sauer, N.: The chromatic number of the product of two 4-chromatic graphs is 4. Combinatorica **5**, 121–126 (1985)
12. Fernau, H.: Minimum dominating set of queens: a trivial programming exercise? Discret. Appl. Math. **158**(4), 308–318 (2010)
13. Frendrup, A., Henning, M.A., Randerath, B., Vestergaard, P.D.: On a conjecture about inverse domination in graphs. Ars Comb. **97A**, 129–143 (2010)
14. Frick, M.: A survey of the path partition conjecture. Discuss. Math. Graph Theory **33**(1), 117–131 (2013)
15. Fricke, G.H., Hedetniemi, S.M., Hedetniemi, S.T., McRae, A.A., Wallis, C.K., Jacobson, M.S., Martin, W.W., Weakley, W.D.: Combinatorial problems on chessboards: a brief survey. Graph Theory Comb. Appl. **I**, 507–528 (1995)
16. Fricke, G.H., Hedetniemi, S.M., Hedetniemi, S.T., Hutson, K.R.: γ-graphs of graphs. Discuss. Math. Graph Theory **31**, 517–531 (2011)
17. Gallai, T.: Über extreme Punkt- und Kantenmengen. Ann. Univ. Sci. Bp. Eotvos Sect. Math. **2**, 133–138 (1959)
18. Garey, M.R., Johnson, D.S.: Computers and Intractability, A Guide to the Theory of NP-completeness. Freeman, New York (1979)
19. Gladwell, M.: Outliers, The Story of Success. Back Bay Books/Little, Brown and Co. (2008)
20. Goncalves, D., Pinlou, A., Rao, M., Thomasse, S.: The domination number of grids. SIAM J. Discret. Math. **25**(3), 1443–1453 (2011)
21. Graham, R.L.: Problem 16. In: Guy, R. et al. (eds.) Combinatorial Structures and Their Applications. Proceedings of the Calgary International Conference on Combinatorial Structures and Their Applications (Calgary, Alberta, 1969), pp. 499–500. Gordon and Breach, New York (1970)

22. Gupta, R.P.: Bounds on the chromatic and achromatic numbers of complementary graphs. In: Tutte, W.T. (ed.) Recent Progress in Combinatorics, Proceeding of the Third Waterloo Conference on Combinatorics, Waterloo, 1968, pp. 229–235. Academic, New York (1969)

23. Gyárfás, A., Lehel, J.: Packing trees of different order into K_n. In: Colloquia Mathematica Societatis János Bolyai 18 Combinatorics. Proceedings of the Fifth Hungarian Colloquium, Keszthely, 1976, pp. 463–469. North-Holland, Amsterdam (1978)

24. Hajós, G.: Über eine Konstruktion nicht n-farberer Graphen. Z. Martin Luther Univ. Halle-Wittenberg **10**, 116–117 (1961)

25. Harary, F.: Graph Theory. Addison-Wesley, Reading (1969)

26. Harary, F., Hedetniemi, S., Prins, G.: An interpolation theorem for graphical homomorphisms. Port. Math. **26**, 453–462 (1967)

27. Hartnell, B.L., Rall, D.F.: Domination in cartesian products: Vizing's conjecture, Chap. 7. In: Haynes, T.W., et al. (eds.) Domination Theory, Advanced Topics, pp. 163–189. Marcel Dekker, New York (1998)

28. Hedetniemi, S.T.: Homomorphisms of graphs and automata. Ph.D. thesis, Technical Report 03105-44-T, University of Michigan (1966)

29. Hedetniemi, S.M., Hedetniemi, S.T., Slater, P.J.: A note on packing two trees into K_n. Ars Comb. **11**, 149–153 (1981)

30. Hedetniemi, S.M., Hedetniemi, S.T., McRae, A.A., Parks, D.A., Telle, J.A.: Iterated colorings of graphs. Discret. Math. **278**, 81–104 (2004)

31. Hobbs, A.M., Bourgeois, B.A., Kasiraj, J.: Packing trees in complete graphs. Discret. Math. **67**(1), 27–42 (1987)

32. Jensen, T.R., Toft, B.: Graph Coloring Problems. Wiley, New York (1995)

33. Krohn, K., Kenneth, R., Mateosian, R., Rhodes, J.: Complexity of ideals in finite semigroups and finite-state machines. Math. Syst. Theory **1**, 59–66 (1967)

34. Kulli, V.R., Sigarkanti, S.C.: Inverse domination in graphs. Natl. Acad. Sci. Lett. **14**, 473–475 (1991)

35. Laborde, J.M., Payan, C., Xuong, N.H.: Independent sets and longest directed paths in digraphs in graphs and other combinatorial topics. Teubner-Texte Math. **59**, 173–177 (1983)

36. Levitin, D.J.: This Is Your Brain on Music: the Science of a Human Obsession, p. 3. Cambridge University Press, Cambridge (1999)

37. Lovász, L.: On decomposition of graphs. Stud. Sci. Math. Hung. **1**, 237–238 (1966)

38. McRae, A.A.: Generalizing NP-completeness proofs for bipartite and chordal graphs. Ph.D. thesis, Department of Computer Science, Clemson University (1994)

39. Ore, O.: Theory of Graphs. American Mathematical Society Colloquium Publications, vol. 38. American Mathematical Society, Providence (1962)

40. Östergard, P.R.J., Weakley, W.D.: Values of the domination numbers of the Queen's graph. Electron. J. Comb. **8**(R29), 19 (2001)

41. Ringel, G.: Färbungsprobleme auf Flächen und Graphen. VEB Deutscher Verlag der Wissenschaften, Berlin (1959)

42. Roditty, Y.: Packing and covering of the complete graph. III. On the tree packing conjecture. Sci. Ser. A Math. Sci. (N.S.) **1**, 81–85 (1988)

43. Sauer, N.: Hedetniemi's conjecture – a survey. Discret. Math. **229**, 261–292 (2001)

44. Sinko, A., Slater, P.J.: Queen's domination using border square and (A, B)-restricted domination. Discret. Math. **308**(20), 4822–4828 (2008)

45. Slater, P.J., Teo, S.K., Yap, H.P.: Packing a tree with a graph of the same size. J. Graph Theory **9**(2), 213–216 (1985)

46. Tardif, C.: Hedetniemi's conjecture 40 years later. Graph Theory Notes N. Y. **LIV**, 46–57 (2008)

47. Vizing, V.G.: The Cartesian product of graphs. Vycisl. Sist. **9**, 30–43 (1963)

48. von Neumann, J.: The general and logical theory of automata. In: Newman, J.R. (ed.) The World of Mathematics, vol. 4, pp. 2070–2098. Simon and Schuster, New York (1956)

49. Zhu, X.: A survey on Hedetniemi's conjecture. Taiwan. J. Math. **2**, 1–24 (1998)

Chapter 9
Chvátal's t_0-Tough Conjecture

Linda Lesniak

Abstract In 1973, Chvátal introduced the concept of "tough graphs" and conjectured that graphs with sufficiently high toughness are hamiltonian. Here we look at some personal perspectives of this conjecture, both those of Chvátal and the author. Furthermore, we present the history of the conjecture and its current status.

9.1 Reminiscences by Vašek Chvátal (October 26, 2014)

The reader who is unfamiliar with the concept of toughness or Chvátal's toughness conjectures is encouraged to read Section 9.2 first.

Chvátal's reminiscences (V. Chvátal, 2014, private communication):

It was a lucky combination of events during my post doctoral year that lead me to the conjecture.

In the fall of 1970, I defended my doctoral thesis. My advisor Crispin Nash-Williams had been interested in hamiltonian graphs among other things; in particular, he coined the term 'forcibly hamiltonian' to mean a sequence such that a graph with this degree sequence must contain a Hamilton cycle. However, rather than steering me toward his own research agenda, he had graciously given me free hand in choosing the topics around which I built the thesis. I remained uninterested in hamiltonian graphs at that time.

This changed when Paul Erdös passed through Waterloo around the time of my defence. In his lecture he mentioned his new generalization of Turán's theorem: If a graph contains no clique on more than r vertices, then it is degree-majorized by a graph containing no such clique for the obvious reason that it is r-partite. I was bowled over by the beauty of Erdös's theorem and wondered at once how its paradigm could be applied to forcibly hamiltonian sequences. Several weeks later, I found the answer: If a graph contains no Hamilton cycle, then it is degree-majorized by a graph containing no such cycle for the obvious reason that the removal of some cut set of k vertices breaks the graph into more than k components. I decided to call graphs without such cut sets 'tough'.

During the same year, Jack Edmonds treated me to private lectures revolving around his notions of good algorithms and good characterizations, notions that he had been promoting since the early 1960's and that later became known as the notions of classes **P** and **NP**. In particular, he expounded his view of theorems asserting that validity of some predicate A is a

L. Lesniak (✉)
Western Michigan University, Kalamazoo, MI, USA
e-mail: lindalesniak@gmail.com

© Springer International Publishing Switzerland 2016
R. Gera et al. (eds.), *Graph Theory*, Problem Books in Mathematics,
DOI 10.1007/978-3-319-31940-7_9

sufficient condition for the validity of another predicate B. Traditionally, such theorems are presented as the implication '$A \rightarrow B$', but Jack preferred to present them as the disjunction '*notA* or B' and he particularly liked these disjunctions when both of their predicates, *notA* and B, had easily verifiable certificates of validity. These tutorials made a lasting impression on me.

Then Herbert Fleischner came from Binghamton to lecture about his fresh proof of the well-known conjecture made independently by Beineke, Nash-Williams, and Plummer several years earlier: The square of every two-connected graph is hamiltonian. I was much impressed by this feat, as was everyone else in the audience. Still, as a recent convert to Jack's doctrine, I could not help noticing that Herbert's theorem did not conform to Jack's paradigm: I did not know any easily verifiable certificate of the predicate "G is not the square of a two-connected graph". This observation made me look for properties of squares of two-connected graphs whose conjunction might make a graph hamiltonian and whose lack had an easily verifiable certificate.

As a warm-up, I set up to prove without recourse to Herbert's theorem that the square of every two-connected graph is tough. This exercise taught me that a stronger statement was true: in order to break the square of a two-connected graph into k components, one has to remove not just k, but at least $2k$ vertices. In turn, the stronger statement suggested a parametric generalization of toughness.

I had known that tough graphs were not necessarily hamiltonian, but I could not find any 2-tough nonhamiltonian graphs; in fact, the largest t for which I could find a t-tough nonhamiltonian graph was $3/2$. So I conjectured that for some t, every t-tough graph is hamiltonian. To provoke an interest in this conjecture, I stated shamelessly that it might be valid for every t greater than $3/2$ and I pointed out that its validity for $t = 2$ would imply Fleischner's Theorem.

It goes without saying that the attention these conjectures received over the last forty years has gratified me very much.

9.2   History of the 2-Tough Conjecture

In his seminal 1973 paper on toughness, Chvátal [12] introduced a new variant for graphs. "It measures in a simple way how tightly various pieces of a graph hold together; therefore, we shall call it toughness. Our central point is to indicate the importance of toughness for the existence of hamiltonian circuits in a graph."

Let $k(G)$ denote the number of components in a graph G. If G is hamiltonian, then clearly $k(G - S) \leq |S|$ for every nonempty proper subset S of $V(G)$. Equivalently, if G is hamiltonian, then

$$\frac{|S|}{k(G - S)} \geq 1$$

for every nonempty proper subset S of $V(G)$.

Chvátal [12] defined a noncomplete graph G to be *t-tough* if

$$\frac{|S|}{k(G - S)} \geq t$$

for every subset S of $V(G)$ with $k(G - S) > 1$. Then the *toughness* $t(G)$ of G is the maximum t for which G is t-tough. For $n \geq 1$, he set $t(K_n) = +\infty$. For example, suppose G is obtained from the complete graph K_n of order $n \geq 3$ by removing an edge $e = uv$. Then $G = K_n - e$, and the only subset of S of G for which $k(G - S) > 1$ is $S = V(G) - \{u, v\}$. Here, $|S| = n - 2$ and $k(G - S) = 2$. Thus,

$$t(G) = \frac{|S|}{k(G - S)} = \frac{n - 2}{2}.$$

Let $\alpha(G)$ denote the (vertex) independence number of a graph G, and let $\kappa(G)$ denote the (vertex) connectivity of G. Chvátal [12] established upper and lower bounds for $t(G)$ in terms of $\alpha(G)$, $\kappa(G)$, and $|V(G)|$.

If G is not complete and S is any subset of $V(G)$ for which $k(G - S) > 1$, then $|S| \geq \kappa(G)$ and $k(G - S) \leq \alpha(G)$. Consequently, G is $(\kappa(G)/\alpha(G))$-tough. It follows, then, from the definition of toughness that

$$t(G) \geq \frac{\kappa(G)}{\alpha(G)}. \tag{9.1}$$

For the upper bounds for $t(G)$, consider first a smallest subset S for which $k(G - S) > 1$. Then $|S| = \kappa(G)$ and $k(G - S) \geq 2$, and so

$$t(G) \leq \frac{\kappa(G)}{2}. \tag{9.2}$$

Similarly, taking S to be the complement of an independent set of $\alpha(G)$ vertices, we obtain

$$t(G) \leq \frac{|V(G)| - \alpha(G)}{\alpha(G)}. \tag{9.3}$$

If G is the complete bipartite graph $K_{m,n}$ with $2 \leq m \leq n$, we have $\kappa(G) = m$, $\alpha(G) = n$, and $|V(G)| = m + n$. Combining the lower bound in (9.1) with the upper bound in (9.3), we see that for $2 \leq m \leq n$, $t(K_{m,n}) = m/n$. Thus, the bounds in (9.1) and (9.3) are sharp.

Recall that a graph is called claw-free if it contains no induced bipartite graph $K_{1,3}$. About ten years after Chvátal obtained his bounds, Matthews and Sumner [20] showed that Chvátal's upper bound $t(G) \leq \kappa(G)/2$ was attained for claw-free graphs.

These results for $t(G)$ in terms of $\alpha(G)$ and $\kappa(G)$ are summarized below.

Theorem 1. *For every noncomplete graph G,*

$$\frac{\kappa(G)}{\alpha(G)} \leq t(G) \leq \frac{\kappa(G)}{2}.$$

These bounds for $t(G)$ are sharp. Furthermore, if G is claw-free, then $t(G) = \frac{\kappa(G)}{2}$.

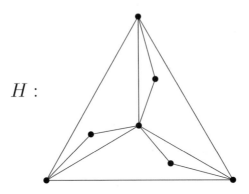

H :

Fig. 9.1 A 1-tough nonhamiltonian graph H

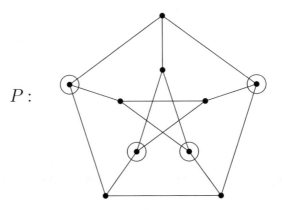

P :

Fig. 9.2 The Petersen graph P has $t(P) \leq \frac{4}{3}$

As we have seen, every hamiltonian graph G is 1-tough since

$$\frac{|S|}{k(G-S)} \geq 1$$

for every nonempty proper subset S of $V(G)$. However, as noted by Chvátal, the converse only holds for graphs with at most six vertices. The graph H in Figure 9.1 was given by Chvátal as an example of a 1-tough nonhamiltonian graph of order 7.

Another example of a 1-tough nonhamiltonian graph is the Petersen graph P. In fact $t(P) = 4/3$. The Petersen graph is shown in Figure 9.2. The set S of four highlighted vertices satisfies $|S|/k(G-S) = 4/3$.

We know that high connectivity in a graph does not imply hamiltonicity. Consider, for example, the complete bipartite graph $K_{m,n}$, with $m < n$. Chvátal's hope was that large enough toughness in a graph G would, in fact, guarantee that G is hamiltonian. In [12], he proposed the following conjecture.

Conjecture 2 (Chvátal's t_0-tough conjecture). *There exists t_0 such that every t_0-tough graph is hamiltonian.*

In [12], Chvátal also constructed an infinite family of $(3/2)$-tough nonhamiltonian graphs. Consequently, he gave the following strengthening of Conjecture 2.

Conjecture 3. *Every t-tough graph with $t > 3/2$ is hamiltonian.*

However, Thomassen (see [5]) showed the existence of infinitely many nonhamiltonian graphs G with $t(G) > 3/2$. Thus, Conjecture 3 was disproved although the original t_0-tough conjecture remained open.

A *k-factor* of a graph G is a k-regular spanning subgraph of G. In particular, every hamiltonian graph has a 2-factor. Since the Petersen graph has a 2-factor but is not hamiltonian, these two concepts are not equivalent.

In [12], Chvátal proposed the following conjecture relating toughness and the existence of k-factors.

Conjecture 4. *Let G be a k-tough graph of order $n \geq k + 1$ and kn even. Then G has a k-factor.*

Conjecture 4 was established in 1985 by Enomoto, Jackson, Katerinis, and Saito [14] and shown to be best possible.

Theorem 5. *Conjecture 4 is true. Furthermore, given $k \geq 1$ and any $\epsilon > 0$, there exists a $(k - \epsilon)$-tough graph of order n with $n \geq k + 1$ and kn even with no k-factor.*

Corollary 6. *Every 2-tough graph has a 2-factor. Furthermore, for any $\epsilon > 0$, there exist infinitely many $(2 - \epsilon)$-tough graphs with no 2-factor.*

Of course, Corollary 6 implies that the smallest possible t_0 for which Chvátal's t_0-tough conjecture is true is $t_0 = 2$. This version became well-known as Chvátal's 2-tough conjecture.

Conjecture 7 (Chvátal's 2-tough conjecture). *Every 2-tough graph is hamiltonian.*

9.3 The 2-Tough Conjecture

Chvátal's 2-tough conjecture was exciting for many reasons. In [12], Chvátal showed that $t(G^2) \geq \kappa(G)$ for every graph G. In particular, if G is a 2-connected graph, then $t(G^2) \geq 2$. Consequently, the truth of the 2-tough conjecture would imply Fleischner's [15] beautiful 1974 result that the square of every 2-connected graph is hamiltonian. Establishing the 2-tough conjecture would also imply the validity of two conjectures made independently in the 1980s. The first, Conjecture 8, is due to Matthews and Sumner [20].

Conjecture 8. *Every 4-connected claw-free graph is hamiltonian.*

The second, Conjecture 9, is due to Thomassen [22].

Conjecture 9. *Every* 4-*connected line graph is hamiltonian.*

Although Conjecture 8 appears stronger than Conjecture 9 (every line graph is claw-free), in 1997, Ryjáček [21] showed that they are equivalent. In order to do so, he described a closure concept for claw-free graphs. This closure was based on adding edges to a graph in such a way that hamiltonicity was not introduced in a nonhamiltonian graph. The addition of edges, of course, always preserves hamiltonicity.

Let G be a claw-free graph. For any vertex v of G, consider the subgraph $G[N(v)]$ of G induced by the neighborhood $N(v)$ of v. If $G[N(v)]$ is connected and not complete, then all edges are added so that, in the resulting graph, the subgraph induced by $N(v)$ is complete. This procedure is repeated until it is impossible to add more edges.

In [21], Ryjáček showed that the closure $cl(G)$ that we obtain this way is well-defined with several important properties.

Theorem 10. *Let G be a claw-free graph. Then,*

1. the closure $cl(G)$ is uniquely determined,
2. $cl(G)$ is hamiltonian if and only if G is hamiltonian,
3. $cl(G)$ is the line graph of a triangle-free graph.

It follows from Theorem 10 that Conjectures 8 and 9 are equivalent. Furthermore, the closure $cl(G)$ of a claw-free graph has come to be called the "Ryjáček closure."

This result of Ryjáček has an interesting story. The story is in some ways indicative of the activity surrounding the 2-tough conjecture in the 25 years following the introduction of the toughness parameter.

The Euler Institute for Discrete Mathematics and its Applications (EIDMA) was a research school based in Eindhoven, the Netherlands, but with participating groups in other cities in the Netherlands as well as in Belgium and Germany. The group in Enschede consisted of faculty at the University of Twente and specialized in graph theory and its applications.

In 1995, this group organized a workshop funded by the Dutch government on the hamiltonicity of 2-tough graphs and related problems. The workshop was held November 19–24, 1995 at the hotel Hölterhof near Enschede. About 20 participants from around the world were invited. I felt honored to be one of them. The following is a quote from the introduction in the technical report on the *EIDMA Workshop on Hamiltonicity of 2-Tough Graphs* [7].

> As the central problem, the 2-tough conjecture was chosen, stating that every 2-tough graph is hamiltonian. If true, the conjecture would imply that 4-connected claw-free graphs are hamiltonian, which in turn would imply that 4-connected line graphs are hamiltonian. We will refer to these conjectures as those of Chvátal, Matthews and Sumner, and Thomassen, respectively, although Chvátal's original conjecture (since disproved by Thomassen) was that t-tough graphs with $t > 3/2$ are hamiltonian, along with the weaker condition (still open) that there exists some t such that every t-tough graph is hamiltonian. His conjectures seem to stem from an analysis of Fleischner's result that squares of 2-connected graphs are hamiltonian.

On Monday three presentations were given by Bill Jackson (London), Jan van den Heuval (Oxford) and Stephen Brandt (Berlin) on respectively claw-free and line graphs, toughness results and eigenvalues. These three topics also roughly give the layout of the material in this report. Mentioned will be results, conjectures, problems, examples, counterexamples and remarks without going too much into details.

On the other days plenary sessions were held in the morning till 10.15 and in the afternoon from 16.00 on. In between and in the evenings the meeting developed into a real workshop. The most important things that happened were the following. Herbert Fleischner suggested on Monday the equivalence of the two conjectures of Matthew and Sumner and Thomassen. During the conference walk through the moors on Tuesday Zdeněk Ryjáček (Pilsen) thought about applying his techniques to this problem and on Wednesday presented his proof to the group, which caused quite some excitement. In the evening the proof was checked by a group of participants. Various consequences were discussed among which a new closure concept called, for obvious reasons, the Ryjáček closure.

On a personal note, the EIDMA workshop on the Hamiltonicity of 2-Tough Graphs was a mathematical experience I will always remember fondly. The location, the hotel Hölterhof, consisted of a beautiful former hunting lodge, where the workshop was held, and small cabins, where the participants stayed. The surrounding woods made this remote location a perfect spot for research and collegiality. Our primary contact with the outside world (remember, these were pre-cell and pre-Wi-Fi days) was the *International Herald Tribune* that my husband bought each day by bicycling the four miles into town. And for the Americans at the workshop, the organizers arranged for a traditional Thanksgiving dinner complete with turkey and all the trimmings.

Returning to the last paragraph from the introduction in the technical report, the following quote is an interesting reflection by the conference organizers J. A. Bondy, H. J. Broersma, C. Hoede, and H. J. Veldman.

> Although the main problem, Chvátal's conjecture, was hardly considered – most participants think the conjecture is false anyway – the outcomes were quite satisfactory. New approaches, such as the one proposed by Brandt, were considered, but the exclamation by Heinz Jung (Berlin) ("Nothing seems to hold!") seems appropriate for the toughness concept. In fact people wondered whether the concept was all that fundamental. The reader may make up his own opinion by reading this report.

My comments? It is worth your time to read the report!

In 2000, about four years after the workshop, three of the participants, D. Bauer, H. J. Broersma, and H. J. Veldman [3], produced the graph G of Figure 9.3, which is the join of F and K_2. They showed that G is 2-tough but not hamiltonian.

To see that G is not hamiltonian, assume to the contrary that C is a hamiltonian cycle of G. At least one of the five pairs $\{s_i, t_i\}$ of vertices, $1 \leq i \leq 5$, contains no vertex that is adjacent to either p or q on C, say $\{s_1, t_1\}$. Then, $s_1u_1, s_1w_1, t_1v_1, t_1x_1 \in E(C)$. Necessarily, $u_1w_1, u_1x_1, v_1x_1, v_1w_1 \notin E(C)$. This, however, implies that $v_1y_1, w_1y_1, u_1z_1, x_1z_1 \in E(C)$, which is impossible.

To see that G is 2-tough, the interested reader is referred to [3].

Furthermore, Bauer, Broersma, and Veldman [3] showed that for every $\epsilon > 0$, there exists a $(9/4 - \epsilon)$-tough graph with no hamiltonian path. It follows that if Chvátal's t_0-tough conjecture is true, that is, there exists t_0 such that every t_0-graph is hamiltonian, then $t_0 \geq 9/4$.

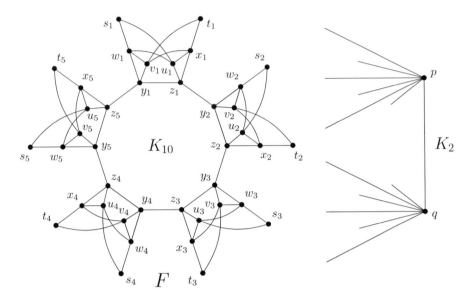

Fig. 9.3 The Bauer-Broersma-Veldman graph, a 2-tough nonhamiltonian graph

The t_0-tough conjecture is still open. However, we now know that it is true for a number of well-studied classes of graphs. We turn to these results in Section 9.4.

9.4   Toughness and Hamiltonicity in Special Classes of Graphs

In this section, we consider sufficient toughness in various classes of graphs that ensures hamiltonicity. Our first such result follows from the well-known theorem of Tutte [23] that 4-connected planar graphs are hamiltonian and the fact that $\kappa(G) \geq 2t(G)$ for every graph G.

Theorem 11. *Let G be a planar graph with $t(G) > 3/2$. Then G is hamiltonian.*

Chvátal's t_0-tough conjecture has been established for many other classes of graphs including, for example, the well-known interval and chordal graphs. The results in this section are presented, with one exception, in chronological order. We also discuss the additional problem of minimizing t_0 once it is known to exist. Since every hamiltonian graph is 1-tough, it follows that in all of these results, $t_0 \geq 1$.

Interval graphs are the intersection graphs of intervals on a line, that is, a graph G is an *interval graph* if there exists a family of intervals that correspond to the vertices of G in such a way that two intervals intersect if and only if the corresponding vertices are adjacent in G. In 1985, Kiel [18] (implicitly) established the following result for interval graphs.

Theorem 12. *Every* 1-*tough interval graph is hamiltonian.*

A graph G is a *chordal graph* if G contains no induced cycle of length four or more. The next result follows from the fact that $t(G) = \frac{\kappa(G)}{2}$ if G is claw-free and the 1986 result of Balikrishnam and Paulraja [1] that 2-connected claw-free chordal graphs are hamiltonian.

Theorem 13. *Every* 1-*tough claw-free chordal graph is hamiltonian.*

A *split graph* is a graph whose vertex set can be partitioned into a clique and an independent set. In 1996, Kratsch, Lehel, and Müller [19] established the minimum toughness guaranteeing hamiltonicity in split graphs.

Theorem 14. *Every* 3/2-*tough split graph is hamiltonian.*

Since Chvátal [12] showed that there exists a $(3/2 - \epsilon)$-tough split graph that is not hamiltonian for every $\epsilon > 0$, Theorem 14 is best possible.

Our next result, due to Deogen, Kratsch, and Steiner [13] in 1997, established sufficient toughness in cocomparability graphs for hamiltonicity. In order to state this result, we first describe the class of cocomparability graphs. A *transitive orientation* of a graph G is an assignment of directions to the edges of G so that the resulting directed graph D is transitive, that is, whenever (u, v) and (v, w) are directed edges of G, then so too is (u, w). A graph that has a transitive orientation is often called a *comparability graph*. A *cocomparability graph* is then a graph whose complement is a comparability graph.

Theorem 15. *Every* 1-*tough cocomparability graph is hamiltonian.*

Theorem 13 addresses sufficient toughness for a claw-free chordal graph to be hamiltonian. The more general problem for chordal graphs was considered by Chen, Jacobson, Kézdy, and Lehel [11] in 1998.

Theorem 16. *Every* 18-*tough chordal graph is hamiltonian.*

It is unlikely that Theorem 16 is best possible. However, Bauer, Broersma, and Veldman [3] showed that for every $\epsilon > 0$, there exists a $(7/4 - \epsilon)$-tough chordal graph with no hamiltonian path. Thus, $t_0 \geq 7/4$ for chordal graphs. In the case of planar chordal graphs, however, Böhme, Harant, and Tkáč [6] established the following 1999 result.

Theorem 17. *Let G be a planar chordal graph with $t(G) > 1$. Then G is hamiltonian*

Böhme, Harant, and Tkáč also showed that for planar chordal graphs, 1-toughness does not ensure hamiltonicity.

Theorems 12, 14, and 16 deal with interval graphs, split graphs, and chordal graphs, respectively. These three classes of graphs have nice characterizations as intersection graphs of connected subgraphs of special classes of trees. A graph G of order n is the *intersection graph* of subgraphs $H_1, \ldots, H_n$ of a graph H if the vertices of G can be put into a one-to-one correspondence with the subgraphs $H_1, \ldots, H_n$ so that two vertices of G are adjacent if and only if the corresponding subgraphs have a common vertex.

It can be shown (see, e.g., [8]) that a graph is an interval graph if and only if it is an intersection graph of subpaths of a path. Similarly, a graph is a split graph if and only if it is an intersection graph of subtrees of a star, that is, a complete bipartite graph $K_{1,t}$. Finally, a graph is a chordal graph if and only if it is an intersection graph of subtrees of a tree. It is clear from these characterizations that all interval graphs and all split graphs are chordal graphs.

In 2007, Kaiser, Král, and Stacho [16] defined a subclass of chordal graphs that is a proper superclass of interval and split graphs. A graph is a *spider* if it is a subdivision of a star. A graph is a *spider graph* if it is an intersection graph of subtrees of a spider. It follows that every interval graph and every split graph is a spider graph, and spider graphs are chordal.

Theorem 18. *Every* $(3/2)$-*tough spider graph is hamiltonian.*

Since every split graph is a spider graph, Theorem 14 is a corollary of Theorem 18, and Theorem 18 is sharp.

In [9], Broersma, Xiong, and Yoshimoto studied another subclass of chordal graphs called k-trees. In order to present their 2007 result, we define k-trees as follows. If the neighborhood of a vertex v in a graph G induces a complete graph of order k, then v is called a k-*simplicial* vertex of G. Let k be a positive integer. The complete graph K_k is the smallest (with respect to order) k-tree, and a graph G of order at least $k + 1$ is a k-*tree* if and only if G contains a k-simplicial vertex v such that $G - v$ is a k-tree. Clearly, 1-trees are just trees.

Theorem 19. *Every* $(k + 1)/3$-*tough* k-*tree is hamiltonian for* $k \geq 2$.

Broersma, Xiong, and Yoshimoto provided infinite classes of nonhamiltonian 1-tough k-trees for each $k \geq 3$. So an open question is to determine the minimum $t_0 = t_0(k)$ for which every t_0-tough k-tree is hamiltonian.

Our next result concerns a superclass of split graphs called $2K_2$-*free graphs*. These are the graphs containing no induced copy of $2K_2$, the graph on four vertices consisting of two independent edges. Clearly, every split graph is $2K_2$-free. Another class of $2K_2$-free graphs consists of the complements of chordal graphs. The 2014 result of Broersma, Patel, and Pyatin [10] gave a sufficient toughness for hamiltonicity in $2K_2$-free graphs.

Theorem 20. *Every* 25-*tough* $2K_2$-*free graph is hamiltonian.*

Since the proof of Theorem 20 relied on the restrictive structure of *triangle-free* $2K_2$-free graphs, the condition of being 25-tough is likely far from best possible.

To close this section, we return to Conjectures 8 and 9. Recall that in 1984, Matthews and Sumner conjectured that every 4-connected claw-free graph is hamiltonian. About the same time, Thomassen [22] conjectured that every 4-connected line graph is hamiltonian. These two conjectures were shown to be equivalent by Ryjáček [21] using the closure concept that came to be known as the Ryjáček closure.

Since every line graph is claw-free and, for a claw-free graph G, we know that $\kappa(G) = 2t(G)$, both Conjectures 8 and 9 can be stated in terms of toughness instead of connectivity. This is true of the remaining results in this section. Since these results were originally presented in terms of connectivity, this is how we shall proceed.

Matthews and Sumner's conjecture, as well as Thomassen's conjecture, remains open. For many years, the best general result related to Thomassen's conjecture was due to Zhan [24] and Jackson (unpublished).

Theorem 21. *Every 7-connected line graph is hamiltonian.*

In fact, the result in [24] showed that every 7-connected line graph G is hamiltonian-connected, that is, for each pair u, v of distinct vertices of G, there is a hamiltonian $u - v$ path in G.

In 2012, Kaiser and Vrána [17] improved Theorem 21 as follows:

Theorem 22. *Every 5-connected line graph with minimum degree at least 6 is hamiltonian.*

Furthermore, using the Ryjáček closure concept, they extended Theorem 22 to claw-free graphs.

Theorem 23. *Every 5-connected claw-free graph with minimum degree 6 is hamiltonian.*

The final result of Kaiser and Vrána [17] included both Theorems 22 and 23 as corollaries.

Theorem 24. *Every 5-connected claw-free graph with minimum degree at least 6 is hamiltonian-connected.*

We close this section by restating Theorem 24 in terms of toughness and an immediate corollary.

Theorem 25. *Every $(5/2)$-tough claw-free graph with minimum degree at least 6 is hamiltonian-connected.*

Corollary 26. *Every 3-tough claw-free graph is hamiltonian.*

9.5 Conclusion

Chvátal's 1973 paper [12] "Tough graphs and hamiltonian circuits" introduced the concept of toughness and posed some intriguing conjectures. As we have seen, the most challenging is still open: there exists t_0 such that every t_0-tough graph is hamiltonian. The conjecture has been shown to be true for many special classes of graphs such as planar graphs, interval graphs, chordal graphs, and claw-free graphs. The most glaring open questions here, in my opinion, are (1) to determine the minimum t_0 that guarantees hamiltonicity in chordal graphs and (2) to determine the minimum t_0 that guarantees hamiltonicity in claw-free graphs. In (1), we know that $7/4 \leq t_0 \leq 18$. In (2), we have $2 \leq t_0 \leq 3$.

But there is more to the toughness story than the t_0-tough conjecture. For example, toughness has been combined with various degree conditions to determine bounds on the length of a longest cycle in a graph. Similar results have been obtained for the existence of factors in a graph. Computational complexity issues have been investigated. For example, it is known that it is **NP**-hard to calculate the toughness of a graph [2]. The beautiful 2006 survey paper [4] "Toughness in Graphs - A Survey" is a must-read for those who are interested in the bigger toughness picture.

Acknowledgment: Many thanks to Shelley Speiss whose many helpful mathematical comments and superb LaTeX'ing truly made this "our" paper. L^2

References

1. Balakrishnan, R., Paulraja, P.: Chordal graphs and some of their derived graphs. Congr. Numer. **53**, 71–74 (1986)
2. Bauer, D., Hakimi, S.L., Schmeichel, E.: Recognizing tough graphs is NP-hard. Discret. Appl. Math. **28**, 191–195 (1990)
3. Bauer, D., Broersma, H.J., Veldman, H.J.: Not every 2-tough graph is Hamiltonian. Discret. Appl. Math. **99**, 317–321 (2000)
4. Bauer, D., Broersma, H., Schmeichel, E.: Toughness in graphs - a survey. Graphs Comb. **22**, 1–35 (2006)
5. Bermond, J.C.: Hamiltonian graphs. In: Beineke, L., Wilson, R.J. (eds.) Selected Topics in Graph Theory, pp. 127–167. Academic, London (1978)
6. Böhme, T., Harant, J., Tkáč, M.: More than one tough chordal planar graphs are Hamiltonian. J. Graph Theory **32**, 405–410 (1999)
7. Bondy, J.A., Broersma, H.J., Hoede, C., Veldman, H.J. (eds.): EIDMA Workshop on Hamiltonicity of 2-tough graphs. Memorandum 1325, University of Twente, Enschede (1996)
8. Brandstädt, A., Le, V., Spinrad, J.: Graph Classes: A Survey. Monographs on Discrete Mathematics and Applications. Society for Industrial and Applied Mathematics, Philadelphia (1999)
9. Broersma, H.J., Xiong, L., Yoshimoto, K.: Toughness and Hamiltonicity in k-trees. Discret. Math. **307**, 832–838 (2007)
10. Broersma, H., Patel, V., Pyatkin, A.: On toughness and hamiltonicity of $2K_2$-free graphs. J. Graph Theory **75**, 244–255 (2014)
11. Chen, G., Jacobson, M.S., Kézdy, A., Lehel, J.: Tough enough chordal graphs are Hamiltonian. Networks **31**, 29–38 (1998)
12. Chvátal, V.: Tough graphs and hamiltonian circuits. Discret. Math. **5**, 215–228 (1973)
13. Deogun, J.S., Kratsch, D., Steiner, G.: 1-Tough cocomparability graphs are hamiltonian. Discret. Math. **170**, 99–106 (1997)
14. Enomoto, H., Jackson, B., Katerinis, P., Saito, A.: Toughness and the existence of factors. J. Graph Theory **9**, 87–95 (1985)
15. Fleischner, H.: The square of every two-connected graph is Hamiltonian. J. Comb. Theory **16B**, 29–34 (1974)
16. Kaiser, T., Král, D., Stacho, L.: Tough spiders. J. Graph Theory **56**, 23–40 (2007)
17. Kaiser, T., Vrána, P.: Hamilton cycles in 5-connected line graphs. Eur. J. Comb. **33**, 924–947 (2012)
18. Kiel, J.: Finding Hamiltonian circuits in interval graphs. Inf. Process. Lett. **20**, 201–206 (1985)
19. Kratsch, D., Lehel, J., Müller, H.: Toughness, hamiltonicity and split graphs. Discret. Math. **150**, 231–245 (1996)

20. Matthews, M., Sumner, D.: Hamiltonian results in $K_{1,3}$-free graphs. J. Graph Theory **8**, 139–146 (1984)
21. Ryjáček, Z.: On a closure concept in claw-free graphs. J. Comb. Theory **70B**, 217–224 (1997)
22. Thomassen, C.: Reflections on graph theory. J. Graph Theory **10**, 309–324 (1986)
23. Tutte, W.T.: A theorem on planar graphs. Trans. Am. Math. Soc. **82**, 99–116 (1956)
24. Zhan, S.: On hamiltonian line graphs and connectivity. Discret. Math. **89**, 89–95 (1991)

Chapter 10
What Do Trees and Hypercubes Have in Common?

Henry Martyn Mulder

Abstract At first sight, trees and hypercubes do not have much in common. They are both connected and bipartite, but these properties are not very distinctive. A closer inspection reveals an interesting common feature. Trees and hypercubes can be constructed using a similar sort of expansion procedure. Now, we can introduce a class of graphs that forms a common generalization of trees and hypercubes: it consists of all those graphs that can be constructed by this expansion procedure. With this generalization in hand, many questions arise. Are there other common properties of trees and hypercubes? And, if so, are these shared by this common generalization? This chapter discusses a very interesting instance of this approach, the case of median graphs.

Mathematics Subject Classification : 05C75, 05C12, 05C05

10.1 Introduction

The following story is historically *incorrect*. It is a work of fiction. Names and characters in this story all exist in the author's imagination, but any resemblance to existing persons is entirely intentional.

The story tells the adventures of the Meta-conjectures that I proposed in 1990. But the origins date back to 1976. Then I had just started as a PhD student in the Math Department at the Vrije Universiteit in Amsterdam. Some time in the spring of 1976, my roommate and fellow PhD student Jan van Mill (yes, the topologist) asked me to work with him on some questions about finite topologies. One of these regarded maximal subbases for a finite topology. A subbase is a family of subsets such that the sets in the basis for the topology of open sets can be obtained by taking intersections of subbase elements. For instance, the open half lines form a subbase

H.M. Mulder (✉)
Econometrisch Instituut, Erasmus Universiteit Rotterdam, P.O. Box 1738,
3000 DR Rotterdam, The Netherlands
e-mail: hmmulder@ese.eur.nl

© Springer International Publishing Switzerland 2016
R. Gera et al. (eds.), *Graph Theory*, Problem Books in Mathematics,
DOI 10.1007/978-3-319-31940-7_10

for the open intervals on the real line, since each open interval is the intersection of two open half lines. Note that, since the complement of an open half line is a closed half line, such a complement is *not* a subbase element. In the finite case, a subbase can contain two complementary sets. The question was what the maximal subbases are that consist of complementary pairs of subsets and satisfy the so-called Helly property: any subfamily of the subbase that consists of pairwise intersecting subbase elements has a nonempty intersection. I called these subbases *maximal Helly copair hypergraphs*.

During the Fifth Hungarian Combinatorial Colloquium in Keszthely, Hungary, June 1976, I took a break and made a walk to see the railway crossing downtown. During that walk, an idea struck me how to construct such a maximal Helly copair hypergraph from a smaller one by an expansion construction. Later, I discussed this with another fellow PhD student Lex Schrijver (yes, the discrete mathematician), and he suggested looking at the underlying graphs. I combined the two ideas and put my result in the technical report series of the Math Department of the Vrije Universiteit [21]. In the spring of 1977, I got rid of the hypergraphs and rewrote the result purely in graph theory terminology [22]. This is the Expansion Theorem below. Although the proof is nontrivial, it is quite natural. The crucial point here is, of course, to get to the idea of expansion. My first proof from Keszthely in hypergraph terminology (without using any graphs) was far from natural and needed thrice as many pages. With hindsight, I am still amazed that I came up with that proof at all; it was nota bene the first theorem that I ever proved in my life. This first proof was never published. I even lost my original notes and cannot reconstruct it anymore, so it is disappearing in the mist of time.

For me, it has always been a fascinating experience: to be *struck by* an idea or to *get* an idea. We do not say "I made an idea," but we say "I got an idea," like it is as a gift from elsewhere. And indeed it is; it comes from our subconscious. I always tell my students the following parable. If you want to solve a problem, then you work on it, you work hard on it, but often the problem resists your efforts. Maybe you have to put it aside, then later work on it again without finding a solution, and so on and so forth. After some time (and in some cases a long time), while you are taking a break, something important happens: your conscious mind is somewhere else, but your subconscious is still at work. And, while you are relaxing, doing something else, like having a drink in the sun at an outdoor cafe or maybe while taking a shower, an idea for a solution pops into your consciousness from your subconscious. And then we can make real progress toward the solution. So I have learned that my most important tool in doing math is my subconscious. But I also warn my students. Your subconscious does not do the necessary work if your consciousness has not put in a big effort. So, if you get your ideas while taking a shower, it won't work to do nothing and just shower 24 hours per day hoping to get productive ideas.

This Expansion Theorem is the basis for the Meta-conjectures that are the focus of this story. Note that I call these *meta*-conjectures. They are not conjectures in the sense that they can be proved or disproved. The statements are on another level and, by applying these to a specific instance, may lead to a "regular" conjecture in the usual sense. It is about a nice and surprising common feature of trees and hypercubes that has led to many new results and may still be quite fruitful in the future to produce new problems and results. It is interesting to note here that, when I was visiting Clemson in 2014 and had given a talk on this topic, Steve Hedetniemi (2014, private communication, Clemson) told me that he had met a conjecture while walking on campus.

10.2 Setting the Stage

Throughout this chapter, $G = (V, E)$ is a finite, simple, connected graph with vertex set V and edge set E. The *order* of G is the number $|V|$ of vertices of G. The distance $d(u, v)$ between two vertices u and v of G is the length of a shortest u, v-path, or u, v-*geodesic*. The *interval* between u and v is the set

$$I_G(u, v) = \{\, x \mid d(u, x) + d(x, v) = d(u, v)\, \},$$

that is, the set of vertices lying on u, v-geodesics. When no confusion arises, we write I instead of I_G. Loosely speaking, it is the set of all vertices *between* u and v in G. In the sequel, we also need the following notation:

$$I(u, v, w) = I(u, v) \cap I(v, w) \cap I(w, u).$$

Anything can happen here, as can be seen in Figure 10.1.

A subset W of V is *convex* if, for any two vertices x and y in W, we have $I(x, y) \subseteq W$. The empty set and V are trivially convex. Note that the intersection of two convex sets is again convex. In abstract convexity theory, a *convexity* on a

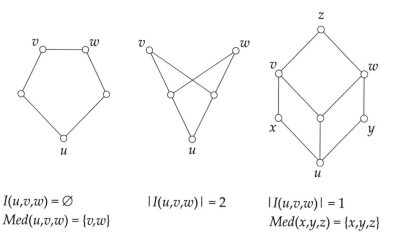

$$I(u,v,w) = \varnothing$$
$$Med(u,v,w) = \{v,w\}$$

$$|I(u,v,w)| = 2$$

$$|I(u,v,w)| = 1$$
$$Med(x,y,z) = \{x,y,z\}$$

Fig. 10.1 $I(u, v, w)$ and $Med(u, v, w)$

finite set V is a family of subsets that contains the empty set and V and is closed under intersection (see [33]). So the convex sets in a connected graph as defined above form a convexity in the sense of abstract convexity theory. A subgraph H of G is *convex* if it is induced by a convex set in G.

A profile π on G of length k is a nonempty sequence $\pi = (x_1, x_2, \ldots, x_k)$ of vertices of V with repetitions allowed. We denote its length by $k = |\pi|$. A subprofile of π is just a subsequence of π. For convenience, we allow a subprofile ρ to be empty and then with length $|\rho| = 0$. The set of all profiles on G is denoted by V^*.

A *consensus function* on G is a function $L: V^* \to 2^V - \{\emptyset\}$, where $2^V - \{\emptyset\}$ is the family of nonempty subsets of V. For convenience, we write $L(x_1, x_2, \ldots, x_k)$ instead of $L((x_1, x_2, \ldots, x_k))$, for any function L defined on profiles. A *median* of a profile $\pi = (x_1, x_2, \ldots, x_k)$ is a vertex x in V minimizing the distance sum $\sum_{i=1}^k d(x, x_i)$. The *median set* $Med(\pi)$ of π is the set of all medians of π. Note that, since G is connected, this defines a consensus function, namely, the *median function* $Med: V^* \to 2^V - \{\emptyset\}$. Trivially, we have $Med(x) = \{x\}$ and $Med(x, y) = I(x, y)$. Moreover, if $I(u, v) \cap I(v, w) \cap I(w, u) \neq \emptyset$, then $Med(u, v, w) = I(u, v) \cap I(v, w) \cap I(w, u)$. Note that $Med(\pi)$ is by definition nonempty, G being connected. So in case $I(u, v, w) = \emptyset$, we have $Med(u, v, w) \neq I(u, v, w)$. See Figure 10.1 for various possibilities.

10.3   Trees and Hypercubes

Trees and hypercubes are well-studied classes in graph theory. They are the main characters in this story. Therefore, we want to make sure that it is clear what they are. To avoid any confusion of what a tree is, we depict in Figure 10.2 a graph that is a tree and one that is not a tree.

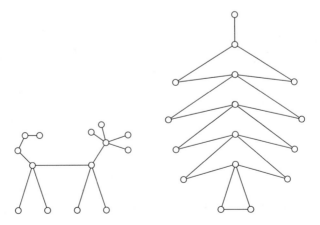

Fig. 10.2 A tree and a non-tree

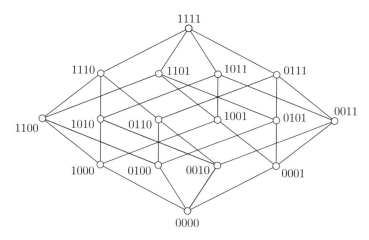

Fig. 10.3 The 4-cube

The *n-dimensional hypercube*, or *n-cube* for short, has the 0,1-vectors of length *n* as vertices. Two vertices are adjacent if, as vectors, they differ in exactly one coordinate. In Figure 10.3, we see the 4-cube. The vertices having 0 as fourth coordinate form a 3-cube, as well as those having a 1 as fourth coordinate. These two 3-cubes are joined by a matching between corresponding vertices of the two 3-cubes.

Equivalently, the *n*-cube can be defined as having the subsets of an *n*-set as vertices, where two vertices are adjacent if their symmetric difference is a singleton. Thus, a hypercube is the Hasse diagram of a Boolean lattice.

In the following observations, the complete graphs K_1 and K_2 are the exceptions: these two graphs are the only graphs that are a tree as well as a hypercube. We call these the *trivial trees*, because on at most two vertices, one cannot have a cycle anyway. But one could say that, for higher orders, trees and hypercubes are almost opposites of each other.

A nontrivial tree of order *k* is cycle-free, irregular, and often even highly irregular: all degrees between 1 and $k − 1$ may occur. The automorphism group is "small." Each automorphism fixes the center as well as the centroid of the tree but usually not much more. The center, as well as the centroid, is either a single vertex or an edge, that is, a K_1 or a K_2. This must be understood in the following sense: the center is mapped on the center and the centroid on the centroid. So, if the center consists of two adjacent vertices, then neither of the vertices needs to be a fixed point, but the "edge" is fixed. This result on automorphism groups is actually one of the oldest results on trees. It was already proved by Camille Jordan in 1869, although for him a "tree" was still an assemblage of curves in the plane (see [15]). The vertex and edge connectivity of any tree is 1. For each value of *k*, there is a tree of order *k*. There is more than one tree of order *k* > 3, and the number of trees grows exponentially as *k* grows.

On the other hand, the order of a hypercube is necessarily a power of 2. For each n, there is a unique n-cube of order 2^n. The 2-cube is a 4-cycle. For larger n, there are many even cycles in the n-cube of many different lengths. The n-cube is regular of degree n. It has a "large" automorphism group; it is even distance transitive. The vertex and edge connectivity is n and hence grows with n. We could go on like this exhibiting even more differences between these two classes of graphs.

Of course trees and hypercubes also share some trivial properties, such as being connected and being bipartite, but these properties are not very distinctive, since they share these with many other classes of graphs. A close inspection of the two classes yields another very interesting and nontrivial property that they also share. As this property needs some explication, we devote the next section to it.

10.4   A Common Construction for Trees and Hypercubes

The simplest way to obtain a tree from a smaller one is by adding a pendant vertex (a vertex of degree 1). We could describe this construction in an elaborate way, and then it becomes an instance of a more general construction to get a tree from a smaller one. We depict this construction in Figure 10.4.

What we do is cover the tree on the left with two subtrees T_1^* and T_2^* that have exactly one vertex in common. To obtain the larger tree, we take two disjoint copies of these subtrees, respectively, T_1 and T_2, as in the tree on the right, and then join the vertices in these subtrees that correspond to the common vertex in T_1^* and T_2^*. We call this an expansion with respect to the covering subtrees T_1^* and T_2^*. Each tree can be obtained by a succession of such expansions from the one vertex graph K_1.

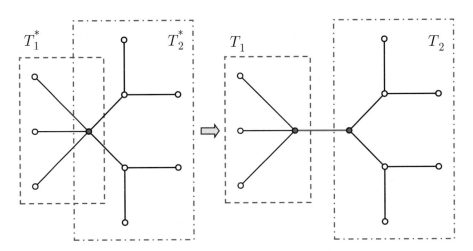

Fig. 10.4 A tree by expansion

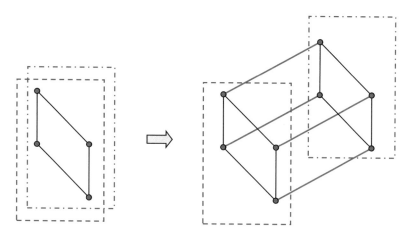

Fig. 10.5 A hypercube by expansion

Note that adding a pendant vertex amounts to taking the expansion with respect to a pair of subtrees with one being the whole tree and the other being the single vertex, where the pendant vertex is to be attached.

Hypercubes can be obtained in a similar manner by expansion, as depicted in Figure 10.5. We cover the n-cube on the left by two subcubes, both of which equal the whole n-cube. We take two disjoint copies of these two subcubes and join respective vertices in the two copies. These new edges form a matching between the two disjoint copies. Thus, we get a hypercube of dimension $n + 1$. Every hypercube can be obtained by a finite sequence of such expansions, starting from the one vertex graph K_1.

This construction is a common property of trees and hypercubes. In both cases, we cover the graph with two subgraphs having something in common. We then take disjoint copies of the two covering subgraphs and join corresponding vertices in these two copies by new edges. At first sight, this might look a bit farfetched as an interesting common feature, but we will see otherwise.

Let us have a closer look at the idea of expansion. In [25], I proposed a broad master plan for expansions. The basic idea is depicted in Figure 10.6.

We need a formal definition. For two graphs $G_1 = (V_1, E_1)$ and $G_2 = (V_2, E_2)$, the *union* $G_1 \cup G_2$ is the graph with vertex set $V_1 \cup V_2$ and edge set $E_1 \cup E_2$, and the *intersection* $G_1 \cap G_2$ is the graph with vertex set $V_1 \cap V_2$ and edge set $E_1 \cap E_2$. We write $G_1 \cap G_2 \neq \emptyset$ when $V_1 \cap V_2 \neq \emptyset$. The graph $G_1 - G_2$ is the subgraph of G_1 induced by the vertices in G_1 but not in G_2 and similarly for $G_2 - G_1$. A *proper cover* of a connected graph G consists of two subgraphs G_1 and G_2 such that $G_1 \cap G_2 \neq \emptyset$ and $G = G_1 \cup G_2$. Note that this implies that there are no edges between $G_1 - G_2$ and $G_2 - G_1$.

Now let G' be a connected graph, and let G'_1, G'_2 be a proper cover of G' with $G'_0 = G'_1 \cap G'_2$. Assume that G'_1 and G'_2 share some property $\mathcal{P}$ and that G'_0 has some property $\mathcal{P}_0$. For $i = 1, 2$, let G_i be an isomorphic copy of G'_i, and let λ_i be the

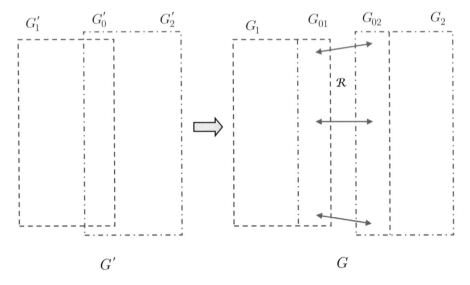

Fig. 10.6 Expansion procedure

isomorphism from G'_i to G_i. We write $G_{0i} = \lambda_i[G'_0]$ and $u_i = \lambda_i(u')$, for u' in G'_0. Let $\mathcal{R}$ be some rule for inserting edges between the subgraphs G_{01} and G_{02}. The $\mathcal{P}, \mathcal{P}_0, \mathcal{R}$-*expansion* of G' with respect to the proper cover G'_1, G'_2, is the graph G obtained as follows: take the disjoint union of G_1 and G_2 and insert edges between G_{01} and G_{02} according to rule $\mathcal{R}$. By varying the properties $\mathcal{P}, \mathcal{P}_0$, and rule $\mathcal{R}$, we get different kinds of expansions, some of which might be interesting for further study. Note that the choice of property $\mathcal{P}_0$ might be restricted by the choice of property $\mathcal{P}$.

Actually, the notion of expansion in [25] was even broader: there a cover with k subgraphs was considered, with $k \geq 2$, and the expansion was defined accordingly. Such an expansion could be called a *k-ary* expansion, where the above type is then a *binary* expansion. For our purposes, here we will restrict ourselves to the binary case. We observe that the well-studied case of quasi-median graphs uses k-ary expansions for arbitrary values of k (see [6, 23]).

What type of expansion would catch the constructions for trees and hypercubes above? For rule $\mathcal{R}$, it seems that we need the following: insert an edge between the vertex in G_{01} and that in G_{02} corresponding to the vertex u', for each vertex u' in G'_0. This amounts to inserting a matching between G_{01} and G_{02} that forms the obvious isomorphism between these two subgraphs. But what to choose for the properties $\mathcal{P}$ and $\mathcal{P}_0$ is not so obvious. There are several possibilities, each of which gives a different type of expansion. Each type of expansion could be used to define a class of graphs: those graphs that can be obtained from K_1 using expansions only of the chosen type. We focus on one type of expansion that yields a nice class of graphs (see the next section). We invite the reader to study other types of expansions.

10.5   The Meta-conjecture and the Strong Meta-conjecture

The expansion that we have in mind is the following. The property P is that of "being convex," so both G_1' and G_2' are convex subgraphs of G', such that G_1', G_2' forms a proper cover of G'. We call such a cover a *convex cover*. Since the intersection of convex subgraphs is a convex subgraph, the property P_0 is also that of "being convex." The rule R of inserting edges is the rule of inserting a matching between corresponding vertices of G_{01} and G_{02}. We call this type of expansion a *convex expansion*. Clearly, trees and hypercubes can be obtained by a succession of convex expansions from K_1.

Now we define a new class of graphs. Recall that this story is historically incorrect. A *median graph* is any graph that can be obtained by a finite sequence of convex expansions from K_1. We will see below why we call such a graph a median graph. Clearly, there are more median graphs than trees and hypercubes (see Figure 10.7 for some small median graphs).

For reasons that are at this point historically obscure, I proposed in 1990 the following "meta-conjectures."

Meta-conjecture: Any (sensible) property that is shared by trees and hypercubes is shared by all median graphs.

Strong Meta-conjecture: Any (sensible) property that is shared by trees and hypercubes *characterizes* median graphs.

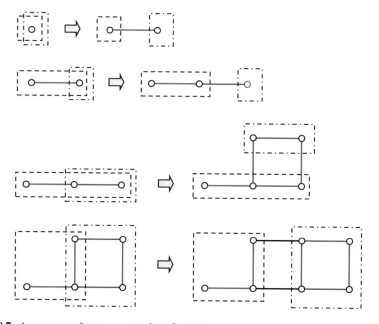

Fig. 10.7 A sequence of convex expansions from K_1

The keyword here is of course "sensible." For instance, in the case of the Meta-conjecture, the property of "being a tree or a hypercube" is trivially shared by trees and hypercubes but not by all median graphs. And in the case of the Strong Meta-conjecture, "being connected" is shared by trees and hypercubes but by no means characterizes median graphs. The focus of this chapter is on "sensible" properties, where the Meta-conjectures actually give us a "real" conjecture and the possibility of a new result. We list the instances chronologically and give some explanation in each case.

10.6 The Expansion Theorem

The first instance dates back to where it all began in 1976 (see the Introduction). Take any three vertices u, v, w in a tree or a hypercube. We always have $|I(u, v, w)| = 1$.

In a tree, the single vertex in $I(u, v, w)$ is the unique vertex lying on the three paths between the pairs among u, v, w. In a hypercube, we take the three 0, 1-vectors and determine the vertex x in the three intervals as follows. For each coordinate, the three vectors vote according to their value in that coordinate. The majority determines the value of the respective coordinate of x. Hence, this property is shared by trees and hypercubes (see Figure 10.8). It turns out that this is a "sensible" property for the Strong Meta-conjecture.

Theorem 1′. *A graph G is a median graph if and only if $|I(u, v, w)| = 1$ for any three vertices u, v, w of G.*

Note that, if $I(u, v, w)$ is nonempty, then it is precisely the median set $Med(\pi)$ of the profile $\pi = (u, v, w)$. When $|I(u, v, w)| = 1$, we call the unique vertex in the intersection of the three intervals the *median* of u, v, w. In view of Theorem 1′, we prefer from now on to call a graph with $|I(u, v, w)| = 1$, for all u, v, w, a

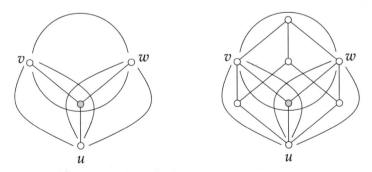

Fig. 10.8 The intersection of three intervals in a tree and a hypercube

median graph. So we *switch terminology*, and then Theorem 1′ should be rephrased
as follows (this represents the factual history better; see [22, 24]).

Theorem 1 (Expansion Theorem [1978]). *A graph G is a median graph if and
only if it can be obtained form K_1 by a finite sequence of convex expansions.*

Median graphs as the graphs having the property $|I(u, v, w)| = 1$, for all u, v, w,
were introduced independently several times. Avann [2] introduced them as *unique
ternary distance graphs* in 1961 relative to distributive semi-lattices. Later Nebeský
[32] introduced them again in 1971 relative to ternary algebras. Finally, Mulder and
Schrijver [22, 31] introduced them in 1978–1979 as underlying graphs of certain
hypergraphs (see the Introduction). From the historical point of view, we would like
to refer the reader to the very interesting application of median graphs in [7].

At first sight, one might think that median graphs are quite esoteric. But in [12]
Imrich, Klavžar and Mulder established a one-to-one correspondence between the
class of connected triangle-free graphs and a special subclass of the class of median
graphs. Since median graphs are triangle-free and connected, this implies that, in
the universe of all graphs, there are as many median graphs as there are connected,
triangle-free graphs.

It is straightforward to prove that the convex expansion of a median graph is
again a median graph. The hard part of the proof of the Expansion Theorem is to
show that a median graph is always the convex expansion of a smaller one. In order
to do this, we need some of the ideas and notation used in this proof. See Figure 10.9
for clarification.

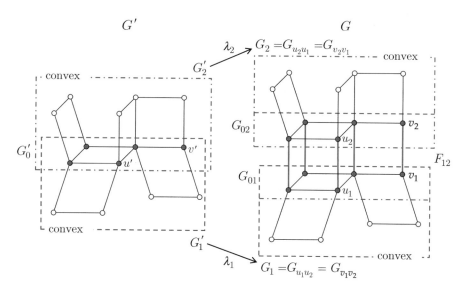

Fig. 10.9 Expansion Theorem

Let G be a median graph as the graph on the right in Figure 10.9. Take an arbitrary edge u_1u_2 in G. Let G_1 be the subgraph consisting of all vertices closer to u_1 than to u_2, and let G_2 be the subgraph consisting of all vertices closer to u_2 than u_1. Since G is bipartite, these two subgraphs partition the vertex set of G. We can prove that these two subgraphs are convex. Let G_{01} be the subgraph of G_1 consisting of the vertices having a neighbor in G_2, and let G_{02} be the subgraph of G_2 consisting of the vertices having a neighbor in G_1. Let F_{12} be the set of edges between G_{01} and G_{02}. We next prove that F_{12} is a matching that induces an isomorphism between G_{01} and G_{02}, as depicted in Figure 10.9. Moreover, G_{01} and G_{02} are convex subgraphs as well. We call G_1, G_2 a *split* of G with *split sides* G_1 and G_2. We will call G_1 and G_2 *opposites* of each other. For the origin of this neologism, see [28].

We next prove that any other edge of F_{12}, say v_1v_2, defines the same split G_1, G_2, that is, G_1 consists of all vertices closer to v_1 than to v_2 and G_2 consists of all vertices closer to v_2 than to v_1. Note, therefore, that we do not have to refer to a specific edge when we consider a split G_1, G_2. We now contract the edges in F_{12} and identify the corresponding vertices in G_{01} and G_{02}, thus producing the graph G' on the left. We can state this more formally by defining the mapping $\kappa : G \to G'$ by setting $\kappa|_{G_1} = \lambda_1^{-1}$ and $\kappa|_{G_2} = \lambda_2^{-1}$. We call κ is the *contraction map*, and we call G' the *contraction* of G with respect to the split G_1, G_2.

The last step of the proof is to show that G' is a median graph with convex cover G_1', G_2'. It is clear then that G is the convex expansion of G' with respect to this cover. Hence, *any* median graph can be obtained this way. It turns out that, in constructing G from K_1 by convex expansions, we may take the expansions in any order.

Note that the hypergraph obtained by taking the vertex set of a split side as a hyper-edge, for all splits, gives us exactly a maximal Helly copair hypergraph in the sense of [21, 31]. The underlying graph is obtained by taking uv as an edge whenever the intersection of all hyper-edges containing u and v is exactly $\{u, v\}$, for distinct u and v.

The Expansion Theorem gives us a very strong tool: on median graphs we can use induction on the number of expansions or, equivalently, the number of splits. In economics (and philosophy), the concept of armchair theorizing exists (see [35]): by sitting in their armchairs and looking at the world, economists can come up with new theories and insights about economics. Buck McMorris introduced me to this concept in mathematics: sitting in our armchair, after proving some heavy-duty theorems, we can let these do the work and come up with nice and new results. This approach is much more solid than that within economics. Most of the results below are an example of the use of armchair theorizing. We use the Expansion Theorem, and the ideas and notation developed in its proof, in combination with the Meta-conjectures. This gives us results that we can prove simply by sitting in our armchair.

10.7 More Applications of the Meta-conjectures

Median graphs appear in different guises in other areas than graph theory. For instance, a median graph is precisely the covering graph of a distributive semi-lattice [23, 28]. Also it is the underlying graph of so-called median algebras [13, 23, 32]. It is also the graph of certain conflict models [8]. Above we have seen that median graphs are the underlying graphs of certain Helly hypergraphs. A survey can be found in [17, 27, 28]. Because these other guises require more details from these other areas, we skip this here. We only observe that also in these cases the Meta-conjectures have shown their value.

In this section, we present more applications of the Meta-conjectures in chronological order.

10.7.1 *Isometric Embedding in Hypercubes*

An *isometric embedding* of a graph G into a graph H is a one-to-one mapping ϕ of the vertex set of G into that of H such that, for any two vertices u and v in G, the distance between u and v in G equals the distance between $\phi(u)$ and $\phi(v)$ in H.

Trivially, any hypercube Q_n can be isometrically embedded into a Q_m with $m \geq n$. It is also an easy exercise to show that a tree T with n vertices can be isometrically embedded into any Q_m with $m \geq n - 1$. Applying the Meta-conjecture, we conjecture that any median graph can be embedded isometrically into a hypercube of sufficiently large dimension. And indeed it can (see [23]). The proof is by induction on the number of splits in the median graph.

Theorem 2 ([1980]). *Let G be a median graph with m splits. Then G can be isometrically embedded in any n-cube with $n \geq m$.*

For the Strong Meta-conjecture, this property is not yet a sensible property. The connected graphs that can be isometrically embedded in a hypercube are precisely the *partial cubes* (see [10, 16]). All even cycles belong trivially to this class. So what "makes sense" in this case? A closer look tells us that we can embed a tree T in a hypercube Q such that the median of any three vertices u, v, w in T is mapped onto the median in Q of the three images in Q of u, v, w. And, yes, this property is sensible for the Strong Meta-conjecture, and we have our second characterization of median graphs after Theorem 1.

Theorem 3 (1980). *A graph G is a median graph if and only if it can be embedded as an isometric subgraph of a hypercube Q such that the median of any three vertices of this embedding is also a vertex of the embedding.*

Otherwise stated, the embedding of G preserves medians. These results were first published in [23] (see also [24]), and [14] for an algorithmic result.

10.7.2   Median Sets

Let G be a connected graph, and let $\pi = (u, v, w)$ be a profile of length 3. By the definition of a median set, we have $Med(\pi) \neq \emptyset$. And if $I(u, v, w) \neq \emptyset$, we also have $Med(\pi) = I(u, v, w)$. In Figure 10.1, we have seen that $I(u, v, w)$ can be empty. So $Med(\pi)$ and $I(u, v, w)$ do not coincide for all profiles $\pi = (u, v, w)$. In trees and hypercubes, they always coincide, but here we have the additional property that these sets are always singleton sets. Combining these two properties into one, we get the following application of the Strong Meta-conjecture of Bandelt and Barthèlemy [4]. A proof of this result was already implicit in [24].

Theorem 4 ([1984]). *A graph G is a median graph if and only if* $|Med(\pi)| = 1$ *for all profiles π of length 3 on G.*

10.7.3   The Majority Counts

Next we focus on the median sets $Med(\pi)$ of profiles π on a connected graph G. For any edge uv, we denote by π_{uv} the subprofile of π of the elements that are closer to u than to v. Recall that a subprofile may be empty. If G is bipartite, then we have

$$|\pi_{uv}| + |\pi_{vu}| = |\pi|.$$

It is straightforward to check that for trees, as well as hypercubes, we have the following property. For any profile π, a vertex x is in $Med(\pi)$ if and only if $|\pi_{xy}| \geq |\pi_{yx}|$ for each neighbor y of x. Loosely speaking, the median set of a profile is always on the majority side of each split. So, by the Meta-conjecture, we get the following result of Bandelt and Barthèlemy [4]. This result is an immediate consequence of the Expansion Theorem, although this was not mentioned in [4].

Theorem 5 ([1984]). *Let G be a median graph, and let π be a profile on G. Then x is in $Med(\pi)$ if and only if $|\pi_{xy}| \geq |\pi_{yx}|$ for each neighbor y of x.*

We can use this idea of the median set being on the majority side even better. In a tree, as well as in a hypercube, we can find the median set of a profile π easily using the following strategy. We start at an arbitrary vertex z and move along edges through the graph. The rule for moving is: when at v and w is a neighbor of v with $|\pi_{wv}| \geq |\pi_{vw}|$, then we move to w. That is, we move to a majority. Two possibilities arise. At some point we cannot move toward another vertex (so there is a strict minority in each direction). In this case, the vertex that we have reached is the unique median of π. Or we are at a vertex v that has a neighbor u such that $|\pi_{uv}| = |\pi_{vu}|$. Now we are allowed to move back and forth between u and v. Again we are in the median set, and it turns out that we can still move freely within the median set, but we cannot get out of it. In order to formulate our results as precisely as possible, we give a formal description of this majority strategy. Note that the way we describe the output reflects the actual origins of this algorithm (see the last paragraph of this section).

Majority Strategy
Input: A connected graph G, a profile π on G, and an *initial vertex* in V.
Output: The set of vertices where signs have been erected.

- Start at the initial vertex.
- If we are at v and w is a neighbor of v with $|\pi_{wv}| \geq \frac{1}{2}|\pi|$, then we *move* to w.
- We move only to a vertex already visited if there is no alternative.
- We stop when:

    (i) we are stuck at a vertex v *or*
    (ii) we have visited vertices at least twice, and, for each vertex v visited

    at least twice and each neighbor w of v, either $|\pi_{wv}| < \frac{1}{2}|\pi|$ or w is also visited at least twice.
- We park and erect a sign at the vertex where we get stuck or at each vertex visited at least twice.

Do we always find the median set using the majority strategy? The answer is no, a simple example suffices. Take the complete graph K_3 with vertices u, v, w and let $\pi = (u, v, w)$. Now, for each edge xy, there is only one vertex closer to y than to x, viz., y itself. So we do not move from x to y. This means that, being at x, we are stuck at x and only find x, whereas $Med(\pi)$ is the whole vertex set. We find *one* median vertex but not all. With the Strong Meta-conjecture at hand, the first equivalence in the following theorem does not come as a surprise (see [26]).

Theorem 6 ([1997]). *Let G be a graph. Then the following statements are equivalent:*

(i) G is a median graph.
(ii) The majority strategy produces $Med(\pi)$ in G, for each profile π.
(iii) The majority strategy produces the same set from any initial position v in G, for each profile.

Statement *(iii)* in the theorem came as a bonus and was not foreseen in any way.

The idea for the majority strategy arose in Louisville, Kentucky, while I was visiting Buck McMorris. We were driving to the University of Louisville along Eastern Parkway. At some stretch, there is a beautiful median on Eastern Parkway, with green grass and large trees. At that time, there were traffic signs along this median that read: "Tow away zone. No parking on the median at any time." I was not working on this problem at all at that time. So again this was a gift from the subconscious.

10.7.4   Retracts of Hypercubes

Let G be a connected graph, and let H be an isometric subgraph of G. We call H a *retract* of G if there exists a *retraction* of G onto H. A retraction is a mapping that maps the vertices of G onto those of H such that restricted to H, this mapping is the identity, and adjacent vertices of G are mapped to a single vertex of H or to adjacent vertices of H. So a retraction fixes every vertex of H and may shrink distances for vertices not in H. Clearly, not every isometric subgraph is a retract.

The identity mapping shows trivially that each hypercube is a retract of the hypercube itself. It is an exercise to show that trees are retracts of hypercubes. Applying the Strong Meta-conjecture, we get the following result of Bandelt [3].

Theorem 7 (1984). *A graph G is a median graph if and only if it is a retract of a hypercube.*

The simplest proof of this theorem uses the Peripheral Expansion Theorem from Section 10.7.6, which is a special case of the Expansion Theorem.

10.7.5   A Fixed Subcube

As stated earlier, Jordan already proved in 1869 that any automorphism of a tree fixes the center and the centroid. The center as well as the centroid of a tree is either a K_1 or a K_2. For hypercubes, the situation is different: the only subgraph that is fixed by all automorphisms (a subgraph that is mapped onto itself) is the hypercube itself. But, looking at it in another way, we see a common feature. Trivially, K_1 and K_2 are subcubes of a tree. So in all cases, a subcube is fixed. It is simple to construct a graph that is not a median graph, where a subcube is fixed by all automorphisms, so the Strong Meta-conjecture is not applicable. But the weaker Meta-conjecture still works, as we see in the following theorem by Bandelt and Van de Vel [5].

Theorem 8 (1987). *Let G be a median graph. Any automorphism of G fixes a subcube of G.*

10.7.6   Peripheral Expansions

Adding a pendant vertex to a tree is the simplest and most obvious way to get a larger tree from a smaller one. This can be phrased as a convex expansion with respect to the convex cover consisting of the smaller tree itself and the single vertex to which the new vertex is to be attached. A little bit more sophisticated, we could also say that the convex cover consists of the whole graph and a convex subgraph. Of course we can formulate the expansion that creates an $(n + 1)$-dimensional cube from an n-cube (see Figure 10.5) in the same more sophisticated manner: the cover consists of the whole graph and a convex subgraph (the whole graph being trivially a convex subgraph). We call such a convex expansion a *peripheral expansion*, following the custom that the set of the pendant vertices of a tree is called the periphery of the tree. Now we have a property that is certainly sensible in the sense of the Strong Meta-conjecture (see [25]).

Theorem 9 (Peripheral Expansion Theorem [1990]). *A graph G is a median graph if and only if it can be obtained from K_1 by successive peripheral expansions.*

This theorem is also quite useful for proving results on median graphs by induction. In the expansion, the side that corresponds to the convex subgraph of the cover is called a *peripheral side*. To find a peripheral side in a median graph, we basically need the Expansion Theorem.

10.7.7   Amalgamation of Hypercubes

Another way to construct a tree is as follows. We start with a finite set of edges or K_2's. Now we glue these together along vertices or K_1's such that no cycle arises. We call this gluing *amalgamation*. The graphs K_1 and K_2 are hypercubes. So we could rephrase this construction: we start with a finite set of hypercubes (in this case all of dimension 1) and amalgamate these along subcubes such that no cycle arises.

Now we have a construction that also applies to hypercubes, even trivially. To get the n-cube, we start with a finite set of hypercubes, viz., a single n-cube. Applying zero amalgamations along subcubes, we get the n-cube. Ah, the Strong Meta-conjecture comes into the picture. But we have to be careful here. In Figure 10.10, we see on the right the 3 by 3 grid, which is a median graph. For the amalgamation construction, we need four 2-cubes. But in the last step, we do not amalgamate along a subcube but along a path that happens to be a convex subgraph. So we rephrase the construction. Instead of amalgamating along a subcube, we amalgamate along

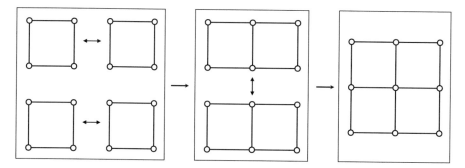

Fig. 10.10 Amalgamation

a convex subgraph. Doing this stepwise, we get for free that the graph is cycle-free in the tree case. So in this case, it took some work to get a sensible property, but it was worth the effort and gives us the next result from [6]. We make a formal definition. Let G_1, G_2 be a convex cover of a graph G. Then we say that G is the *convex amalgamation* of G_1 and G_2.

Theorem 10 (1994). *A graph G is a median graph if and only if it can be obtained from a set of hypercubes by convex amalgamations.*

10.8 Axiomatics of Consensus Problems

Finding the median set of a profile is a so-called location problem. Such a problem can sometimes be phrased as a consensus problem. One wants to reach consensus among agents or clients in a rational way. This is modeled using a consensus function (see [9]). The input of the function consists of certain information about the agents, and the output concerns the issue about which a consensus should be reached. The rationality of the process is guaranteed by the fact that the consensus function satisfies certain "rational" rules or *consensus axioms*. Such axioms should be appealing and simple. Of course this depends on the consensus function at hand. A function that has nice properties might indeed be characterized by such axioms, but a function that behaves badly might need more complicated or less appealing axioms. The study of the axiomatics of consensus functions was initiated by the economist K. Arrow in his seminal paper [1] of 1951. In a location problem, the input is the location of the clients: a profile consisting of vertices. Holzman [11] was the first to study location problems as a consensus problem. The median function was first characterized axiomatically by Vohra [34] on tree networks (the continuous variant of a tree). The discrete case was first dealt with by McMorris, Mulder, and Roberts [18]. They rephrased Vohra's axioms as follows. In these axioms, L is a consensus function on a connected graph $G = (V, E)$.

(A) **Anonymity:** For any profile $\pi = (x_1, x_2, \ldots, x_k)$ on V and any permutation σ of $\{1, 2, \ldots, k\}$, we have $L(\pi) = L(\pi^\sigma)$, where $\pi^\sigma = (x_{\sigma(1)}, x_{\sigma(2)}, \ldots, x_{\sigma(k)})$.
    The consensus function cannot distinguish among the group of clients.
(B) **Betweenness:** $L(u, v) = I(u, v)$, for $u, v \in V$.
    All locations between exactly two preferred locations are equally good.
(C) **Consistency:** If $L(\pi) \cap L(\rho) \neq \emptyset$ for profiles π and ρ, then
    $L(\pi\rho) = L(\pi) \cap L(\rho)$.
    Loosely speaking, if two profiles agree on some output x, then x is in the output of the concatenation as well.

It is straightforward to check that the median function satisfies these axioms on all connected graphs. So the question arises what extra axioms are needed to characterize the median function on any class of graphs. A *cube-free median graph* is a median graph that does not contain a 3-cube. Trivially, any tree is a cube-free median graph. In [18], it was proved that on cube-free median graphs, a location function $L: V^* \to 2^V - \{\emptyset\}$ is the median function if and only if it satisfies (A), (B), and (C). In particular this holds on trees. In [18] it was suggested, between the lines, that on arbitrary median graphs, a fourth axiom was needed, and a result involving an extra axiom was given. But basically, it was an open problem whether such a fourth axiom was actually necessary. In 2011 Mulder and Novick [29] were able to show that on hypercubes, the median function is characterized by the three basic axioms (A), (B), and (C). Their proof used techniques that were very specific for hypercubes. But, with the Meta-conjecture at hand, it became a necessity to search for a proof that on all median graphs, the median function is the unique function $L: V^* \to 2^V - \{\emptyset\}$ satisfying the three basic axioms (A), (B), and (C). And again the Meta-conjecture showed its worth: this result was proved in [30], but the proof was nontrivial.

Theorem 11 (2013). *Let L be a consensus function on a median graph G. Then $L = Med$ if and only if L satisfies (A), (B), and (C).*

In the literature, simple examples are available to show that (A) and (B) do not imply (C) and that (A) and (C) do not imply (B). But, surprisingly, it turns out to be far from trivial to show that (A) is independent from (B) and (C). Using the rich structure of median graphs, a heavy-duty example is constructed in [20] that shows independence on any median graph with at least two vertices.

In [19], another axiomatic characterization of the median function on median graphs was given. It involves the next two axioms and consistency.

(F) **Faithfulness:** $L(x) = \{x\}$, for all $x \in V$.

If there is only one client, then the preferred location of this client is returned by the consensus function.

$(Cond)$ $\frac{1}{2}$**-Condorcet:** $u \in L(\pi)$ if and only if $v \in L(\pi)$, for each profile π on G and any edge uv of G with $|\pi_{uv}| = |\pi_{vu}|$.

The consensus function does not make a distinction between neighbors that are equally preferred.

Theorem 12 (2000). *Let L be a consensus function on the median graph G. Then $L = Med$ if and only if L satisfies (F), (C), and $(Cond)$.*

One consequence of Theorems 11 and 12 is that the three axioms (C), $(Cond)$, and (F) imply (A) and (B) on median graphs. In [20], all possible dependencies and independencies on median graphs between the above five axioms are determined. There is one rather surprising result. On hypercubes, axioms (C) and $(Cond)$ already imply (A), so axiom (F) is *not* needed for anonymity. With the Meta-conjecture in hand, the next question immediately forces itself upon us: do axioms (C) and $(Cond)$ imply (A) on trees as well? And consequently, what on median graphs? Another heavy-duty example in [20] shows that (A) is independent from (C) and $(Cond)$ on all nontrivial trees (with at least three vertices), hence on all median graphs that are not a hypercube. So in this case, we really need (F) as an additional axiom to force anonymity. Although this is a negative example, it still underscores the Meta-conjecture.

10.9   Concluding Remarks

It seems to me that we have shown that the two Meta-conjectures are rather powerful tools for obtaining interesting and nontrivial results about median graphs. In all cases, the Expansion Theorem turned out to play a crucial role in the proofs, another fact that underscores the Meta-conjectures. And the story does not end here. The reader is invited to find new applications of one of the two Meta-conjectures.

There is another interesting type of expansion that gives a nice common generalization of trees and hypercubes. If we take as properties $\mathcal{P}$ and $\mathcal{P}_0$ the property of "being an isometric subgraph," then we get the class of partial cubes when again starting with K_1 (see [10, 16]). But so far there are no Meta-conjectures in this case. Other avenues of research might present themselves when we try to find another type of expansion that gives a productive common generalization of trees and hypercubes.

10.10   Where I Began

This section is historically correct. As I said in the Introduction, this story began in 1976. The Expansion Theorem was published in 1978 [22], and median graphs were first studied extensively in my PhD thesis [24]. On the cover of my thesis, I had a picture of a beautiful windmill called "De Herder" (see Figure 10.11). So it plays a role in the origins of the Meta-conjectures. The photograph I used for it has yellowed by age as you can see. When you go by train from Schiphol Airport to Leiden, you will see the mill on your right shortly before you arrive at Leiden Station. This windmill plays a crucial role in my life. The old Dutch word for miller is mulder. During the French and Napoleontic occupation (1795–1813), French laws were introduced that necessitated every Dutch family to take a family name. Until that time, many people did not have one and usually were just called by their first name or their profession. My great-great-great-great-grandfather was a miller and owned the windmill in the picture. When he had to take a family name, he took his profession as his name, so he became Mulder (as many other people did, who were not family at all, the windmill being the main source of industrial power in the Netherlands at that time). So, in a sense, I "began" at this windmill, and thus this windmill is part of the prehistory as well as the actual history of the Meta-conjectures and therefore deserves its place here.

Fig. 10.11   Windmill "De Herder"

References

1. Arrow, K.J.: Social Choice and Individual Values. Cowles Commission Monograph, vol. 12. Wiley, New York (1951)
2. Avann, S.P.: Metric ternary distributive semi-lattices. Proc. Am. Math. Soc. **12**, 407–414 (1961)
3. Bandelt, H.J.: Retracts of hypercubes. J. Graph Theory **8**, 501–510 (1984)
4. Bandelt, H.J., Barthélemy, J.-P.: Median sets in median graphs. Discret. Appl. Math. **8**, 131–142 (1984)
5. Bandelt, H.J., van de Vel, M.: A fixed cube theorem for median graphs. Discret. Math. **67**, 129–137 (1987)
6. Bandelt, H.J., Mulder, H.M., Wilkeit, E.: Quasi-median graphs and algebras. J. Graph Theory **18**, 681–703 (1994)
7. Barthélemy, J.-P.: From copair hypergraphs to median graphs with latent vertices. Discret. Math. **76**, 9–28 (1989)
8. Barthélemy, J.-P., Constantin, J.: Median graphs, parallelism and posets. Discret. Math. **111**, 49–63 (1993)
9. Day, W.H.E., McMorris, F.R.: Axiomatic Consensus Theory in Group Choice and Biomathematics. Society for Industrial and Applied Mathematics, Philadelphia (2003)
10. Djoković, D.: Distance preserving subgraphs of hypercubes. J. Comb. Theory Ser. B **14**, 263–267 (1973)

11. Holzmann, R.: An axiomatic approach to location on networks. Math. Oper. Res. **15**, 553–563 (1990)
12. Imrich, W., Klavžar, S., Mulder, H.M.: Median graphs and triangle-free graphs. SIAM J. Discret. Math. **12**, 111–118 (1999)
13. Isbell, J.R.: Median algebra. Trans. Am. Math. Soc. **260**, 319–362 (1980)
14. Jha, P.K., Slutzki, G.: Convex-expansion algorithms for recognizing and isometric embedding of median graphs. Ars Comb. **34**, 75–92 (1992)
15. Jordan, C.: Sur les assemblages de lignes. J. Reine Angew. Math. **70**, 193–200 (1869)
16. Klavžar, S.: Hunting for cubic partial cubes. In: Changat, M., et al. (eds.) Convexity in Discrete Structures. Ramanujan Mathematical Society Lecture Notes Series, vol. 5, pp. 87–95. Ramanujan Mathematical Society, Mysore (2008)
17. Klavžar, S., Mulder, H.M.: Median graphs: characterizations, location theory, and related structures. J. Comb. Math. Comb. Comput. **30**, 103–127 (1999)
18. McMorris, F.R., Mulder, H.M., Roberts, F.S.: The median procedure on median graphs. Discret. Appl. Math. **84**, 165–181 (1998)
19. McMorris, F.R., Mulder, H.M., Powers, R.C.: The median function on median graphs and semilattices. Discret. Appl. Math. **101**, 221–230 (2000)
20. McMorris, F.R., Mulder, H.M., Novick, B., Powers, R.C.: Five axioms for location functions on median graphs. Discret. Math. Algoritm. Appl. **7**(2), 30 pp. (2015). 1550013
21. Mulder, H.M.: Maximal Helly Copair systems. Technical Report WS 34, Vrije Universiteit, Amsterdam (1976)
22. Mulder, H.M.: The structure of median graphs. Discret. Math. **24**, 197–204 (1978)
23. Mulder, H.M.: *n*-Cubes and median graphs. J. Graph Theory **4**, 107–110 (1980)
24. Mulder, H.M.: The Interval Function of a Graph. Mathematical Centre Tracts, vol. 132. Mathematisch Centrum, Amsterdam (1980)
25. Mulder, H.M.: The expansion procedure for graphs. In: Bodendiek, R., (ed.) Contemporary Methods in Graph Theory, pp. 459–477. B.I.-Wissenschaftsverlag, Mannheim/Wien/Zürich (1990)
26. Mulder, H.M.: The majority strategy on graphs. Discret. Appl. Math. **80**, 97–105 (1997)
27. Mulder, H.M.: Transit functions on graphs (and posets). In: Changat, M., et al. (eds.) Convexity in Discrete Structures. Ramanujan Mathematical Society Lecture Notes Series, vol. 5, pp. 117–130. Ramanujan Mathematical Society, Mysore (2008)
28. Mulder, H.M.: Median graphs. A structure theory, In: Kaul, H., Mulder, H.M. (eds.) Advances in Interdisciplinary Discrete Applied Mathematics. Interdisciplinary Mathematical Sciences, vol. 11, pp. 93–125. World Scientific, Singapore (2010)
29. Mulder, H.M., Novick, B.: An axiomatization of the median procedure on the *n*-cube. Discret. Appl. Math. **159**, 939–944 (2011)
30. Mulder, H.M., Novick, B.A.: A tight axiomatization of the median procedure on median graphs. Discret. Appl. Math. **161**, 838–846 (2013)
31. Mulder, H.M., Schrijver, A.: Median graphs and Helly hypergraphs. Discret. Math. **25**, 41–50 (1979)
32. Nebeský, L.: Median graphs. Comment. Math. Univ. Carol. **12**, 317–325 (1971)
33. van de Vel, M.: Theory of Convex Structures. North Holland, Amsterdam (1993)
34. Vohra, R.V.: An axiomatic characterization of some locations in trees. Eur. J. Oper. Res. **90**, 78–84 (1996)
35. Wikipedia, Armchair Theorizing

Chapter 11
Two Chromatic Conjectures: One for Vertices and One for Edges

P. Mark Kayll

Abstract Erdős, Faber, and Lovász conjectured that a pairwise edge-disjoint union of n copies of the complete graph K_n has chromatic number n. This seeming parlour puzzle has eluded proof for more than four decades, despite the attack by a few of this era's more powerful combinatorial minds. Regarding edges, the list-colouring conjecture asserts, loosely, that list colouring is no more difficult than ordinary edge colouring. Probably first proposed by Vizing, this notorious conjecture—also having garnered the attention of leading combinatorialists—has itself defied proof for forty years. Like any good mature conjecture, both of these have spawned interesting mathematics vainly threatening their resolution. This chapter considers some of the related partial results in concert with the conjectures themselves.

Mathematics Subject Classification 2010: Primary 05-02,05C15; Secondary 05C65, 05-03, 01A65, 01A70

Introduction

When I prepared the lecture [58] on which this chapter is based—and later as I produced this written account—I remembered a belief that Alan Mekler (1947–1992) once shared with me. A prodigious and gregarious young professor at Simon Fraser University, Alan forever shaped my thinking about early mathematical influences. We got to know each other even though I never had the privilege of taking one of his courses. Later when I was a graduate student at Rutgers University, Alan

This work was partially supported by a grant from the Simons Foundation (#279367 to Mark Kayll).

P.M. Kayll (✉)
Department of Mathematical Sciences, University of Montana, 32 Campus Drive, Missoula, MT 59812-0864, USA
e-mail: mark.kayll@umontana.edu

occasionally visited New Brunswick. Once we enjoyed dinner together talking math and reminiscing over SFU days. During those years, Alan played several roles in my life: mathematical exemplar, personal connection to my hometown and SFU, and friend.

He believed in a sort of mathematical cycle of life. As Alan described it, we're especially impressionable while in graduate school. The open problems we learn then are the ones that endure most strongly as our professional lives unfold. Incrementally, these problems morph into theorems, bringing us contentment and closure later in our careers, regardless of the solvers.

I could see the truth of Alan's belief through his eyes, in part because of his excitement about the imminent appearance of his monograph [24] with Paul Eklof. But as a grad student hearing these ideas, I didn't have the experience to measure them myself. Now in mid-career, I think they help explain my choices of 'favourite conjectures', both of which I learned at Rutgers. Initially, I resisted hearkening back to those student days, but Alan's axiom gave me permission—even encouragement—for doing exactly this.

Sadly, Alan died before I had another opportunity to dine with him. Because his long-ago shared observation influenced my choices for this piece, I remember and acknowledge him here.

A Word on Definitions Two of my favourite graph theory conjectures concern colouring, both the vertex and edge variants. Readers who know the definitions of chromatic number χ, list-chromatic number χ_L, chromatic index χ', and list-chromatic index χ'_L can skip the Appendix, where these terms and others used in this chapter are catalogued. Mirroring the lecture, I'll dive right in here, too.

11.1 The List-Colouring Conjecture

The Acknowledgements by Erdős et al. [31] begin:

> It got started when we tried to solve Jeff Dinitz's problem.

"It" means their introduction and seminal study of the graph invariant χ_L, which was begun independently by Vizing [84]. Mildly paraphrased, [31] continues by posing the following question raised by Dinitz at the Southeastern Conference on Combinatorics, Graph Theory, and Computing in Boca Raton, April 1979:

> Given an $n \times n$ array of n-sets, is it always possible to choose one element from each set, keeping the chosen elements distinct in every row and distinct in every column?

The Dinitz conjecture—the assertion that his question has an affirmative answer—had been circulating since the late 1970s; see [48, p. 202] or [51, p. 320]. On his home page [22], Dinitz indicates that he posed it to Erdős in 1979. Now a consequence of a theorem of Galvin [33], it can be stated as $\chi'_L(K_{n,n}) = n$. During 1978–1994, Dinitz' conjecture stood as perhaps the highest-profile unproved case of the List-Colouring Conjecture 1.2, to which we turn shortly. First we consider

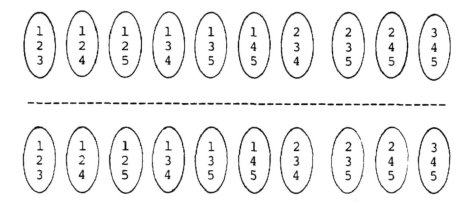

Fig. 11.1 List assignment to $V(K_{10,10})$ exhausting $\binom{[5]}{3}$ on both parts (from [31])

vertex list colouring because the conjectured general behaviour of χ'_L contrasts so strikingly with the actual behaviour of χ_L.

Figure 11.1, copied from [31], depicts a list assignment to the vertices of $K_{10,10}$. In both the top (X) and bottom (Y) parts, each 3-set in $\binom{[2 \cdot 3 - 1]}{3} = \binom{[5]}{3}$ appears as the list on some vertex. Let us suppose that $K_{10,10}$ admits a proper list colouring $\sigma : V \to [5]$ respecting this assignment. By symmetry between X and Y, we may assume that σ assigns at most $2 (= 3 - 1)$ colours ℓ and ℓ' to X; for if σ assigns at least 3 colours to both of X and Y, then some colour is used on both parts of the bipartition, so σ is improper. But now some vertex $x \in X$ has the list $[5] \smallsetminus \{\ell, \ell'\}$, which implies that x failed to receive a legal colour under σ. This contradiction shows that there can exist no such proper list colouring of $V(K_{10,10})$, with the given assignment of 3-sets, and hence that $\chi_L(K_{10,10}) > 3$.

Erdős et al. [31] observed that χ_L can be forced above every prescribed integer $t \geq 1$ for a sufficiently large complete bipartite graph. Given t, let $n := \binom{2t-1}{t}$ and set a colour palette $S := \{1, 2, \ldots, 2t - 1\}$. If $\{X, Y\}$ is a bipartition of $K_{n,n}$, then, as a colour list, assign each t-set in $\binom{S}{t}$ to a vertex of X and a vertex of Y. Using 't' here in the role of '3' in the preceding paragraph gives the following result.

Theorem 1.1 ([31]). If $n = \binom{2t-1}{t}$, then $\chi_L(K_{n,n}) > t$.

Of course, for the ordinary chromatic number, we have $\chi(K_{n,n}) = 2$, so even as commonplace a family as complete bipartite graphs may witness

$$\frac{\chi_L}{\chi} \to \infty \quad (\text{as } n \to \infty). \tag{11.1}$$

But perhaps not for line graphs. The list-colouring conjecture says that for them, the trivial inequality $\chi_L \geq \chi$ holds with equality:

Conjecture 1.2 (LCC). *If G is a multigraph, then $\chi'_L(G) = \chi'(G)$.*

Though Conjecture 1.2 is not an Erdős prize problem, resolving it would no doubt bring fame and glory beyond all conceivable bounds, at least within *some* circles.

The basic relations

$$\Delta \le \chi' \le \chi'_L \le 2\Delta - 1 \qquad (11.2)$$

show that the ratio χ'_L / χ' is sandwiched between 1 and 2, so the divergence phenomenon (11.1) cannot occur when colouring edges. When pondering Conjecture 1.2, it's worthwhile keeping in mind the two seminal bounds on χ' in terms of Δ. These are

$$\chi'(G) \le \left\lfloor \frac{3\Delta(G)}{2} \right\rfloor \text{ for multigraphs } G \qquad (11.3)$$

(due to Shannon [80]) and

$$\chi'(G) \le \Delta(G) + 1 \text{ for simple graphs } G \qquad (11.4)$$

(due to Vizing [83]). Both bounds are sharp. For (11.3), this is seen by taking G to have three vertices, each pair joined by either $\lfloor \Delta/2 \rfloor$ or $\lceil \Delta/2 \rceil$ parallel edges (as needed to achieve maximum degree Δ). Now each pair of these $\lfloor 3\Delta/2 \rfloor$ edges shares a common endpoint, so the edges must all receive distinct colours. For (11.4), this is seen by taking G to be any nonempty regular simple graph of odd order. If G is Δ-regular and has $2k + 1$ vertices, then it has $\Delta(2k + 1)/2$ edges. Each colour class of an edge colouring has size at most k, so with G being nonempty, we have $\chi' \ge \Delta(2k + 1)/2k > \Delta$.

Conjecture 1.2 has been attributed to numerous mathematicians, including Vizing [84], Gupta, Erdős [28], Albertson and Tucker, and Albertson and Collins. I pieced this list together from [1, 18, 19, 37, 48, 51] and refer readers to these sources for more details. In 1985, Bollobás and Harris [8] were the first authors to state the conjecture in print. However, Vizing should probably be credited for its earliest public formulation, in September 1975, during a conference problem session in Odessa, Ukraine; see reference [11] in [60]. Albertson (on the list above) also shared this information with Kahn (see [51]).

11.1.1 *Progress Towards the List-Colouring Conjecture*

Results shedding light on the LCC fall into (at least) three categories: those producing upper bounds on χ'_L—typically in terms of Δ—for either multigraphs or simple graphs, those obtaining an asymptotic estimate $\chi'_L \sim \chi'$ as these invariants grow large, and those achieving $\chi'_L = \chi'$ exactly for specific classes of graphs. We tabulate a sample of theorems of the last type in Table 11.1 on the following page. Considering (11.16) in this chapter's Appendix, it should come as no surprise that

Table 11.1 A sample of graph classes satisfying $\chi'_L = \chi'$, 1995–2014

Graph class (brief description)	Reference	Publication year
Bipartite multigraphs	[33]	1995
Δ-regular Δ-edge chromatic planar multigraphs	[25]	1996
Complete graphs of odd order	[39]	1997
Planar simple graphs with $\Delta \geq 12$	[12]	1997
Multigraphs with perfect line graphs	[70]	1999
(Correction to preceding)	[71]	2003
Multicircuits	[90]	1999
Series-parallel (aka K_4-minor free) simple graphs	[49]	1999
Outerplanar (aka $K_{2,3}$- and K_4-minor free) simple graphs	[89]	2001
Near-outerplanar (aka $K_{2,3}$- or K_4-minor free) simple graphs	[41]	2006
Planar, simple, contains no C_4, and $\Delta \geq 7$	[44]	2006
Euler char $\varepsilon \geq 0$, simple, contains no C_5, and $\Delta \geq 11$	[88]	2007
$\varepsilon \geq 0$, simple, contains neither C_4 nor C_5, and $\Delta \geq 7$	[88]	2007
Planar, simple, contains neither C_4 nor C_6, and $\Delta \geq 6$	[63]	2008
$\varepsilon < 0$, simple, and $\Delta \geq \sqrt{25 - 24\varepsilon} + 10$	[91]	2008
$\varepsilon \geq 0$, simple, no two 3-cycles sharing edge, and $\Delta \geq 9$	[21]	2009
Planar, simple, no two 4-cycles sharing vertex, and $\Delta \geq 8$	[64]	2009
Planar, simple, no C_3 sharing edge with a C_4, and $\Delta \geq 8$	[61]	2011
1-planar simple graphs with $\Delta \geq 21$	[92]	2012
Pseudo-outerplanar simple graphs with $\Delta \geq 5$	[82]	2014
Complete graphs of even, successor of a prime, order	[78]	2014

the LCC is often studied in conjunction with conjectures about χ'' or χ''_L; see, e.g., [12, 21, 23, 37, 41, 44, 60, 61, 63, 64, 89, 91, 92]. To limit our scope, we shall not discuss total colouring as we weave together the three sorts of results on the LCC, mostly chronologically.

The survey [1] points out that when $\Delta > 2$, the trivial upper bound for χ'_L in (11.2) can be dropped to $2\Delta - 2$ using a Brooks-type theorem for χ_L, proved independently in [31] and [84]. The first significant improvement was made by Bollobás and Harris [8], who showed that simple graphs satisfy $\chi'_L \leq 11\Delta/6 + o(\Delta)$. Hind [42]—a subsequent student of Bollobás—proved that $\chi'_L \leq 9\Delta/5$ for multigraphs and $\chi'_L \leq 5\Delta/3$ for triangle-free multigraphs. About the same time, Chetwynd and Häggkvist [18] established the bound $\chi'_L \leq 9\Delta/5$ for triangle-free simple graphs. Although their article followed Hind's stronger results, these appeared exclusively in his doctoral dissertation and were not widely known at the time. Soon thereafter, he and Bollobás [9] published a further improvement for simple graphs: $\chi'_L \leq 7\Delta/4 + o(\Delta)$.

Next, Kostochka [60] showed that (necessarily simple) graphs of sufficiently large girth (roughly, girth at least $\Omega(\Delta \log \Delta)$) satisfy

$$\chi'_L \leq \Delta + 1. \tag{11.5}$$

As in [8, 9, 18], Kostochka used a clever recolouring procedure along the lines of the proof of Vizing's theorem [83]. That all simple graphs satisfy (11.5) has come to be called Vizing's conjecture, from [84]. Vizing's theorem (11.4) shows that this conjecture would follow from the LCC. Besides Kostochka, several other authors (e.g. [11, 17, 23, 85]) have established the intermediate (11.5) for various classes of graphs, but we make no attempt to survey this line of inquiry towards the LCC.

In a separate but concurrent advance, Kahn applied powerful probabilistic arguments—including the so-called 'incremental random' method—first to obtain

$$\chi'_L \leq \Delta + o(\Delta) \text{ as } \Delta \to \infty \qquad (11.6)$$

for simple graphs. This was announced in [51] as part of the proceedings of a 1991 meeting and presented in full detail in [52]. Together with (11.2), the conclusion (11.6) shows that simple graphs satisfy

$$\chi'_L \sim \chi' \qquad (11.7)$$

as either of these parameters (or Δ) grows large; i.e., simple graphs satisfy Conjecture 1.2 asymptotically. Kahn [55] later proved (11.7) for multigraphs as well. This oversimplifies these two Kahn articles because we're viewing the results through an 'LCC lens'. In [52], he actually proved a hypergraph result that specializes to give (11.6) for simple graphs. And in [55], he actually proved that

$$\chi'_L \sim \chi'^* \text{ as } \chi'^* \to \infty \qquad (11.8)$$

for multigraphs, which, together with his intermediate theorem [53] that such graphs satisfy $\chi' \sim \chi'^*$ as $\chi'^* \to \infty$, yields (11.7). Kahn's results in [53, 55] both rely on the (approximate) stochastic independence properties of 'hard-core' distributions on the set of matchings of a graph—first proved in [56]—in passing from fractional to ordinary edge or list-edge colourings.

Let us rewind to the early 1990s. Alon [1], Häggkvist and Chetwynd [37], and Kostochka [60] noted the state of the LCC art at that time. The conjecture had been proven only for forests, graphs with $\Delta \leq 2$, snarks [40], graphs with no cycles of length exceeding three [34], complete bipartite graphs $K_{r,s}$ with $7r \leq 2s$ [36], planar graphs with $\Delta \geq 14$ [11], and a few small cases (viz. K_4, $K_{2,s}$, $K_{3,3}$, $K_{4,4}$, and $K_{6,6}$). As it turns out, some important new cases of Conjecture 1.2 were about to fall. First, using tools in [3], Janssen [47] established that $\chi'_L = \chi'$ for complete bipartite graphs $K_{n-i,n}$ with unequal parts. Quickly on the heels of Janssen's breakthrough, Galvin [33] proved Conjecture 1.2 for all bipartite multigraphs, which, in particular, settled the Dinitz conjecture in the affirmative. Perhaps surprisingly, Galvin's attack did not pick up where Janssen's left off. He combined the 'Bondy-Boppana-Siegel lemma' (observed in [3]) with a result of Maffray [65] to give an unexpectedly simple and elementary proof.

These developments provided both exhilaration and frustration to combinatorics grad students, especially those who'd wrestled with Conjecture 1.2. I remember having proved some simple cases, c. 1990, eventually becoming stuck and changing research direction. Then came Janssen's result, followed closely by a photocopied manuscript in Galvin's hand—see Figure 11.2—making the rounds at Hill Center, Rutgers. My friend and fellow student Paul O'Donnell gave me a copy, and within the week, I was presenting it in Mike Saks' office. This was early 1994, my final year as a student.

   Almost concurrently with Galvin's announcement, Häggkvist and Janssen circulated a manuscript [38] of their own, in which they applied Janssen's methods in [47] to prove that

$$\chi'_L \leq \Delta + O\left(\Delta^{2/3}\sqrt{\log \Delta}\right) \tag{11.9}$$

for simple bipartite graphs. Though Galvin's result eclipsed this before it would be published, the authors had already announced that their ideas could be extended to achieve (11.9) for all simple graphs and hence to establish another, now explicit, upper bound for χ'_L as in (11.6). Häggkvist and Janssen also proved that complete graphs of order n satisfy

$$\chi'_L(K_n) \leq n, \tag{11.10}$$

which yields Conjecture 1.2 when n is odd. Their results eventually appeared in [39].

   The 'incomplete survey' segment of my lecture [58] ended here, but I take the opportunity next to present the highlights from [39] to the present. Table 11.1 on p. 175 gives the highlights of the highlights during 1995–2014.

11.1.2   A Miscellany of LCC-related Results

Though we aren't trying for complete or detailed coverage, our sample in this section should paint a representative picture of the LCC landscape from the mid-1990s until this writing.

   As it served Janssen [38, 39, 47], the main tool in [3] served Ellingham and Goddyn [25] in their LCC investigation. They used it to relate the edge choosability of a multigraph G to certain coefficients in the 'graph monomial' of G's line graph. They verified the LCC for various families of 1-factorable graphs, including Δ-regular Δ-edge chromatic planar multigraphs. Eventually, Schauz [77] developed an algebraic framework—even more general than Alon and Tarsi's—from which Ellingham and Goddyn's results also follow; see the end of this section for a more striking consequence of Schauz's machinery.

   Galvin's method in [33] ignited a spark that led Borodin et al. [12] to generalize both Galvin's theorem and Shannon's theorem (11.3). They proved that if edges $e = \{x, y\}$ of certain multigraphs G are assigned lists L_e, then G admits an L-edge colouring under either of the following sets of hypotheses: (i) G is bipartite

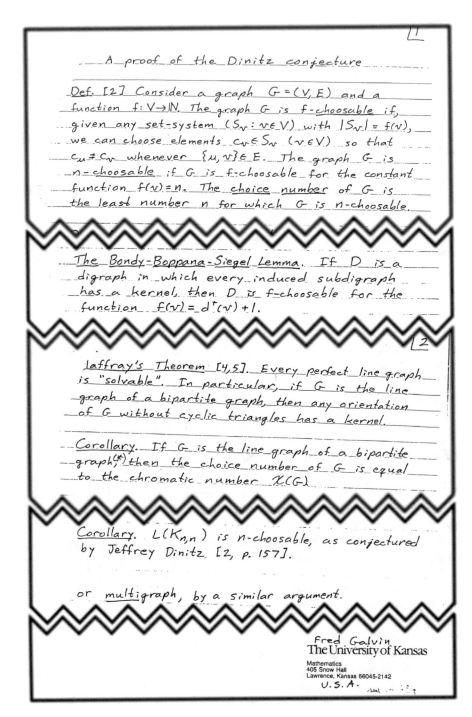

A proof of the Dinitz conjecture

Def. [2] Consider a graph $G = (V, E)$ and a function $f: V \to \mathbb{N}$. The graph G is f-choosable if, given any set-system $(S_v : v \in V)$ with $|S_v| = f(v)$, we can choose elements $c_v \in S_v$ $(v \in V)$ so that $c_u \neq c_v$ whenever $\{u, v\} \in E$. The graph G is n-choosable if G is f-choosable for the constant function $f(v) = n$. The choice number of G is the least number n for which G is n-choosable.

The Bondy-Boppana-Siegel Lemma. If D is a digraph in which every induced subdigraph has a kernel, then D is f-choosable for the function $f(v) = d^+(v) + 1$.

Jaffray's Theorem [4,5]. Every perfect line graph is "solvable". In particular, if G is the line graph of a bipartite graph, then any orientation of G without cyclic triangles has a kernel.

Corollary. If G is the line graph of a bipartite graph,(*) then the choice number of G is equal to the chromatic number $\chi(G)$.

Corollary. $L(K_{n,n})$ is n-choosable, as conjectured by Jeffrey Dinitz [2, p. 157].

or multigraph, by a similar argument.

Fred Galvin
The University of Kansas

Mathematics
405 Snow Hall
Lawrence, Kansas 66045-2142
U.S.A.

Fig. 11.2 Excerpts from Galvin's handwritten manuscript [1994]

with each $|L_e| \geq \max\{\deg(x), \deg(y)\}$; or (ii) G is arbitrary and each $|L_e| \geq \max\{\deg(x), \deg(y)\} + \lfloor\min\{\deg(x), \deg(y)\}/2\rfloor$. In particular, (i) implies that $\chi'_L \leq \Delta$ in the bipartite case (Galvin's theorem), while (ii) implies that $\chi'_L \leq \lfloor 3\Delta/2 \rfloor$ in general (a list analogue of (11.3)). These authors also settled the LCC for simple planar graphs (indeed, those with nonnegative Euler characteristic) with $\Delta \geq 12$. In addition, they gave other sufficient conditions in terms of Δ and the 'maximum average degree' of G in order that $\chi'_L = \Delta$.

Another generalization of Galvin's theorem was obtained by Peterson and Woodall, who in [70] (and [71]) established that multigraphs with perfect line graphs satisfy the LCC. Galvin's result follows because bipartite multigraphs have this property. A *multicircuit* is a multigraph whose underlying simple graph is a cycle. In [90], Woodall proved the LCC first for multicircuits and, then, building on his results with Peterson [70], for any multigraph in which every block has one of four properties: (i) is bipartite, (ii) is a multicircuit, (iii) has at most four vertices, or (iv) has its underlying simple graph of the form $K_{1,1,p}$.

The same year (1999), Juvan et al. [49] confirmed the LCC for simple *series-parallel* graphs, i.e., for (simple) K_4-minor-free graphs. Precisely stated, their result asserts that for every integer $k \geq 3$, for every such graph G with $\Delta \leq k$, and for every list assignment $\{L_e : e \in E(G)\}$ with each $|L_e| \geq k$, there exists an L-edge colouring of G. This implies a later result of Wang and Lih [89], who proved, among other related theorems, the LCC for simple outerplanar graphs. The implication is perhaps most easily seen by recalling the well-known characterization of outerplanar graphs as those that are both $K_{2,3}$- and K_4-minor free.

In a pair [86, 87] of papers from 2005 and 2007, Wang and Huang investigated the LCC for Cartesian products of n-cycles C_n and order-m paths P_m. In the first article, they confirmed the conjecture for products $C_n \,\square\, C_m$ with both of n, m odd. As Galvin's theorem handles the case with both of n, m even, we see that the LCC holds for these products whenever n and m have the same parity. In the second, they confirmed the LCC for products $C_n \,\square\, P_m$ when $m \geq 3$; indeed, they showed that the latter graphs all satisfy $\chi'_L = \chi' = 4$.

A graph is *near outerplanar* if it is either $K_{2,3}$- or K_4-minor free. Hetherington and Woodall [41] used an efficient approach to proving the LCC for simple near-outerplanar graphs; viz., they proved it for (simple) $(\overline{K_2} \vee (K_1 \cup K_2))$-minor-free graphs. To see that this takes care of near-outerplanar graphs, notice that the join $\overline{K_2} \vee (K_1 \cup K_2)$ can be viewed both as $K_{2,3}$ plus an edge connecting two degree-two vertices and as K_4 with one subdivided edge. We should point out that the LCC verification is a straightforward consequence of the main results of this paper, whose larger scope included total list-colouring near-outerplanar graphs.

Starting in the mid-2000s, numerous authors discovered combinations of forbidden cycle and minimum-Δ conditions sufficient for simple graphs to satisfy the LCC. These account for eight entries in Table 11.1; we cover them in slightly more detail here. Graphs in this paragraph are always simple. Hou et al. [44] considered planar graphs containing no cycle with length between 4 and k (inclusive). They confirmed the LCC whenever the pair (k, Δ) is entrywise at least any member of the set $\{(4, 7), (5, 6), (8, 5), (14, 4)\}$. Using [12] together with their own tools, Wang and Lih [88] proved several results of a similar flavour, now for graphs embedded in any surface of nonnegative Euler characteristic ε. (Aside from the Euclidean plane, this adds the projective plane, the Klein bottle, and the torus to the allowable host surfaces.) For these graphs G, they showed that

$$\chi'_L = \chi' \ (= \Delta) \tag{11.11}$$

under any one of the following conditions: (i) G contains neither 4- nor 5-cycles and $\Delta \geq 7$; (ii) G contains no 4-cycle and $\Delta \geq 9$; (iii) G contains no two 3-cycles sharing a vertex and $\Delta \geq 9$; (iv) G contains no two 3-cycles sharing an edge and $\Delta \geq 11$; or (v) G contains no 5-cycle and $\Delta \geq 11$. For planar graphs, again, Liu et al. [63] also established (11.11) and so confirmed the LCC for graphs containing neither 4- nor 6-cycles and satisfying $\Delta \geq 6$ or containing neither 5- nor 6-cycles and satisfying $\Delta \geq 7$. Next we mention a result of Wu and Wang [91], who considered graphs embeddable on a surface with ε negative. For such graphs, they extended a result from [12] by proving that (11.11) holds whenever $\Delta \geq \sqrt{25 - 24\varepsilon} + 10$. The next two articles return to the case of nonnegative ε. For these graphs G, Cranston [21] obtained (11.11) whenever G does not contain two 3-cycles sharing an edge and $\Delta \geq 9$; this covers just one theorem among several in his paper concerning both total and edge list colouring. Liu et al. [64] reached the same conclusion (also among other results) for graphs with $\varepsilon \geq 0$, not containing two 4-cycles sharing a vertex and having $\Delta \geq 8$. Li and Xu [61] published the most recent result of this flavour chosen for this chapter. They established (11.11) for planar graphs not containing a 3-cycle sharing an edge with a 4-cycle and satisfying $\Delta \geq 8$. Like many of the theorems mentioned in this paragraph, this result was proved using 'discharging', a technique renowned for its effectiveness on the 4-colour problem.

The LCC has also been verified for certain graphs enjoying one of a couple of other variations of planarity. A graph is 1-*planar* if it can be drawn in the plane so that each edge is crossed by at most one other edge. Zhang et al. [92] proved that simple 1-planar graphs with $\Delta \geq 16$ satisfy Vizing's conjecture (11.5), while those with $\Delta \geq 21$ satisfy $\chi'_L \leq \Delta$, hence (11.11). A graph is *pseudo-outerplanar* if each of its blocks has a plane embedding so that its vertices lie on a fixed circle and its edges lie inside the disc of this circle, with each of them crossing at most one of the others. Tian and Zhang [82] confirmed the LCC for simple pseudo-outerplanar graphs with $\Delta \geq 5$.

In light of the early progress [39] on the LCC for complete graphs (see (11.10)), it is vexing that the conjecture remains open for most of the even-order cases. Perhaps it's not surprising because the full even-order case implies the odd: just notice that due to $K_{2m-1} \subseteq K_{2m}$, if the LCC holds for K_{2m}, then

$$\chi'_L(K_{2m-1}) \leq \chi'_L(K_{2m}) = \chi'(K_{2m}) = 2m - 1,$$

which gives the LCC for odd n. For even order, the first nontrivial case is K_4, which satisfies the LCC by a result in [25] (because K_4 is 3-regular, 3-edge chromatic, and planar). An elementary proof of this case was also given by Cariolaro and Lih [14].

A few years later, in [15], Cariolaro collaborated with three others (including his father) to prove the LCC for K_6. Similar tools and a more sophisticated approach eventually led to the next result by one of those three, Schauz, who in [78] showed that $\chi'_L(K_{p+1}) = p$ for all odd primes p (hence confirming the LCC for these graphs). Schauz's proof uses a version of the 'Combinatorial Nullstellensatz' (see [2]) introduced by himself in [77]. His beautiful theorem offers a perfect point to conclude our discussion of the LCC.

11.2 The Erdős-Faber-Lovász Conjecture

Now we turn to a well-known Erdős prize problem, for which he[0] offered $500 for either a proof or a counterexample. For the last twenty years of his life, Erdős considered it one of his favourite combinatorial problems—see, e.g., [27, 29, 30].

One appealing formulation, due to Haddad and Tardif [35], supposes n committees, no two sharing more than one common member and each with n members. They hold their meetings in a common boardroom containing n distinguishable chairs. The Erdős-Faber-Lovász (EFL) conjecture asserts that it is possible for each person to select a 'persistent' chair, that is, one to be used during each meeting of every committee to which she/he belongs. Looking back, years after its original formulation (not the one above), Faber [32] restated the problem succinctly:

> given n sets, no two of which meet more than once and each with n elements, color the elements with n colors so that each set contains all the colors. (11.12)

As Erdős [26] remarked following one of the earliest published statements of the EFL conjecture: "It clearly fails if we have $n + 1$ sets." (Thanks to Alex Soifer [81, p. 363] for pointing out this reference.)

The conjecture's genesis provides a quintessential example of Pólya's adage (from [72]): "If you can't solve a problem, then there is an easier problem you can't solve: find it." During a 1972 meeting [6] at the Ohio State University (Columbus), Vance Faber, together with Paul Erdős and László Lovász, began discussing a variety of linear hypergraph problems. A few weeks later, they reconvened at a tea party in Faber's apartment in Boulder, CO, where he was a postdoc at the National Center for Atmospheric Research. The difficulties posed by these problems led the three mathematicians to create the more elementary 'n sets problem' (11.12). In [32], Faber relates that the three initially figured the conjecture would be easy to resolve and planned to gather the next day to write the proof. Instead, in the spirit espoused by Pólya, the new problem itself became a source for numerous 'easier problems', continuing to this day.

---

[0]Erdős' tradition of offering cash rewards for certain of his favourite problems lives on, now underwritten by Chung and Graham [19] in his honour.

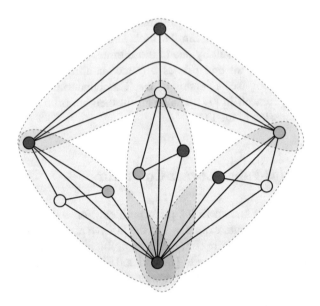

Fig. 11.3 An EFL graph with $n = 4$ (Public domain image by David Eppstein, from http://en. wikipedia.org/wiki/File:Erdős-Faber-Lovász_conjecture.svg; usage permission is granted under the GNU Free Documentation License.)

We present a sample of EFL-related results. More thorough treatments can be found, e.g., in the surveys [51] (up to 1994) and [75] (to 2007).

Another natural formulation casts the EFL assertion in terms of graphs (with a moniker borrowed from [75]):

Conjecture 2.1 (EFL: graphical). *If, for $1 \leq i \leq n$, the n-cliques $K_n(i)$ on vertex sets V_i satisfy $|V_i \cap V_j| \leq 1$ for $i \neq j$, then the union $G := \bigcup_{i=1}^{n} K_n(i)$ has $\chi(G) = n$.*

Graphs G appearing in the statement of Conjecture 2.1 are called *EFL graphs*. It's worth noting that these graphs may contain n-cliques besides the obvious ones. In Figure 11.3, the upper yellow, leftmost (magenta), rightmost (green), and lowest (red) vertices form a 4-clique that is not one of the building block K_4's. Noticing this phenomenon, one might wonder whether G can contain cliques even larger than the $K_n(i)$'s.

To settle this question, consider a clique H in G. Such an H gives rise to a hypergraph $\mathcal{H}$ with vertex set $\{K_n(i)\}_{i=1}^{n}$ by a construction called *dualizing*: for each vertex x of H, we introduce an edge A_x of $\mathcal{H}$ via $A_x := \{K_n(i): x \in V_i\}$. (Viewed as vertices in $\mathcal{H}$, some of the $K_n(i)$'s may lie in none of $\mathcal{H}$'s edges, but this doesn't interfere with our discussion.) As H is a clique in G, the hypergraph $\mathcal{H}$ is intersecting, but because the original cliques $K_n(i)$ are pairwise edge-disjoint, we have $|A_x \cap A_y| = 1$ whenever x and y are distinct vertices of H. This is exactly the set-up for a 1948 result of de Bruijn and Erdős:

Theorem 2.2 ([13]). *If $\mathcal{H}$ is a hypergraph of order n, with $|A \cap B| = 1$ for all distinct $A, B \in \mathcal{H}$, then $|\mathcal{H}| \leq n$ (with equality only when $\mathcal{H}$ is a [possibly degenerate] projective plane or is a star with a loop at the central vertex).*

Let us correct a propagating bibliographic error that crept in to the lecture version [58] of this chapter; cf. [51, 54, 75]. The citation for Theorem 2.2 is indeed [13] and not a 1946 de Bruijn solo article where he gave us the 'universal cycles' now carrying his name.

Back to the $\mathcal{H}$ dualizing H. As a consequence of Theorem 2.2, we have $|\mathcal{H}| \leq n$, whence $|V(H)| \leq n$, which shows that the clique number $\omega(G)$ is at most n. Of course, now $\omega(G) = n$.

The device used to explore the cliques in G also leads to a more commonly studied version of the EFL conjecture. First notice that vertices lying in exactly one of the $K_n(i)$'s defining G play no essential role in determining $\chi(G)$: if G' is obtained from G by deleting these vertices, then $\chi(G) \leq n$ if and only if $\chi(G') \leq n$. Because the former relation here is equivalent to $\chi(G) = n$, confirming Conjecture 2.1 amounts to proving the latter. If we dualize G', as we did for cliques H above, then we obtain a hypergraph $\mathcal{H}$, again with vertex set $\{K_n(i)\}_{i=1}^{n}$. Now $\mathcal{H}$ may fail to be intersecting, but at least it's simple. By construction, vertex-colouring G' is equivalent to edge-colouring $\mathcal{H}$, so we arrive at a hypergraph colouring variant of Conjecture 2.1:

Conjecture 2.3 (EFL: hypergraphical). *If $\mathcal{H}$ is a simple hypergraph of order n, then $\chi'(\mathcal{H}) \leq n$.*

Hindman [43] first noticed the reduction just described and the equivalence of Conjectures 2.1 and 2.3.

Theorem 2.2 not only reveals the clique number of the EFL graphs but also settles the EFL conjecture for intersecting hypergraphs. For if $\mathcal{H}$ is both simple and intersecting, then it satisfies the de Bruijn-Erdős hypothesis, and so $\chi'(\mathcal{H}) \leq |\mathcal{H}| \leq n$.

We've already encountered several instances where Conjecture 2.3 is sharp; to wit, (i) $\mathcal{H}$ is a (possibly degenerate) projective plane, or (ii) $\mathcal{H}$ is a complete graph of odd order $n \geq 3$. The cases (i) here align with the sharp cases in Theorem 2.2, from which we omit the looped star example only because it is not simple. The cases (ii) here are among the sharp cases in Vizing's theorem (11.4) because $K_{2\ell+1}$ is a regular simple graph of odd order with $\Delta = 2\ell$. Other authors (e.g. [51, 75]) have noted that minor modifications in case (ii) also yield equality in Conjecture 2.3; we add some flesh to this observation with just one example. If $\ell \geq 1$, then any graph G obtained from $K_{2\ell+1}$ by deletion of fewer than ℓ edges is a sharp EFL instance. For suppose that $D\ (< \ell)$ counts the deleted edges. With G of odd order, it satisfies $\nu \leq (n-1)/2 = \ell$, and because of the universal bound

$$\chi' \geq \frac{m}{\nu}\,, \tag{11.13}$$

we obtain

$$\chi'(G) \geq \frac{\ell(2\ell + 1) - D}{\ell} = 2\ell + 1 - \frac{D}{\ell} > 2\ell,$$

so that $\chi'(G) \geq 2\ell + 1$. Of course, Vizing's theorem gives the reverse bound, confirming our claim about G.

While we're on the topic of Vizing's theorem, we ought to point out that it implies Conjecture 2.3 when $\mathcal{H} = G$ is a (simple) graph. In fact, it gives a stronger bound unless $\Delta(G) = n - 1$. But in this case, we may as well view G as a subgraph of K_n and colour all of $E(K_n)$ with at most n colours (which possibility is well known and easier to prove than Vizing's theorem; cf. [5]). Restricting this colouring to $E(G)$ gives the required EFL bound, $\chi'(G) \leq n$, without having to resort to the full power of Vizing's theorem. See [54] for a generalization of Conjecture 2.3 which, in the case of graphs, specializes to Vizing's theorem exactly.

11.2.1 Progress Towards the EFL Conjecture

Basic counting shows that in a simple hypergraph $\mathcal{H}$, an edge A of size $k \, (\geq 2)$ intersects at most

$$\frac{k(n - k)}{k - 1} \tag{11.14}$$

other edges of size at least k. Indeed, since $\mathcal{H}$ is simple, each such edge has a single vertex in common with A, and for each of the k vertices x in A, each of the $n - k$ vertices not in A appears in at most one edge with x. When $k \geq 2$, the expression (11.14) is at most $2n - 4$, so $\chi'(\mathcal{H}) \leq 2n - 3$ follows by greedily edge-colouring $\mathcal{H}$ in nonincreasing order of edge sizes. If the minimum such size is at least three, then (11.14) yields an upper bound on $\chi'(\mathcal{H})$ of roughly $3n/2$. Independently, Mitchem [66] and Chang and Lawler [16] established the same bound, namely, $\chi'(\mathcal{H}) \leq \lceil 3n/2 - 2 \rceil$, for all simple hypergraphs. Much more recently, Sánchez-Arroyo [76] used the upper bound (11.14), in its dual form, to prove Conjecture 2.3 under the added hypothesis that $\mathcal{H}$'s minimum edge size exceeds $\sqrt{n}$ (more on this after Theorem 2.7).

Ten years after the EFL conjecture's first appearance, Seymour [79] published the following significant result.

Theorem 2.4. *If $\mathcal{H}$ is a simple hypergraph of order n, then $\nu(\mathcal{H}) \geq |\mathcal{H}|/n$, with equality only if $\mathcal{H}$ is either (i) a (possibly degenerate) projective plane or (ii) a complete graph of odd order.*

Because of (11.13), Theorem 2.4 follows from the hypergraphical EFL conjecture. In the same paper where he proved this theorem, Seymour conjectured a fractional analogue of Conjecture 2.3 which he later confirmed in an article [57] with Kahn:

Theorem 2.5 (EFL: fractional). *If $\mathcal{H}$ is a simple hypergraph of order n, then $\chi'^*(\mathcal{H}) \leq n$.*

Coincidentally, I took Paul Seymour's structural graph theory course the autumn preceding their submission of the manuscript [57]. Because he used Jeff Kahn's office during that semester, office hours would interrupt their collaboration. Once or twice I wondered about their chalkboard hieroglyphics—now I think I know what they were about.

By LP duality, Theorem 2.5 is equivalent to the statement that for a simple hypergraph $\mathcal{H}$, every function $w \colon \mathcal{H} \to \mathbb{R}^+$ that is feasible for problem (11.18) (in this chapter's Appendix) satisfies

$$\sum_{A \in \mathcal{H}} w(A) \leq n.$$

Noticing the feasibility of $w \equiv 1/\nu(\mathcal{H})$, we see that Theorem 2.4 follows from Theorem 2.5, which (because of $\chi'^* \leq \chi'$) places the latter result between the former and the EFL conjecture itself. In proving Theorem 2.5, Kahn and Seymour used Motzkin's lemma (from [68]), which contributed to a simpler argument than the one originally used by Seymour in proving Theorem 2.4. Though these are no doubt beautiful theorems, 20 plus years ago, Kahn [51] lamented that "there is presently little reason to think that these results are tending toward a proof of [Conjecture 2.3]."

In a contemporaneous article [50], Kahn proved that the EFL conjecture is asymptotically correct:

Theorem 2.6 (EFL: asymptotic). *Simple hypergraphs $\mathcal{H}$ of order n satisfy $\chi'(\mathcal{H}) < n + o(n)$.*

In 'epsilontics', this means that for every $\varepsilon > 0$, if $\mathcal{H}$ is a simple hypergraph with sufficiently large order n, then $\chi'(\mathcal{H}) < n + \varepsilon n$. For this outstanding result, Erdős "immediately gave him [Kahn] a consolation prize of \$100", a snippet of folklore believed widely among Rutgers combinatorics students at the time and confirmed in this quote from [30]. In [51], Kahn sketches a second proof of Theorem 2.6 using (a weak version of) the main result in [52].

Faber himself made one of the more recent contributions to the EFL literature.

Theorem 2.7 ([32]). *Every simple, Δ-regular, r-uniform hypergraph $\mathcal{H}$ of order n satisfies the following: (i) if $\Delta \leq r+1$, then $\chi'(\mathcal{H}) \leq n$; and (ii) for some (universal) constant C, if $r \geq C$ and $n \geq Cr^2$, then $\chi'(\mathcal{H}) \leq n$.*

The proof relies on a theorem of Alon et al. [4].

Sánchez-Arroyo's result mentioned at the start of this section shows that we may assume here that $n \geq r^2$. Combining this with Theorem 2.7 narrows the range of values of n, relative to a finite set of choices for r, for which EFL counterexamples

Photo courtesy of V. Faber

Fig. 11.4 Vance Faber (January 2014) investigating his 'Faber-Gator conjecture'

can still exist when $\mathcal{H}$ is uniform and regular. Given Δ, potential counterexamples are restricted by $r < \Delta - 1$, and a choice with $r \geq C$ bounds n within the finite 'Goldilocks' range $r^2 \leq n < Cr^2$.

Approaching the 2014 Joint Meetings, I figured that the audience of my lecture [58] might appreciate a recent photograph of Professor Faber. After he graciously supplied the one in Figure 11.4—taken on vacation in the Florida Everglades—I used the photo to mark the lecture's endpoint. Here it signals this chapter's wind-down.

11.2.2 A Mini-Miscellany of EFL-Related Results

Augmented by the excellent surveys [51, 75] mentioned before Conjecture 2.1, the preceding section contributes to a reasonably complete catalogue of EFL-related results to 2007. Here we cite a few newer articles together with an earlier one that slipped through the surveys' cracks. Again we make no completeness claim; the point is to highlight some recent EFL progress.

Perhaps because it didn't appear in a combinatorics journal, the article [7] somehow evaded coverage in either of the EFL surveys. Beutelspacher et al. proved the EFL conjecture for projective geometries. In this case, Conjecture 2.3 asserts that $\chi'(PG(d,q)) \leq q^d + q^{d-1} + \cdots + q + 1 =: \Theta_d$. These authors established a more stringent bound for such set systems, namely, for $d \geq 3$, that $(\Theta_{d-1} \leq) \chi'(PG(d,q)) \leq 2\Theta_{d-1} + 2\Theta_{d-2}$. They also showed that the lower bound here is sharp when $q = 2$ or $d = 2^i - 1$ for an integer $i \geq 2$ and provided a table of best-known values of $\chi'(PG(d,q))$ when $d \leq 11$.

Using MathSciNet®, we found only a little more than a handful of EFL-related articles since 2007: [46, 74, 59, 67, 69, 62, 73]. Preprints of both [46] and [59] were cited and discussed in the latest survey [75]; we include them here only to update the citations.

Romero and Sánchez-Arroyo [74] proved the EFL conjecture (in the n sets formulation (11.12)) for the cases when the edges of $\mathcal{H}$ can be labelled with the integers $1, 2, \ldots, n$ so that for every vertex x of $\mathcal{H}$, the labels of the edges containing x can be split into at most two discrete intervals of the form $\{i, i+1, \ldots, i+k\}$. As singletons take this form, their result extends the case in which every vertex is in at most two edges; in dual form, this is the case settled by Vizing's theorem (11.4) (cf. the paragraph preceding Sect. 11.2.1).

Mitchem and Schmidt [67] revisited an early result of the first author [66] and confirmed Conjecture 2.1 for two new classes of EFL graphs. For such a graph G with a vertex x, they call the number of $K_n(i)$'s containing x the *special degree* of x. Then G is *r-uniform* if every one of its vertices has special degree 1 or r (corresponding, if G were dualized, to an r-uniform hypergraph $\mathcal{H}$ in Conjecture 2.3). The theorem from [66]—proved more simply by Mitchem and Schmidt—verifies Conjecture 2.1 for EFL graphs in which each $K_n(i)$ contains at most one vertex of special degree exceeding two. Their newly verified classes are both r-uniform EFL graphs, the first satisfying $3 \leq r \leq n \leq r(r-1)(r-2)+1$ and the second with $r \geq 3$ and carrying additional structure inherited from a 'resolvable transversal design' associated with a certain subgraph of G.

Paul and Germina [69] confirmed Conjecture 2.3 when $\Delta(\mathcal{H}) \leq \sqrt{n + \sqrt{n} + 1}$. By counting, it follows from Sánchez-Arroyo's [76] minimum edge-size hypothesis (noted in the first paragraph of Sect. 11.2.1) that the hypergraphs considered by him satisfy the Paul-Germina hypothesis on $\Delta(\mathcal{H})$ and consequently that their result generalizes [76].

A *b-colouring* of a graph G is a proper colouring of $V(G)$ in every colour class of which there is a vertex having neighbours in all the other colour classes; the *b-chromatic number* $\chi_b(G)$ is the maximum integer k for which G admits a b-colouring using k colours. These concepts were introduced in [45], and Lin and Chang [62] recently discovered a connection between them and the EFL conjecture. They constructed a family $\mathcal{B}_n$ of bipartite graphs with the following property: if Conjecture 2.1 holds, then every $H \in \mathcal{B}_n$ satisfies $\chi_b(H) \in \{n, n-1\}$. Based on this theorem, they proposed a weakening of Conjecture 2.1: if $H \in \mathcal{B}_n$, then $\chi_b(H) \in \{n, n-1\}$. They confirmed this conjecture for a certain class of graphs.

The most recent article on the EFL conjecture appeared in 2014, the year before this writing. Extending Hindman's study in [43] of small cases (up to $n = 10$), Romero and Alonso-Pecina [73] have now confirmed Conjecture 2.1 for $n \leq 12$. Luckily, for ending our EFL discussion and starters on new research, that leaves 13.

Appendix: Notation and Terminology

Our purpose here is to fix the notation and terminology appearing in this chapter and not to provide an exhaustive definition list for related nomenclature. Omissions of a combinatorial or graph-theoretic nature may be found in [10], while linear programming omissions can be rectified using [20].

Sets We denote the sets of real and nonnegative real numbers by $\mathbb{R}$ and $\mathbb{R}^+$, respectively. If n is a positive integer, then $[n]$ means the set $\{1, 2, \ldots, n\}$. If S is a set and k a nonnegative integer, then $\binom{S}{k}$ denotes the set of all k-element subsets of S.

Hypergraphs and Graphs A *hypergraph* consists of a finite set V of *vertices*, together with a finite multiset $\mathcal{H}$ of subsets of V; elements of $\mathcal{H}$ are called *edges*. We follow a common practice and use $\mathcal{H}$ to refer both to a hypergraph and its edge set. The *order* of a hypergraph is the cardinality of V and is usually denoted by n, while m is reserved for the *size* $|\mathcal{H}|$ of $\mathcal{H}$. Most hypergraphs here are *simple*, meaning they contain no singleton edges and any two distinct edges intersect in at most one vertex. The *degree* of a vertex is the number of edges containing it, and $\Delta(\mathcal{H})$ denotes the maximum degree in $\mathcal{H}$. A hypergraph is *regular* if every vertex has degree Δ and in this event is called Δ-*regular*. (We generally omit the argument from a hypergraph invariant when there's no danger of ambiguity.) A hypergraph $\mathcal{H}$ is *uniform* if every edge $A \in \mathcal{H}$ contains the same number r of vertices and in this event is called r-*uniform*. We call $\mathcal{H}$ *intersecting* when $A \cap B \neq \emptyset$ for every pair A, B of edges in $\mathcal{H}$. One natural example of a hypergraph enjoying all of these last three properties is a projective plane $\mathcal{P}$ of 'order' q (for an integer $q \geq 2$). Of course, as a hypergraph, such a $\mathcal{P}$ has order and size $q^2 + q + 1$ and is $(q + 1)$-regular and $(q + 1)$-uniform, and pairwise edge intersections are all singletons. A *degenerate projective plane* (sometimes called a 'near pencil' in the literature) is a hypergraph on the vertex set $[n]$ with edge set $\{\{1, n\}, \{2, n\}, \ldots, \{n - 1, n\}, \{1, 2, \ldots, n - 1\}\}$.

A *multigraph* $G = (V, E)$ is a 2-uniform hypergraph with vertex set V and edge set E; this definition, though not quite standard, conveniently disallows G to contain loops, which get in the way of careful definitions of both vertex and edge colouring. A *simple graph* is a multigraph that is simple in the hypergraph sense. We sometimes use the generic 'graph' when there is no reason to be specific regarding multigraph or simple graph. The maximum number of vertices in a clique of G is denoted by $\omega(G)$.

Colouring For a graph $G = (V, E)$ and a positive integer k, a *k-colouring* of G is a function $\sigma : V \rightarrow [k]$ such that

$$\sigma(x) \neq \sigma(y) \text{ whenever } \{x, y\} \in E. \tag{11.15}$$

Such functions are usually called 'proper colourings', but we never consider improper colourings and thus dispense with the adjective. The least k for which G admits a k-colouring is G's *chromatic number* $\chi(G)$. A *k-list assignment* L is a function that assigns to each vertex x of G a k-set (or 'list') L_x (of natural numbers, say). Given such an L, an *L-colouring* of G is a function $\sigma : V \rightarrow \bigcup_{x \in V} L_x$ such that $\sigma(x) \in L_x$ for each $x \in V$ and the usual colouring condition (11.15) is satisfied. The *list-chromatic number* $\chi_L(G)$ is the least integer k for which G admits an L-colouring for every k-list assignment L. Because one such L has each $L_x = [k]$, we always have $\chi_L \geq \chi$.

Both of χ, χ_L have edge and 'total' analogues. The *chromatic index* $\chi'(G)$ of G can be defined as the chromatic number of the line graph of G and likewise for the *list-chromatic index* $\chi'_L(G)$. The *total graph* of $G = (V, E)$ has vertex set $V \cup E$ and an edge joining every pair of its vertices corresponding to an incident or adjacent pair of objects (vertices or edges) in G. The *total chromatic number* $\chi''(G)$ is the chromatic number of the total graph of G, while the *total list-chromatic number* $\chi''_L(G)$ is defined analogously to $\chi'_L(G)$. It's an exercise to prove that these invariants always satisfy

$$(\Delta + 1 \leq) \ \chi'' \leq \chi''_L \leq \chi'_L + 2 \ (\leq 2\Delta + 1). \tag{11.16}$$

When we consider χ' for hypergraphs $\mathcal{H}$, it's useful to have in mind the connection with matchings. A *matching* in $\mathcal{H}$ is a set of pairwise disjoint edges of $\mathcal{H}$, and we write $\mathcal{M}$ for the set of matchings of $\mathcal{H}$. We denote by $\nu(\mathcal{H})$ the maximum size of a matching in $\mathcal{H}$, i.e., $\max\{|M| : M \in \mathcal{M}\}$. Now $\chi'(\mathcal{H})$ is the least size of a subset of $\mathcal{M}$ whose union is $\mathcal{H}$. This formulation may be cast in linear programming terms. First we define the *fractional chromatic index* $\chi'^*(\mathcal{H})$ as the optimal value of the LP (in the nonnegative orthant of $\mathbb{R}^{\mathcal{M}}$):

$$\min \ \sum_{M \in \mathcal{M}} x(M)$$

$$\text{subject to } \sum_{A \in M \in \mathcal{M}} x(M) \geq 1 \text{ for each } A \in \mathcal{H}. \tag{11.17}$$

Notice that any optimal solution $x \in \mathbb{R}^{\mathcal{M}}$ to the LP (11.17), under the extra constraint that x have integer entries, must have $\{0, 1\}$-entries. Thus $\chi'(\mathcal{H})$ is the optimal value of this integer LP, whose linear relaxation (11.17) defines $\chi'^*(\mathcal{H})$.

We also have one occasion to refer to the LP dual of problem (11.17) (in the nonnegative orthant of $\mathbb{R}^{\mathcal{H}}$):

$$
\begin{aligned}
\max \quad & \sum_{A \in \mathcal{H}} w(A) \\
\text{subject to} \quad & \sum_{A \in M} w(A) \leq 1 \text{ for each } M \in \mathcal{M}.
\end{aligned}
\tag{11.18}
$$

In (11.8)—see Sect. 11.1.1—it would have been natural to write χ'^{*}_{L} in place of χ'^{*}, and indeed, we could have done so because these two invariants turn out to be the same; see, e.g., [55].

Acknowledgements Special thanks to the editors (Ralucca Gera, Stephen Hedetniemi, Craig Larson) for organizing the lectures, for making this volume happen, and for their patience and kindness with the deadlines. Extra thanks to the anonymous referee for the careful reading and a constructive report. Loving thanks to Jennifer Walworth for the editorial advice.

References

1. Alon, N.: Restricted colorings of graphs. In: Surveys in Combinatorics, 1993 (Keele). London Mathematical Society Lecture Note Series, vol. 187, pp. 1–33. Cambridge University Press, Cambridge (1993). MR 1239230 (94g:05033)
2. Alon, N.: Combinatorial Nullstellensatz. Comb. Probab. Comput. **8**(1–2), 7–29 (1999). Recent trends in combinatorics (Mátraháza, 1995). MR 1684621 (2000b:05001)
3. Alon, N., Tarsi, M.: Colorings and orientations of graphs. Combinatorica **12**(2), 125–134 (1992). MR 1179249 (93h:05067)
4. Alon, N., Krivelevich, M., Sudakov, B.: Coloring graphs with sparse neighborhoods. J. Comb. Theory Ser. B **77**(1), 73–82 (1999). MR 1710532 (2001a:05054)
5. Alspach, B.: The wonderful Walecki construction. Bull. Inst. Comb. Appl. **52**, 7–20 (2008). MR 2394738
6. Berge, C. (ed.): Hypergraph Seminar. Lecture Notes in Mathematics, vol. 411. Springer, Berlin/New York (1974). Dedicated to Professor Arnold Ross. MR 0349451 (50 #1945)
7. Beutelspacher, A., Jungnickel, D., Vanstone, S.A.: On the chromatic index of a finite projective space. Geom. Dedicata. **32**(3), 313–318 (1989). MR 1038405 (90m:05032)
8. Bollobás, B., Harris, A.J.: List-colourings of graphs. Graphs Comb. **1**(2), 115–127 (1985). MR 951773 (89e:05086)
9. Bollobás, B., Hind, H.R.: A new upper bound for the list chromatic number. Discret. Math. **74**(1–2), 65–75 (1989). Graph colouring and variations. MR 989123 (90g:05078)
10. Bondy, J.A., Murty, U.S.R.: Graph Theory. Graduate Texts in Mathematics, vol. 244. Springer, New York (2008). MR 2368647 (2009c:05001)
11. Borodin, O.V.: A generalization of Kotzig's theorem and prescribed edge coloring of planar graphs. Mat. Zametki **48**(6), 22–28 (1990). 160. MR 1102617 (92e:05046)
12. Borodin, O.V., Kostochka, A.V., Woodall, D.R.: List edge and list total colourings of multigraphs. J. Comb. Theory Ser. B **71**(2), 184–204 (1997). MR 1483474 (99d:05028)
13. de Bruijn, N.G., Erdős, P.: On a combinatorial problem. Ned. Akad. Wet. Proc. **51**, 1277–1279 (1948). = Indagationes Math. **10**, 421–423 (1948). MR 0028289 (10,424a)
14. Cariolaro, D., Lih, K.-W.: The edge-choosability of the tetrahedron. Math. Gaz. **92**(525), 543–546 (2008)

15. Cariolaro, D., Cariolaro, G., Schauz, U., Sun, X.: The list-chromatic index of K_6. Discret. Math. **322**, 15–18 (2014). MR 3164031

16. Chang, W.I., Lawler, E.L.: Edge coloring of hypergraphs and a conjecture of Erdős, Faber, Lovász. Combinatorica **8**(3), 293–295 (1988). MR 963120 (90a:05141)

17. Chen, Y., Zhu, W., Wang, W.: Edge choosability of planar graphs without 5-cycles with a chord. Discret. Math. **309**(8), 2233–2238 (2009). MR 2510350 (2010h:05252)

18. Chetwynd, A., Häggkvist, R.: A note on list-colorings. J. Graph Theory **13**(1), 87–95 (1989). MR 982870 (90a:05081)

19. Chung, F., Graham, R.: Erdős on Graphs. AK Peters, Wellesley (1998). His legacy of unsolved problems. MR 1601954 (99b:05031)

20. Chvátal, V.: Linear Programming. A Series of Books in the Mathematical Sciences. W. H. Freeman and Company, New York (1983). MR 717219 (86g:90062)

21. Cranston, D.W.: Edge-choosability and total-choosability of planar graphs with no adjacent 3-cycles. Discuss. Math. Graph Theory **29**(1), 163–178 (2009). MR 2548793 (2011a:05099)

22. Dinitz, J.H.: Home Page for Jeff Dinitz. Available at www.emba.uvm.edu/~jdinitz/ (2015). [Online; accessed 28 Mar 2015]

23. Dong, A., Liu, G., Li, G.: List edge and list total colorings of planar graphs without 6-cycles with chord. Bull. Korean Math. Soc. **49**(2), 359–365 (2012). MR 2934486

24. Eklof, P.C., Mekler, A.H.: Almost Free Modules. North-Holland Mathematical Library, vol. 46. North-Holland Publishing Co., Amsterdam (1990). Set-theoretic methods. MR 1055083 (92e:20001)

25. Ellingham, M.N., Goddyn, L.: List edge colourings of some 1-factorable multigraphs. Combinatorica **16**(3), 343–352 (1996). MR 1417345 (98a:05068)

26. Erdős, P.: Problems and results on finite and infinite combinatorial analysis. In: Infinite and Finite Sets (Colloq., Keszthely, 1973; dedicated to P. Erdős on his 60th birthday), vol. I, pp. 403–424. North-Holland, Amsterdam (1975). Colloq. Math. Soc. János Bolyai, vol. 10. MR 0389607 (52 #10438)

27. Erdős, P.: Problems and results in graph theory and combinatorial analysis. In: Graph Theory and Related Topics (Proceedings of the Conference, University of Waterloo, Waterloo, Ont., 1977), pp. 153–163. Academic Press, New York/London (1979). MR 538043 (81a:05034)

28. Erdős, P.: Some old and new problems in various branches of combinatorics. In: Proceedings of the Tenth Southeastern Conference on Combinatorics, Graph Theory and Computing (Florida Atlantic University, Boca Raton, 1979). Congressus Numerantium, vol. XXIII–XXIV, pp. 19–37. Utilitas Mathematica, Winnipeg (1979). MR 561032 (81f:05001)

29. Erdős, P.: On the combinatorial problems which I would most like to see solved. Combinatorica **1**(1), 25–42 (1981). MR 602413 (82k:05001)

30. Erdős, P.: Some of my favorite problems and results. In: The Mathematics of Paul Erdős, I. Algorithms and Combinatorics, vol. 13, pp. 47–67. Springer, Berlin (1997). MR 1425174 (98e:11002)

31. Erdős, P., Rubin, A.L., Taylor, H.: Choosability in graphs. In: Proceedings of the West Coast Conference on Combinatorics, Graph Theory and Computing (Humboldt State University, Arcata, 1979). Congressus Numerantium, vol. XXVI, pp. 125–157. Utilitas Mathematica, Winnipeg (1980). MR 593902 (82f:05038)

32. Faber, V.: The Erdős-Faber-Lovász conjecture—the uniform regular case. J. Comb. **1**(2), 113–120 (2010). MR 2732509 (2012c:05221)

33. Galvin, F.: The list chromatic index of a bipartite multigraph. J. Comb. Theory Ser. B **63**(1), 153–158 (1995). MR 1309363 (95m:05101)

34. Gutner, S.: M.Sc. thesis, Tel Aviv University, Israel (1992)

35. Haddad, L., Tardif, C.: A clone-theoretic formulation of the Erdős-Faber-Lovász conjecture. Discuss. Math. Graph Theory **24**(3), 545–549 (2004). MR 2120637 (2005h:05076)

36. Häggkvist, R.: Towards a solution of the Dinitz problem? Discret. Math. **75**(1–3), 247–251 (1989). Graph theory and combinatorics (Cambridge, 1988). MR 1001399 (90f:05022)

37. Häggkvist, R., Chetwynd, A.: Some upper bounds on the total and list chromatic numbers of multigraphs. J. Graph Theory **16**(5), 503–516 (1992). MR 1185013 (93i:05060)

38. Häggkvist, R., Janssen, J.: On the list-chromatic index of bipartite graphs. Technical Report 15, Department of Mathematics, University of Umeå, 24 pp. (1993)

39. Häggkvist, R., Janssen, J.: New bounds on the list-chromatic index of the complete graph and other simple graphs. Comb. Probab. Comput. **6**(3), 295–313 (1997). MR 1464567 (98i:05076)

40. Harris, A.J.: Problems and conjectures in extremal graph theory. Ph.D. thesis, University of Cambridge (1985)

41. Hetherington, T.J., Woodall, D.R.: Edge and total choosability of near-outerplanar graphs. Electron. J. Comb. **13**(1), Research Paper 98, 7 pp. (2006). MR 2274313 (2007h:05058)

42. Hind, H.R.F.: Restricted edge-colourings. Ph.D. thesis, Peterhouse College, University of Cambridge (1988)

43. Hindman, N.: On a conjecture of Erdős, Faber, and Lovász about n-colorings. Can. J. Math. **33**(3), 563–570 (1981). MR 627643 (82j:05058)

44. Hou, J., Liu, G., Cai, J.: List edge and list total colorings of planar graphs without 4-cycles. Theor. Comput. Sci. **369**(1–3), 250–255 (2006). MR 2277573 (2007j:05071)

45. Irving, R.W., Manlove, D.F.: The b-chromatic number of a graph. Discret. Appl. Math. **91**(1–3), 127–141 (1999). MR 1670155 (2000a:05079)

46. Jackson, B., Sethuraman, G., Whitehead, C.: A note on the Erdős-Farber-Lovász conjecture. Discret. Math. **307**(7–8), 911–915 (2007). MR 2297176 (2008a:05091)

47. Janssen, J.C.M.: The Dinitz problem solved for rectangles. Bull. Am. Math. Soc. (N.S.) **29**(2), 243–249 (1993). MR 1215310 (94b:05032)

48. Jensen, T.R., Toft, B.: Graph Coloring Problems. Wiley-Interscience Series in Discrete Mathematics and Optimization. Wiley, New York (1995). A Wiley-Interscience Publication. MR 1304254 (95h:05067)

49. Juvan, M., Mohar, B., Thomas, R.: List edge-colorings of series-parallel graphs. Electron. J. Comb. **6**, Research Paper 42, 6 pp. (1999). MR 1728012 (2000h:05081)

50. Kahn, J.: Coloring nearly-disjoint hypergraphs with $n + o(n)$ colors. J. Comb. Theory Ser. A **59**(1), 31–39 (1992). MR 1141320 (93b:05127)

51. Kahn, J.: Recent results on some not-so-recent hypergraph matching and covering problems. In: Extremal Problems for Finite Sets (Visegrád, 1991). Bolyai Society Mathematical Studies, vol. 3, pp. 305–353. János Bolyai Mathematical Society, Budapest (1994). MR 1319170 (96a:05108)

52. Kahn, J.: Asymptotically good list-colorings. J. Comb. Theory Ser. A **73**(1), 1–59 (1996). MR 1367606 (96j:05001)

53. Kahn, J.: Asymptotics of the chromatic index for multigraphs. J. Comb. Theory Ser. B **68**(2), 233–254 (1996). MR 1417799 (97g:05078)

54. Kahn, J.: On some hypergraph problems of Paul Erdős and the asymptotics of matchings, covers and colorings. In: The Mathematics of Paul Erdős, I. Algorithms and Combinatorics, vol. 13, pp. 345–371. Springer, Berlin (1997). MR 1425195 (97m:05193)

55. Kahn, J.: Asymptotics of the list-chromatic index for multigraphs. Random Struct. Algorithms **17**(2), 117–156 (2000). MR 1774747 (2001f:05066)

56. Kahn, J., Kayll, P.M.: On the stochastic independence properties of hard-core distributions. Combinatorica **17**(3), 369–391 (1997). MR 1606040 (99e:60034)

57. Kahn, J., Seymour, P.D.: A fractional version of the Erdős-Faber-Lovász conjecture. Combinatorica **12**(2), 155–160 (1992). MR 1179253 (93g:05108)

58. Kayll, P.M.: Two chromatic conjectures: one for vertices, one for edges. Joint Mathematics Meetings: AMS Special Session on My Favorite Graph Theory Conjectures, Baltimore MD, 17 Jan 2014. Beamer available at faculty.nps.edu/rgera/Conjectures/ JointMeetings-2014/mark-JMM-2014.pdf

59. Klein, H., Margraf, M.: A remark on the conjecture of Erdős, Faber and Lovász. J. Geom. **88**(1–2), 116–119 (2008). MR 2398480 (2009a:05072)

60. Kostochka, A.V.: List edge chromatic number of graphs with large girth. Discret. Math. **101**(1–3), 189–201 (1992). Special volume to mark the centennial of Julius Petersen's "Die Theorie der regulären Graphs", Part II. MR 1172377 (93e:05033)

61. Li, R., Xu, B.: Edge choosability and total choosability of planar graphs with no 3-cycles adjacent 4-cycles. Discret. Math. **311**(20), 2158–2163 (2011). MR 2825660 (2012h:05115)
62. Lin, W.-H., Chang, G.J.: *b*-coloring of tight bipartite graphs and the Erdős-Faber-Lovász conjecture. Discret. Appl. Math. **161**(7–8), 1060–1066 (2013). MR 3030590
63. Liu, B., Hou, J., Liu, G.: List edge and list total colorings of planar graphs without short cycles. Inf. Process. Lett. **108**(6), 347–351 (2008). MR 2458420 (2009g:05059)
64. Liu, B., Hou, J., Wu, J., Liu, G.: Total colorings and list total colorings of planar graphs without intersecting 4-cycles. Discret. Math. **309**(20), 6035–6043 (2009). MR 2552636 (2010k:05092)
65. Maffray, F.: Kernels in perfect line-graphs. J. Comb. Theory Ser. B **55**(1), 1–8 (1992). MR 1159851 (93i:05061)
66. Mitchem, J.: On *n*-coloring certain finite set systems. Ars Comb. **5**, 207–212 (1978). MR 0505583 (58 #21667)
67. Mitchem, J., Schmidt, R.L.: On the Erdős-Faber-Lovász conjecture. Ars Comb. **97**, 497–505 (2010). MR 2743755 (2011i:05195)
68. Motzkin, Th.: The lines and planes connecting the points of a finite set. Trans. Am. Math. Soc. **70**, 451–464 (1951). MR 0041447 (12,849c)
69. Paul, V., Germina, K.A.: On edge coloring of hypergraphs and Erdos-Faber-Lovász conjecture. Discret. Math. Algorithms Appl. **4**(1), 1250003, 5 pp. (2012). MR 2913089
70. Peterson, D., Woodall, D.R.: Edge-choosability in line-perfect multigraphs. Discret. Math. **202**(1–3), 191–199 (1999). MR 1694489 (2000a:05086)
71. Peterson, D., Woodall, D.R.: Erratum: "Edge-choosability in line-perfect multigraphs" [Discret. Math. **202**(1–3), 191–199 (1999). MR1694489 (2000a:05086)]. Discret. Math. **260**(1–3), 323–326 (2003). MR 1948402 (2003m:05076)
72. Pólya, G.: How to Solve it. Princeton Science Library. Princeton University Press, Princeton (2004). A new aspect of mathematical method, Expanded version of the 1988 edition, with a new foreword by John H. Conway. MR 2183670 (2006f:00007)
73. Romero, D., Alonso-Pecina, F.: The Erdős-Faber-Lovász conjecture is true for $n \leq 12$. Discret. Math. Algorithms Appl. **6**(3), 1450039, 5 pp. (2014). MR 3217845
74. Romero, D., Sánchez-Arroyo, A.: Adding evidence to the Erdős-Faber-Lovász conjecture. Ars Comb. **85**, 71–84 (2007). MR 2359282 (2008m:05119)
75. Romero, D., Sánchez-Arroyo, A.: Advances on the Erdős-Faber-Lovász conjecture. In: Combinatorics, Complexity, and Chance. Oxford Lecture Series in Mathematics and its Applications, vol. 34, pp. 272–284. Oxford University Press, Oxford (2007). MR 2314574 (2008f:05128)
76. Sánchez-Arroyo, A.: The Erdős-Faber-Lovász conjecture for dense hypergraphs. Discret. Math. **308**(5–6), 991–992 (2008). MR 2378934 (2008m:05121)
77. Schauz, U.: Algebraically solvable problems: describing polynomials as equivalent to explicit solutions. Electron. J. Comb. **15**(1), Research Paper 10, 35 pp. (2008). MR 2368915 (2009c:41009)
78. Schauz, U.: Proof of the list edge coloring conjecture for complete graphs of prime degree. Electron. J. Comb. **21**(3), Paper 3.43, 17 pp. (2014). MR 3262280
79. Seymour, P.D.: Packing nearly disjoint sets. Combinatorica **2**(1), 91–97 (1982). MR 671149 (83m:05044)
80. Shannon, C.E.: A theorem on coloring the lines of a network. J. Math. Phys. **28**, 148–151 (1949). MR 0030203 (10,728g)
81. Soifer, A.: The Colorado Mathematical Olympiad and Further Explorations. Springer, New York (2011)
82. Tian, J., Zhang, X.: Pseudo-outerplanar graphs and chromatic conjectures. Ars Comb. **114**, 353–361 (2014). MR 3203677
83. Vizing, V.G.: On an estimate of the chromatic class of a *p*-graph. Diskret. Analiz No. **3**, 25–30 (1964). MR 0180505 (31 #4740)
84. Vizing, V.G.: Coloring the vertices of a graph in prescribed colors. Diskret. Analiz No. 29 Metody Diskret. Anal. v Teorii Kodov i Shem **101**, 3–10 (1976). MR 0498216 (58 #16371)
85. Wang, W.: Edge choosability of planar graphs without short cycles. Sci. China Ser. A **48**(11), 1531–1544 (2005). MR 2203603 (2006m:05099)

86. Wang, G., Huang, Q.: The edge choosability of $C_n \times C_m$. Graph Theory Notes N.Y. **49**, 11–13 (2005). MR 2202294 (2006h:05086)

87. Wang, G., Huang, Q.: The edge choosability of $C_n \times P_m$. Ars Comb. **83**, 161–167 (2007). MR 2305755 (2008c:05073)

88. Wang, W.-F., Lih, K.-W.: On the sizes of graphs embeddable in surfaces of nonnegative Euler characteristic and their applications to edge choosability. Eur. J. Comb. **28**(1), 111–120 (2007). MR 2261807 (2007i:05072)

89. Weifan, W., Lih, K.-W.: Choosability, edge choosability, and total choosability of outerplane graphs. Eur. J. Comb. **22**(1), 71–78 (2001). MR 1808085 (2002g:05091)

90. Woodall, D.R.: Edge-choosability of multicircuits. Discret. Math. **202**(1–3), 271–277 (1999). MR 1694465 (2000a:05090)

91. Wu, J., Wang, P.: List-edge and list-total colorings of graphs embedded on hyperbolic surfaces. Discret. Math. **308**(24), 6210–6215 (2008). MR 2464909 (2009k:05087)

92. Zhang, X., Wu, J., Liu, G.: List edge and list total coloring of 1-planar graphs. Front. Math. China **7**(5), 1005–1018 (2012). MR 2965951

Chapter 12
Some Conjectures and Questions in Chromatic Topological Graph Theory

Joan P. Hutchinson

Abstract We present a conjecture and eight open questions in areas of coloring graphs on the plane, on nonplanar surfaces, and on multiple planes. These unsolved problems relate to classical graph coloring and to list coloring for general embedded graphs and also for planar great-circle graphs and for locally planar graphs.

Mathematics Subject Classification 2010: Primary 05C15; Secondary 05C10

12.1 Introduction

Much of chromatic graph theory was first stimulated by the four-color conjecture; much of recent research in this area has been further stimulated by the proof of the four-color theorem. The conjecture, posed in 1852 by a student, Francis Guthrie, asks if the regions of every planar map can be colored with four colors, one color assigned to each region, so that every pair of regions that share a border receives different colors. See Jensen and Toft's comprehensive book on graph coloring problems [35, §1.2] for the early history of this conjecture (Figure 12.1).

In the dual, graph theory version, the conjecture asks if every planar graph can have one of four colors assigned to each vertex so that adjacent vertices receive different colors. After many false attempts, the four-color theorem was first proved in 1976 by Appel and Haken [6] with a computer-assisted proof, which was revolutionary in its computer technique and monumental in its length and depth.

The four-color theorem has since been reproven at least twice, first by Robertson, Sanders, Seymour, and Thomas in 1996 [51] and later by Gonthier in 2008 [27], both proofs assisted by computer searches. Needless to say, a more elementary

---

In Memory of Dan Archdeacon, 1954–2015, and Albert Nijenhuis, 1926-2015.

J.P. Hutchinson (✉)
Department of Mathematics, Statistics, and Computer Science,
Macalester College, St. Paul, MN 55105, USA
e-mail: hutchinson@macalester.edu

© Springer International Publishing Switzerland 2016
R. Gera et al. (eds.), *Graph Theory*, Problem Books in Mathematics,
DOI 10.1007/978-3-319-31940-7_12

Fig. 12.1 A 4-coloring of all the 3,093 counties in the contiguous US states [58]

(straightforward, purely theoretical) proof without the use of a computer would be most desirable, but the search for such a proof is not a task to be undertaken lightly.

In other words, beware of the "four-color disease." On the other hand, the four-color problem has created many colorful and engaging related conjectures, whose solutions are likely more tractable and for which partial results are possibly more easily obtained.

12.2   Some Challenging Open Problems

Here are four direct offspring of the four-color problem that lead to one solved and three unsolved chromatic problems, Conjecture 1 and Questions 2 and 3. Questions 4–9, in subsequent sections, pose newer variations.

First, since a (finite) graph can be drawn without edge crossings in the plane if and only if it can be so drawn on the surface of a sphere, one can consider other objects on which to draw graphs; classically these are the closed and bounded, two-dimensional surfaces without boundary. These surfaces are classified for orientable surfaces as the sphere plus handles or, more descriptively, as the sphere, the torus, the double torus, ..., the k-holed torus, or equivalently the sphere plus k handles, for some nonnegative integer k.

For nonorientable surfaces these are the sphere plus crosscaps (each crosscap being a disk with opposite boundary points identified; see Figure 12.6) or the projective plane, the Klein bottle, ..., and the sphere plus j crosscaps for some $j > 0$. For an excellent, comprehensive book, consult Mohar and Thomassen's *Graphs on Surfaces* [42], where a completely combinatorial explanation is given of surfaces, handles, crosscaps, and the graphs that embed on surfaces.

For some beautiful depictions of nonorientable surfaces, consult Ferguson [21].

Definition 1. The *Euler genus* g of a surface S is defined to be $g = 2k$ if S is orientable and is (homeomorphic to) the sphere plus $k \geq 0$ handles and $g = j$ if S is nonorientable and is the sphere plus $j > 0$ crosscaps.

Definition 2. A graph *embeds* on a surface if it can be drawn on that surface without edge crossings. The *genus of a graph* is the least genus of a surface on which the graph embeds. A graph of genus 0 is called a *planar graph*.

Theorem 1 (Euler–Poincaré Formula). *If G embeds on a surface of Euler genus $g \geq 0$ with every face a 2-cell (i.e., contractible to a point on the surface), then*

$$v - e + f = 2 - g,$$

where v is the number of vertices and e is the number of edges of G, which is embedded with f faces.

Definition 3. For $k > 0$, a graph is said to be *k-colored* if one of k colors is assigned to each vertex so that every pair of adjacent vertices receives different colors. A graph is *k-chromatic* if k is the least integer for which the graph can be k-colored. A graph is said to be *k-critical* if it is k-chromatic, but every subgraph can be $(k-1)$-colored.

In part because of frustration with the four-color conjecture, Heawood in 1890, using the Euler–Poincaré formula, determined an upper bound on the number of colors needed for graphs that embed on a surface of Euler genus $g > 0$.

Theorem 2 (Heawood [30]). *If G embeds on a surface of Euler genus $g > 0$, then it can be colored in at most $\left\lfloor \frac{7 + \sqrt{24g+1}}{2} \right\rfloor$ colors.*

Thus a graph on the projective plane $(g = 1)$ needs at most six colors, a graph on the torus or on the Klein bottle $(g = 2)$ needs at most seven colors, a graph on the double torus $(g = 4)$ needs at most eight colors, and a graph on the 17-holed torus or on the sphere plus 34 crosscaps needs at most $\left\lfloor \frac{7 + \sqrt{817}}{2} \right\rfloor = 17$ colors.

Are so many colors ever needed? Figure 12.2 shows two maps where seven and eight colors are needed. The dual graphs are the complete graphs K_7 and K_8 (see Definition 5), which need seven and eight colors, respectively. Leading a band of graph colorers, in 1968 Ringel and Youngs [49] proved that Heawood's upper bound can always be achieved except on the Klein bottle, for which Franklin [22] had proved earlier that only six colors are ever needed (Figure 12.3).

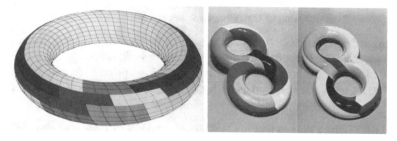

Fig. 12.2 Seven mutually adjacent countries on the torus (left) and eight mutually adjacent countries on the double torus (right, with both sides shown)

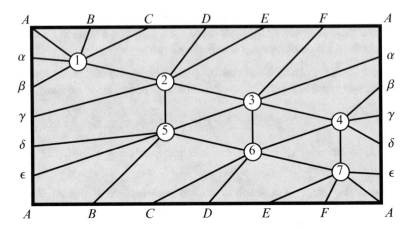

Fig. 12.3 K_7 embedded on the torus, represented as a rectangle with top and bottom identified and left and right identified

Definition 4. The *chromatic number of a surface* is the maximum chromatic number among all graphs that embed on that surface. (In other words, it is the least k for which every graph on that surface can be k-colored.)

Theorem 3 (Ringel–Youngs Map-Color Theorem [49]). *The chromatic number of a surface of Euler genus $g > 0$ is given by $H(g) = \left\lfloor \frac{7+\sqrt{24g+1}}{2} \right\rfloor$ except for the Klein bottle, where only six colors are needed. In particular, for every $g > 0$, the complete graph on $H(g)$ vertices, $K_{H(g)}$, embeds on the surface of Euler genus g.*

The map-color theorem was proved without the use of a computer and is comprehensively and clearly written up by Ringel in [48]; see also Gross and Tucker [28] for an alternative approach and proof. Note also that although the substitution $g = 0$ into the Heawood bound enticingly gives the answer 4, the Ringel–Youngs proof is valid for all surfaces except the sphere since it requires $g > 0$. Much work and many conjectures have been developed from this remarkable result; see Section 12.4 below. After this pioneering result, the search for a solution to the four-color problem continued.

A second direction was developed from the four-color problem in 1943 by Hadwiger [29], in part to develop the chromatic theory of graphs and to escape the topological constraints of the plane or sphere.

Definition 5. A graph G is said to *contract* to a graph H if by a sequence of vertex deletions, edge deletions, and *edge contractions*, in which two adjacent vertices are identified, G can be transformed to H. The complete graph on k vertices, K_k, consists of k vertices with every pair adjacent. A graph that does not contract to K_k is said to be K_k *-minor-free* (Figure 12.4).

CONJECTURE 1 (Hadwiger's Conjecture or HC). Every k-chromatic graph contracts to K_k. Equivalently a graph that is K_k-minor-free can be $(k-1)$-colored.

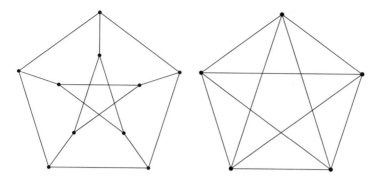

Fig. 12.4 The Petersen graph (left) contracts to K_5 (right) with five edge contractions

One can easily check that the Petersen graph can be 3-colored and is 3-chromatic even though it contracts to K_5. This conjecture does not give a necessary and sufficient condition for being k-chromatic.

For a history of work on this conjecture, see [35, Ch. 6]. For some expository write-ups of recent research, see Toft [54], Chudnovsky [14], and Bondy [9]. Hadwiger's conjecture was first proven for $k \leq 4$; see [35, §6.1]. In 1937 Wagner [57] proved that the case of $k = 5$ is equivalent to the four-color conjecture and so was validated in 1976. The strongest result to date is the proof that HC holds for $k = 6$ by Robertson et al. [50] with a computer-assisted proof that reduces the problem to the four-color theorem. In a different approach, Bollobas, Catlin, and Erdős [8] used a probabilistic proof to prove the striking result that "almost all" graphs satisfy HC.

There are many recent results on special cases and variations on this conjecture, and many, many people have been, and are, working on these problems. We suggest a deeper study of related, recent work before tackling this problem; for example, see the recently published *Topics in Chromatic Graph Theory* [7, Ch. 4]. In summary, despite assaults on many fronts of HC, it has resisted full solution so far. Stay tuned.

A third direction concerns the coloring of graphs on the disjoint union of spheres (or planes), known as Ringel's Earth–Moon problem and also as the problem of the chromatic number of graphs of thickness t; see [35, pp. 36–37] and Ringel [47].

In the former problem, suppose the Moon were colonized. How many colors would be needed to properly color the countries of the Earth and of the Moon so that every country and its lunar colony receive the same color? The maximum number of colors needed lies between 9 and 12, inclusively (see below), but despite many years of interest, no further progress has been made on narrowing these bounds. Some have suggested that this problem is as hard as two or three four-color theorems!

Stated in dual, graph theory terms, what is the maximum number of colors needed for a graph that is the edge union of two planar graphs? Or the edge union of more than two planar graphs? Here is the precise definition.

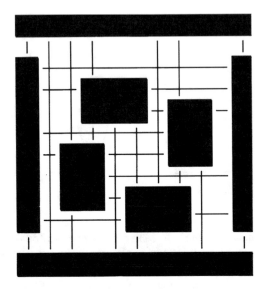

Fig. 12.5 K_8 shown to have thickness 2

Definition 6. A graph that is the edge union of t, but no fewer, planar graphs, each on the same set of vertices, is said to have *thickness t*. Let χ_t denote the chromatic number of thickness t graphs; that is, χ_t is the maximum number of colors needed for a graph of thickness t.

Another way to explain the thickness parameter is, given a graph G with n vertices, to make t copies of the n-vertex set of G and then assign each edge of G to one of the t copies so that t planar graphs are created. G has thickness t if t is the smallest for which this assignment can be made.

In Figure 12.5 let each black rectangle represent a vertex and each horizontal or vertical line an edge. The resulting graph is K_8 with its edges divided into two planar subgraphs, formed from the horizontal and the vertical edges.

Ringel's initial work [47] on the Earth-Moon problem showed that $8 \leq \chi_2 \leq 12$. Since K_8 has thickness 2, it follows that $8 \leq \chi_2$, but Sulanke [24] brought to light the graph $K_{11} - C_5$ (where C_n denotes a simple cycle of n vertices). $K_{11} - C_5$ is the same as the Cartesian product of K_6 with C_5 (the disjoint union of these two graphs with all possible edges joining the two). It is an interesting challenge to divide this graph into two planar subgraphs, as Sulanke did. Its chromatic number is easier to determine: K_6 needs six colors, C_5 needs three, and thus the product of these graphs needs nine. Sulanke's graph shows that nine colors are sometimes needed for thickness-2 graphs and thus that $9 \leq \chi_2 \leq 12$.

Euler's formula arguments (for planar graphs) show that for $t > 2$, $6t - 2 \leq \chi_t \leq 6t$. For these bounds also, no tightening has been found.

For example, $K_{11} - C_5$ is a 9-critical (see Definition 3) thickness-2 graph and was the only known graph with these two properties for many years. In Boutin et al. [11] and Gethner and Sulanke [26], many more 9-critical, thickness-2 graphs are determined; these lead to an infinite family of such graphs. In these papers bounds

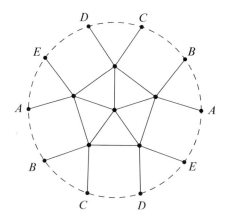

Fig. 12.6 K_6 embedded on the projective plane, represented as a disk with antipodal points identified

on the chromatic number of some families of thickness-2 graphs are obtained, and the authors propose some areas in which to search for a 10-chromatic thickness-2 graph. In summary the ranges for χ_t, $t \geq 2$, remain stubbornly open.

QUESTION 2. Can the known bounds on χ_t be improved? And ultimately, what is the value of χ_t for $t \geq 2$?

For this section's final open question, we mention Albertson's four-color problem [2]; see [35, §3.3].

QUESTION 3. Given $g > 0$, is there a constant $M(g)$ such that for every graph G that embeds on a surface of Euler genus g, all but at most $M(g)$ vertices of G can be 4-colored? In particular for the torus, is $M(2) = 3$?

There is evidence for an affirmative answer. For example, every 6-chromatic graph on the projective plane contains K_6, proved by Albertson and Hutchinson [3], and every 7-chromatic graph on the torus contains K_7, proved by Ungar and Dirac [17]; see [35, §1.2]. In addition both K_6 and K_7 embed with all faces three sided on their respective surfaces; see Figures 12.3 and 12.6. For $i = 1, 2$, a coloring of K_{5+i} with colors $\{1, 2, \ldots, 5 + i\}$, embedded as in Figures 12.6 and 12.3, extends to a coloring of the interior of each three-sided face with colors $\{1, 2, 3, 4\}$ (by the four-color theorem!) so that except for $i + 1$ vertices, the graph can be 4-colored.

A list of all unavoidable graphs in 6-chromatic graphs on the Klein bottle has also been obtained with more complex proofs by Kawarabayashi et al. [36] and by Chenette et al. [13].

Note that an affirmative solution to Albertson's problem would imply the four-color theorem, for if G were a 5-chromatic planar graph, then an infinite number of copies of G would embed on the torus, contradicting the existence of $M(2)$. Again do not undertake the full scope of this problem lightly! But some variations are approachable, and one such has recently been solved by Nakamoto and Ozeki [43].

12.3   Some More Recent Open Problems on Coloring and "List Coloring" of Planar Graphs

We move now to some (possibly) more tractable problems, which either have received less scrutiny or are more recently conjectured.

Definition 7. A *plane graph* is a planar graph embedded in the plane. An *outerplanar graph* is a plane graph drawn with all vertices on one (typically the outer) face. A *triangulation* is an embedded graph with all faces three sided (recall Definition 2). A face of an embedded graph is said to be *even sided* if it is bounded by an even number of edges. (An edge with both sides lying on one face contributes two to the count of edges of that face.)

The "easiest" coloring result on embedded graphs is the result that every plane graph can be 2-colored if and only if every face is even sided (because an even-faced graph is necessarily bipartite).

No characterization of 3-colorable or 3-chromatic plane graphs is known; even the recognition problem for the latter class of graphs is NP-complete; see Garey and Johnson [25]. But some classes of 3-colorable graphs have been studied; see [35, Ch. 2]. Here are two sample results, the first from 1898.

Theorem 4 (Heawood [31]). *A plane triangulation can be 3-colored if and only if every vertex has even degree.*

Theorem 5 (Król [37]). *A planar graph can be 3-colored if and only if it is a subgraph of a plane triangulation with all vertices of even degree.*

Though Król's condition sounds like a characterization, it can be very hard to determine whether it holds for a given planar graph.

Here is a not-so-well-known 3-color problem.

Definition 8. Imagine a sphere on which are drawn some great circles, no three meeting at a point. From this drawing the *great-circle graph* is the 4-regular plane graph with a vertex for each intersection point of two great circles and an edge for each subarc of a great circle that joins two intersection points (Figure 12.7).

Theorem 6 (Brooks [12]). *If a connected graph G has maximum degree Δ, then it can be Δ-colored unless G is $K_{\Delta+1}$ or $\Delta = 2$ and G is an odd cycle.*

By Brooks' theorem, great-circle graphs can be 4-colored.

QUESTION 4 (Felsner et al. [20]). Is every great-circle graph 3-colorable?

It's "easy" to show that any hemisphere of a great-circle graph can be 3-colored by considering a projection of the arcs on a hemisphere onto straight lines in the plane and using a sweep-line proof since the plane graph and every subgraph have a vertex of degree at most 2; see §17.7 of [58]. But can two hemispheres so 3-colored be meshed to obtain one 3-coloring of the whole graph? Though

Fig. 12.7 A great-circle graph on the sphere with its vertices 3-colored [58]

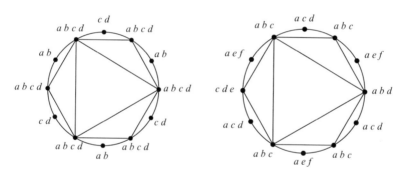

Fig. 12.8 A non-list-colorable planar graph (left) and a list-colorable variation (right)

this question is not well known among graph colorers, it has intrigued a number of combinatorial geometers; see ⟨www.openproblemgarden.org/category/graph_theory⟩ for this problem and many others too.

Many variations on graph coloring have been introduced over the years. Here we focus on the powerful concept of "list coloring" with first a look at its relevance to planar graphs; this concept was first introduced by Vizing [55] and by Erdős, Rubin, and Taylor [19].

Definition 9. Suppose each vertex v of a graph G has been given a nonempty list of colors $L(v)$. When G has a proper coloring with each vertex receiving a color from its list, G is said to be *L-list colorable* (or just *list colorable* when the lists L are clear). If G is list colorable whenever each vertex receives a list of size at least $k > 0$, G is said to be *k-list colorable* (or *k-choosable*). G is said to be *k-list chromatic* if k is the least integer for which G is k-list colorable (Figure 12.8).

(To convince yourself that the left graph of Figure 12.8 cannot be list colored, pick a color for a vertex of degree 6. If you picked a or b, try extending the coloring counterclockwise; if c or d, try extending clockwise.)

When all lists are the same, say $L(v) = \{1, 2, \ldots, k\}$ for each vertex v, then G is L-list colorable if and only if it is k-colorable. Thus the k-list-chromatic number is at least as large as the chromatic number. Perhaps surprisingly the former can be strictly larger than the latter. For an example with a wonderful proof, see the following theorem; in fact the proof is so wonderful that it is included in a book of exemplary proofs by Aigner and Ziegler [1].

Theorem 7 (Thomassen [53]). *Every planar graph is 5-list colorable.*

Voigt [56] and others have found planar graphs that cannot be 4-list colored. In [19] 2-list-colorable graphs are characterized. See also [35, §2.13] for work in this area.

QUESTION 5 (Meta). What can be said about planar graphs that are 4-list colorable? What can be said that distinguishes 3-, 4-, and 5-list-colorable planar graphs?

Mahdian and Mahmoodian [39] have introduced ideas and results on graphs that are uniquely list colorable.

Definition 10. A graph G is *uniquely k-list colorable*, for $k \geq 2$, when there exists a collection of lists with $|L(v)| = k$ for each vertex v of G for which there is a unique L-list coloring. G has property $M(k)$ if whenever L gives lists of size k to the vertices of G and G has an L-list coloring, then G has more than one L-coloring. Let $m(G)$ denote the least k for which $M(k)$ holds for G.

Thus G is uniquely k-list colorable if and only if it does not have the property $M(k)$. In [39] the authors characterize all graphs that can be uniquely 2-list colored; an example, which can be generalized to more, is shown in Figure 12.9. From Theorem 7, it follows (from the proof) that $m(G) \leq 5$ for each planar graph G. The authors ask the next question.

QUESTION 6. [40] Is there a planar graph G for which $m(G) = 5$? QUESTION 6 has now been answered in the negative; see [18].

We pose one final question about list-coloring planar graphs, one of many that are open, even though we know the best-possible list-coloring bound for planar graphs.

Definition 11. A *Gallai tree* [23] is a graph whose every *block* (a maximal 2-connected subgraph) is an odd cycle or a complete graph (Figures 12.10).

Theorem 8 (Vizing [55], Borodin [10], Erdős et al. [19]). *A graph G can be L-list colored when* $|L(v)| \geq \deg(v)$ *for every v in G provided that G is not a Gallai tree.*

Since we also know that planar graphs are 5-list colorable, the following seems quite reasonable.

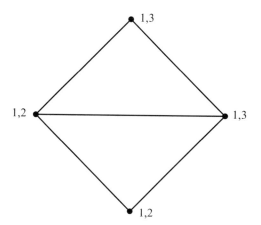

Fig. 12.9 A uniquely 2-list-colorable graph

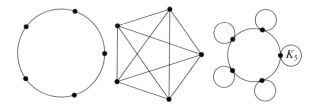

Fig. 12.10 Three examples of Gallai trees: a 5-cycle, K_5, and a 5-cycle with a copy of K_5 attached at each vertex

QUESTION 7. A.B. Richter (personal communication). If $G \neq K_4$ is planar, 3-connected, and, for each vertex v, $|L(v)| \geq \min\{\deg(v), k\}$, is G L-list colorable when $k = 6$?

Notice that by the first example in Figure 12.8, for 2-connected, planar graphs, the answer is "no" when $k = 4$. The answer is shown to be "yes" for outerplanar graphs $G \neq K_3$ (see Definition 7) that are merely 2-connected when $k = 5$ by Hutchinson [33]. Since K_5 is not planar, no planar graph can contract to K_5 and so is K_5-minor-free (see Definition 5). The best-known (partial) results on Question 7 appear in Cranston et al. [15]. There they ask for which pairs $\{r, k\}$ is a graph G L-list colorable when G is K_5-minor-free, r-connected, not a Gallai tree, and $|L(v)| \geq \min\{\deg(v), k\}$ for each vertex v of G. Before working on this, be sure to consult their paper.

There is a large and growing body of research on list coloring of general graphs; for a start see [35, §§1.9, 7.1, 7.2] and [7, Ch. 6]. Then do some Internet searches to look for the latest papers—this is a popular area!

12.4   The Power of Locally Planar Graphs
on Nonplanar Surfaces

Returning to coloring graphs on surfaces, a very fruitful point of view is the following. Let $\mathcal{P}$ be a graph property. Suppose there is a constant $k > 3$ such that every
graph that is embedded on a nonplanar surface with every noncontractible cycle of
length at least k satisfies $\mathcal{P}$. (A *noncontractible cycle* is one that cannot be contracted
to a point on the surface.) Then we say that *locally planar* graphs satisfy $\mathcal{P}$.
Informally, we call an embedded graph locally planar if all noncontractible cycles
are "suitably long," depending on the problem at hand (Figures 12.11 and 12.12).

Locally planar graphs are also said to have large *edge width* (see [42]). The
concept of local planarity was introduced by Albertson and Stromquist [5]. The
central problem studies the extent to which these embedded graphs act like planar
graphs.

The next three results illustrate the planar-like properties of locally planar graphs.

Theorem 9 (Thomassen [52]). *On all nonplanar, orientable surfaces, locally
planar graphs are 5-colorable.*

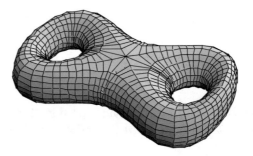

Fig. 12.11 A graph on the double torus with all noncontractible cycles of length at least 10 [58]

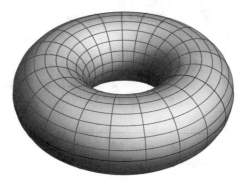

Fig. 12.12 A locally planar graph on the torus with all faces even sided [58]

In this proof, a graph embedded on a surface of Euler genus g with all noncontractible cycles of length at least $64\left(2^{7g}\right)$ is shown to be 5-colorable; the cycle-length bound is not claimed to be optimal. Another proof of Thm. 9 is given by Albertson and Hutchinson [4]. Whereas Thomassen does not assume the four-color theorem in [52], with that assumption the authors in [4] supply an easier proof provided all noncontractible cycles have length at least $64\ (2^g - 1)$.

Theorem 10 (Hutchinson [32]). *On all nonplanar, orientable surfaces, locally planar graphs with all faces even sided are 3-colorable.*

The same result does not hold on nonorientable surfaces, but the maximum chromatic numbers for locally planar graphs, embedded with all faces even sided, have been determined in Mohar and Seymour [41] and in Nakamoto, Negami, and Ota [44].

QUESTION 8 (Meta). What can be said about which locally planar graphs can be 4-colored? Or what can be said to distinguish the 3-, 4-, and 5-colorable locally planar graphs?

By the map-color theorem, Theorem 3, we know that $K_{H(g)}$ embeds on a surface of genus g for $g > 0$. There are infinitely many cases when $K_{H(g)}$ embeds as a triangulation; we see two such examples in Figures 12.3 and 12.6. When $H(g)$ is odd, these embedded graphs have all vertices of even degree, namely, $H(g) - 1$. Thus Heawood's theorem of Theorem 4 does not hold on surfaces. Instead we asked to what extent that theorem holds for locally planar graphs and found the following answer.

Theorem 11 (Hutchinson, Richter and Seymour [34]). *On all nonplanar orientable surfaces, locally planar triangulations with all vertices of even degree are 4-colorable.*

The more general problem about properties of embedded 4-chromatic graphs is asked in [42, Prob. 8.4.10] and studied in Kündgen and Thomassen [38].

What happens when you ask about list-coloring graphs on surfaces and in particular about list-coloring locally planar graphs? First, for all nonplanar surfaces, it is not hard to see that the Heawood bound of Theorem 2 gives the best-possible list-coloring bound for graphs on surfaces of Euler genus $g > 0$ [35, §1.9]; thus not much changes with the introduction of list coloring. However, there is an important strengthening of Theorem 9.

Theorem 12 (DeVos et al. [16], Postle and Thomas [46]). *Locally planar graphs on every nonplanar surface can be 5-list colored.*

Recall that a locally planar graph embedded on an orientable surface with all faces even sided can be 3-colored.

QUESTION 9. Kawarabayashi (personal communication) has asked whether a locally planar graph, embedded with all faces even sided on an orientable surface, can be 4-list colored or even 3-list colored.

Since the proofs for 5-list-coloring locally planar graphs on surfaces are quite hard, a solution to this final question may also be very challenging.

In conclusion, graph coloring is a very popular subject. This article is in no way intended to be comprehensive; instead I have included my favorite open problems and questions. Another comprehensive list of topological problems that were the favorites of Dan Archdeacon can be found at ⟨www.cems.uvm.edu/~darchdea/problems/problems.html⟩, but note that this was last updated in 2003. For one of the latest collections of open problems, see [7, Ch. 15]. Albert Nijenhuis, together with Herbert S. Wilf, did pioneering work on combinatorial algorithms [45], an approach that has developed into a fruitful area of chromatic and in topological graph theory.

References

1. Aigner, M., Ziegler, G.M.: Proofs from the Book. Springer, Berlin (1991)
2. Albertson, M.O.: Open Problem 2. In: Chartrand, G., et al. (eds.) The Theory and Applications of Graphs, p. 609. Wiley, New York (1981)
3. Albertson, M.O., Hutchinson, J.P.: The three excluded cases of Dirac's map-color theorem. Ann. N. Y. Acad. Sci. **319**, 7–17 (1979)
4. Albertson, M.O., Hutchinson, J.P.: Extending precolorings of subgraphs of locally planar graphs. Eur. J. Comb. **25**, 863–871 (2004); also ArXiv: 1602.06985v3
5. Albertson, M.O., Stromquist, W.: Locally planar toroidal graphs are 5-colorable. Proc. Am. Math. Soc. **84**, 449–456 (1982)
6. Appel, K., Haken, W.: Every planar map is four colorable. Bull. Am. Math. Soc. **82**, 711–712 (1976)
7. Beineke, L.W., Wilson, R.J. (eds.): Topics in Chromatic Graph Theory. Cambridge University Press, Cambridge (2015)
8. Bollobas, B., Catlin, P.A., Erdős, P.: Hadwiger's conjecture is true for almost every graph. Eur. J. Comb. **1**, 195–199 (1980)
9. Bondy, A.: Beautiful conjectures in graph theory. Eur. J. Comb. **37**, 4–23 (2014)
10. Borodin, O.V.: Problems of colouring and covering the vertex set of a graph by induced subgraphs (in Russian). Ph.D. thesis, Novosibirsk State University, Novosibirsk (1979)
11. Boutin, D., Gethner, E., Sulanke, T.: Thickness-two graphs Part one: new nine-critical graphs, permuted layer graphs, and Catlin's graphs. J. Graph Theory **57**, 198–214 (2008)
12. Brooks, R.L.: On colouring the nodes of a network. Proc. Camb. Philol. Soc. **37**, 194–197 (1941)
13. Chenette, N., Postle, L., Streib, N., Thomas, R., Yerger, C.: Five-coloring graphs on the Klein bottle. J. Comb. Theory Ser. B **102**, 1067–1098 (2012)
14. Chudnovsky, M.: Hadwiger's conjecture and seagull packing. Not. Am. Math. Soc. **57**, 733–736 (2010)
15. Cranston, D.W., Pruchnewski, A., Tuza, Z., Voigt, M.: List-colorings of K_5-minor-free graphs with special list assignments. J. Graph Theory **71**, 18–30 (2012)
16. DeVos, M., Kawarabayashi, K., Mohar, B.: Locally planar graphs are 5-choosable. J. Comb. Theory Ser. B **98**, 1215–1232 (2008)
17. Dirac, G.A.: Short proof of the map colour theorem. Can. J. Math. **9**, 225–226 (1957)
18. Eshahchi, Ch., Ghebleh, M., Hajiabolhassan, H.: Some concepts in list coloring. J. Comb. Math. Comb. Comput. 41, 151–160 (2002)
19. Erdős, P., Rubin, A.L., Taylor, H.: Choosability in graphs. In: Proceedings of the West Coast Conference on Combinatorics, Graph Theory, and Computing, Arcada, 1979. Congressus Numerantium, vol. 26, pp. 125–157. Utilitas Mathematica, Winnipeg (1997)

20. Felsner, S., Hurtado, F., Noy, M., Streinu, I.: Hamiltonicity and colorings of arrangement graphs. Discret. Appl. Math. **154**, 2470–2483 (2006)
21. Ferguson, H.: Mathematics in Stone and Bronze. Meridian Creative Group, Erie (1994)
22. Franklin, P.: A six color problem. J. Math. Phys. **13**, 363–369 (1934)
23. Gallai, T.: Kritische graphen. I. Magyar Tud. Akad. Math. Kutató Int. Közl. **8**, 165–192 (1963)
24. Gardner, M.: Mathematical games. Sci. Am. **242**, 14–21 (1980)
25. Garey, M.R., Johnson, D.S.: Computers and Intractability: a Guide to the Theory of NP-Completeness. W.H. Freeman and Co., New York (1979)
26. Gethner, E., Sulanke, T.: Thickness-two graphs Part two: more new nine-critical graphs, independence ratio, doubled planar graphs, and singly and doubly outerplanar graphs. Graphs Comb. **25**, 197–217 (2009)
27. Gonthier, G.: Formal proof—the four-color theorem. Not. Am. Math. Soc. **55**, 1382–1393 (2008)
28. Gross, J.L., Tucker, T.W.: Topological Graph Theory. Wiley, New York (1987)
29. Hadwiger, H.: Über eine Klassifikation der Streckenkomplexe. Vierteljschr. Naturforsch. Gesellsch. Zürich **88**, 133–142 (1943)
30. Heawood, P.J.: Map colour theorem. Q. J. Pure Appl. Math. **24**, 332–333 (1890)
31. Heawood, P.J.: On the four-colour map theorem. Q. J. Pure Appl. Math. **29**, 270–285 (1898)
32. Hutchinson, J.P.: Three-coloring graphs embedded with all faces even-sided. J. Comb. Theory Ser. B **65**, 139–155 (1995)
33. Hutchinson, J.P.: On list-coloring outerplanar graphs. J. Graph Theory **59**, 59–74 (2008)
34. Hutchinson, J.P., Richter, A.B., Seymour, P.: Colouring Eulerian triangulations. J. Comb. Theory Ser. B **84**, 225–239 (2002)
35. Jensen, T.R., Toft, B.: Graph Coloring Problems. Wiley, New York (1995)
36. Kawarabayashi, K., Král, D., Kynčl, J., Lidický, B.: 6-critical graphs on the Klein bottle. SIAM J. Discret. Math. **23**, 372–383 (2009)
37. Król, M.: On a necessary and sufficient condition of 3-colorability of planar graphs, I. Prace Nauk. Inst. Mat. Fiz. Teoret PWr. **6**, 37–40 (1972). II, **9**, 49–54 (1973)
38. Kündgen, A., Thomassen, C.: Spanning quadrangulations of triangulated surfaces. Abhandlungen aus dem Mathematischen Seminar der Universität Hamburg. (to appear)
39. Mahdian, M., Mahmoodian, E.S.: A characterization of uniquely 2-list colorable graphs. Ars Comb. **51**, 295–305 (1999)
40. Mahmoodian, E.S., Mahdian, M.: On the uniquely list colorable graphs. Ars Comb. **59**, 307–318 (2001)
41. Mohar, B., Seymour, P.: Coloring locally bipartite graphs on surfaces. J. Comb. Theory Ser. B **84**, 301–310 (2002)
42. Mohar, B., Thomassen, C.: Graphs on Surfaces. The Johns Hopkins University Press, Baltimore (2001)
43. Nakamoto, A., Ozeki, K.: Coloring of locally planar graphs with one color class small. Dan Archdeacon Memorial Volume, (submitted 2015)
44. Nakamoto, A., Negami, S., Ota, K.: Chromatic numbers and cycle parities of quadrangulations on nonorientable closed surfaces. Discret. Math. **185**, 211–218 (2004)
45. Nijenhuis, A., Wilf, H.S.: Combinatorial Algorithms, 2nd edn. Academic, Orlando (1978). also ⟨www.math.upenn.edu/~wilf/website/CombAlgDownld.html⟩
46. Postle, L., Thomas, R.: 5-list-coloring graphs on surfaces, L. Postle Ph.D. Dissertation, Georgia Institute of Technology (2012); also Hyperbolic families and coloring graphs on surfaces. Manuscript (2012)
47. Ringel, G.: Färbungsprobleme auf Flächen und Graphen. VEB Deutscher Verlag der Wissenschaften, Berlin (1959)
48. Ringel, G.: Map Color Theorem. Springer, New York (1974)
49. Ringel, G., Youngs, J.W.T.: Solution of the Heawood map-color problem. Proc. Natl. Acad. Sci. USA **60**, 438–445 (1968)
50. Robertson, N., Seymour, P., Thomas, R.: Hadwiger's conjecture for K_6-free graphs. Combinatorica **13**, 279–361 (1993)

51. Robertson, N., Sanders, D.P., Seymour, P., Thomas, R.: A new proof of the four-colour theorem. Electron. Res. Announc. Am. Math. Soc. **2**, 17–25 (1996)
52. Thomassen, C.: Five-coloring maps on surfaces. J. Comb. Theory Ser. B **59**, 89–105 (1993)
53. Thomassen, C.: Every planar graph is 5-choosable. J. Comb. Theory Ser. B **62**, 180–181 (1994)
54. Toft, B.: Survey of Hadwiger's conjecture. Congr. Numer. **115**, 241–252 (1996)
55. Vizing, V.G.: Vertex colorings with given colors (in Russian). Metody Diskret. Analiz **29**, 3–10 (1976)
56. Voigt, M.: List colourings of planar graphs. Discret. Math. **120**, 215–219 (1993)
57. Wagner, K.: Über eine Eigenschaft der ebenen Komplexe. Math. Ann. **114**, 570–590 (1937)
58. Wagon, S.: Mathematica in Action, 3rd edn. Springer, New York (2010)

Chapter 13
Turán's Brick Factory Problem: The Status of the Conjectures of Zarankiewicz and Hill

László A. Székely

Abstract In this chapter, we explore the history and the status of the Zarankiewicz crossing number conjecture and the Hill crossing number conjecture, on drawing complete bipartite and complete graphs in the plane with a minimum number of edge crossings. We discuss analogous problems on other surfaces and in different models of drawing.

Mathematics Subject Classification 2010: 05C10, 52C10

13.1 Origins

The concept of plane graphs (graphs embedded into the plane without edge crossings) goes well back into the nineteenth century, even to Cauchy [16], as flattening the skeleton of polyhedra. It turns out that K_5 and $K_{3,3}$ cannot be drawn as plane graphs as they have too many edges (see formula (13.6)). In 1934, Hanani (then Chojnacki) [17] proved a stronger statement: if either of these graphs is reasonably drawn in the plane (see Section 13.2), then two edges will cross an odd number of times. It turns out that these two graphs essentially exhibit all the obstacles for a plane drawing: in 1930, Kuratowski [44] showed that if a graph cannot be drawn crossing-free in the plane, then it contains a subdivision of K_5 or $K_{3,3}$, and in 1937, Wagner [75] showed that such a graph has K_5 or $K_{3,3}$ as minor.

The impossibility of drawing $K_{3,3}$ without a crossing has been well known in recreational mathematics. In 1917, Dudeney's book [24] contained the following Problem 251: "WATER, GAS, AND ELECTRICITY. There are some half-dozen

This research was supported in part by the NSF DMS contract 1300547.

L.A. Székely (✉)
Department of Mathematics, University of South Carolina,
LeConte Bldg, Columbia, SC 29208, USA
e-mail: laszlo@mailbox.sc.edu; Szekely@math.sc.edu

© Springer International Publishing Switzerland 2016
R. Gera et al. (eds.), *Graph Theory*, Problem Books in Mathematics,
DOI 10.1007/978-3-319-31940-7_13

puzzles, as old as the hills, that are perpetually cropping up, and there is hardly a month in the year that does not bring inquiries as to their solution. Occasionally one of these, that one had thought was an extinct volcano, bursts into eruption in a surprising manner. I have received an extraordinary number of letters respecting the ancient puzzle that I have called "Water, Gas, and Electricity." It is much older than electric lighting, or even gas, but the new dress brings it up to date. The puzzle is to lay on water, gas, and electricity, from W, G, and E, to each of the three houses, A, B, and C, without any pipe crossing another. Take your pencil and draw lines showing how this should be done. You will soon find yourself landed in difficulties." Kullman [43] cites versions of this problem as the houses and wells problem, the Corsican vendetta problem, or the Persian caliph's problem. In 1970, I heard the problem as three doghouses and three wells from my polytechnics teacher, who was unlikely to have any background in graph theory.

At this point, Paul Turán changed the question from whether crossings must happen in any drawing of the graph to what is the minimum number of crossings over all drawings. This happened while he was in a forced labor camp in WWII. He describes the moment in his foreword to the first issue of the *Journal of Graph Theory* [72]: "There were some kilns where the bricks were made and some open storage yards where the bricks were stored. All the kilns were connected by rail with all storage yards. ... the trouble was only at crossings. The trucks generally jumped the rails there, and the bricks fell out of them; in short this caused a lot of trouble and loss of time ... the idea occurred to me that this loss of time could have been minimized if the number of crossings of the rails had been minimized. But what is the minimum number of crossings?" In modern terminology, he asked what is the crossing number of the complete bipartite graph $K_{m,n}$. This problem has been known as *Turán's Brick Factory Problem*.

Independently, in 1934, sociologists started drawing graphs in the plane as sociograms, see Moreno's book [51]. In 1944, Bronfenbrenner [14] recommended drawing sociograms with the least amount of crossings and so did Moreno in 1953 the second edition of [51]. (David Eppstein and Marcus Schaefer [64] discovered these relevant references.) To give more credit to sociology, I mention here that Sándor Szalai, who is often called the "father of Hungarian sociology," recognized the graph Ramsey theorem from his sociogram data: many people exhibit large cliques or large anti-cliques and asked Paul Turán whether mathematicians know about this.

The most thorough account on crossing numbers is the encyclopedic survey of Schaefer [64]. I follow the notation in [64] for different kinds of crossing numbers. The online bibliography of crossing numbers [74] is another useful resource. This paper does not discuss the general lower bound techniques (see [66]) for crossing numbers as Leighton's Lemma (or Crossing Lemma), bisection width, and graph embedding, since these do not yield the best known results for the conjectures of Zarankiewicz and Hill. The paper does not get into the relevance of crossing numbers for VLSI (i.e., chip design for computers), although this research direction led to the general lower bound techniques above in Leighton's work [46]. We also avoid the applications of the crossing number method to discrete geometry [68] that resulted in considerable progress in several fields, from incidence geometry to number theory. I am indebted to [10] and [64] for using the results of their thorough research on the origins of Turán's Brick Factory Problem. I also use ideas from my earlier surveys on crossing numbers [66, 69–71].

13.2 Conjectures

In this paper, the term *graph* means a finite simple graph, that is, an undirected graph having no loops or multiple edges. A *drawing* of a graph G on the plane places the vertices of G into distinct points on the plane and then, for every edge uv in G, draws a continuous simple curve in the plane connecting the two points corresponding to u and v, in such a way that no curve has a vertex point as an internal point.

The *crossing number* $cr(G)$ of a graph G is the minimum number of intersection points among the interiors of the curves representing edges, over all possible drawings of the graph, where no three edges have a common interior point. (It is easy to see that the latter condition can be dropped without changing the value of the crossing number, if we change the objective to minimize $\sum_{\{e,f\}edges\ e\neq f} |int(e) \cap int(f)|$, where $int(e)$ denotes the interior of the curve corresponding to edge e. This alternative approach is necessary if we do not want to give up otherwise beautiful straight-line drawings, where multiple edges have common internal points.) It is also easy to see that a drawing of a graph G realizing the crossing number $cr(G)$ must have the following two properties, otherwise G can be redrawn with fewer crossings:

(i) Pairs of edges sharing the same endpoint do not cross,
(ii) Any two edges intersect at most once.

If crossing number problems are posed similarly for the sphere instead of the plane, stereographic projection shows that the corresponding planar and spheric crossing numbers are equal, and so is the crossing number on any oriented surface of genus 0, using continuous deformation.

In order to draw the complete bipartite graph $K_{n,m}$ with a minimum number of crossings, Zarankiewicz placed $\lfloor n/2 \rfloor$ vertices to positive positions on the x-axis, $\lceil n/2 \rceil$ vertices to negative positions on the x-axis, $\lfloor m/2 \rfloor$ vertices to positive positions on the y-axis, and $\lceil m/2 \rceil$ vertices to negative positions on the y-axis and drew nm edges in straight-line segments to obtain a drawing of $K_{n,m}$ (see Figure 13.1). It is not hard to check that the following formula gives the number of crossings in the Zarankiewicz drawing:

$$Z(n, m) = \left\lfloor \frac{n}{2} \right\rfloor \left\lfloor \frac{n-1}{2} \right\rfloor \left\lfloor \frac{m}{2} \right\rfloor \left\lfloor \frac{m-1}{2} \right\rfloor. \tag{13.1}$$

Zarankiewicz [77] and Urbaník [73] independently claimed and published that $cr(K_{n,m})$ was equal to (13.1), their result was cited and used in follow-up papers, and even the proof was reprinted in the book by Busacker and Saaty [15]. In 1965

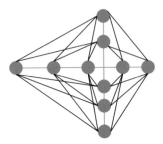

Fig. 13.1 The Zarankiewicz drawing of $K_{5,5}$

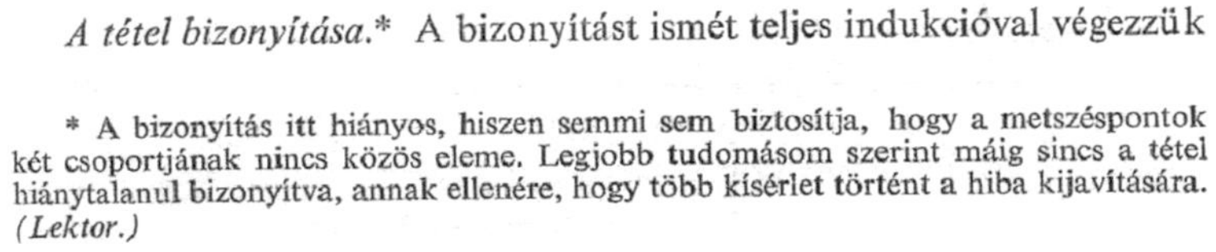

A tétel bizonyítása.* A bizonyítást ismét teljes indukcióval végezzük

* A bizonyítás itt hiányos, hiszen semmi sem biztosítja, hogy a metszéspontok két csoportjának nincs közös eleme. Legjobb tudomásom szerint máig sincs a tétel hiánytalanul bizonyítva, annak ellenére, hogy több kísérlet történt a hiba kijavítására. *(Lektor.)*

Fig. 13.2 The footnote from the publisher's reader—one of the earliest references in print to the problems with the Zarankiewicz theorem. "There is a gap in the proof, as nothing guarantees that the two sets of intersection points are disjoint. As far as I know, the gap has not been fixed, notwithstanding several attempts. (The publisher's reader)"

Kainen and in 1966 Ringel discovered a gap in the argument. Richard Guy deserves much credit for clarifying the situation in [29] and [30]. (As a high school student in the mathematics program of Fazekas Gimnázium in Budapest, I read the Hungarian translation of [15], which still reprinted the incorrect proof, with a footnote added by the publisher's reader, Katona: this proof has a gap (Figure 13.2). I was wondering how this can happen.) The still open Zarankiewicz crossing number conjecture postulates that

$$cr(K_{n,m}) = Z(n, m). \tag{13.2}$$

As $Z(n, 1) = Z(n, 2) = 0$, the first instance of (13.2) that requires a proof is $n \geq 3$ $m = 3$. The original induction proof of Zarankiewicz actually works in this instance. The base case is $n = 3$: $cr(K_{3,3}) = 1 = Z(3, 3)$, as $K_{3,3}$ is nonplanar. Assume that u and v are two distinct vertices from the n-element partite set. Consider and fix a drawing of $K_{n,3}$. Assume first that in this drawing edges with endvertex u never cross edges with endvertex v. If w is a third vertex from the n-element partite set, all edges going out from u, v, w make a drawing of a $K_{3,3}$ and a crossing in it. For different w's it is a different crossing, a total of $n - 2$ crossings. After the removal of u and v from the drawing, we must see at least $cr(K_{n-2,3})$ crossings, and those differ from the previous $n - 2$ crossings. Using the hypothesis and doing some algebra we observe at least

$$Z(n - 2, 3) + n - 2 = Z(n, 3)$$

crossings in the drawing. If the assumption above fails, then for every u and v, there are two edges with endvertices u and v that cross, providing $\binom{n}{2} > Z(n, 3)$ crossings.

In 1970, Kleitman [39] showed that (13.2) holds for $m \leq 6$ (the last proof to an instance of the conjecture without computer). He also proved that the smallest counterexample to the Zarankiewicz's conjecture must occur for odd n and m, using the counting argument to be discussed in Section 13.4. Furthermore, he showed that any two drawings of $K_{2n+1,2m+1}$ with no two tangential edges and no pairs of crossing adjacent edges have the same number of crossings modulo 2. This allows computing the parity of $cr(K_{2n+1,2m+1})$. Namely, $cr(K_{2n+1,4m+1})$ is even, while $cr(K_{4n+3,4m+3})$ is odd. The argument behind this observation follows the crossing number, when an edge is pulled over a vertex, a generic step to move from one drawing to another.

In 1993, Woodall [76] used elaborate computer search to show that (13.2) holds for $K_{7,7}$ and $K_{7,9}$, leaving $K_{7,11}$ and $K_{9,9}$ the smallest unsettled instances of the conjecture (13.2). The best bounds for them are as follows. We know $cr(K_{7,10}) = Z(7, 10) = 180$. The counting argument (see Section 13.4) for copies of $K_{7,10}$ in $K_{7,11}$ yields $220 \leq cr(K_{7,11})$. As $cr(K_{7,11})$ is odd, we conclude

$$221 \leq cr(K_{7,11}) \leq Z(7, 11) = 225.$$

Even the result (13.14) did not improve on this lower bound. On the other hand, the bound (13.15) provides

$$242 \leq cr(K_{9,9}) \leq Z(9, 9) = 256.$$

From 1958, Anthony Hill, a British artist and amateur mathematician, looked for the crossing number of K_n and formulated a conjecture based on his drawings. His conjecture, which postulates that

$$cr(K_n) = H(n), \tag{13.3}$$

where

$$H(n) = \frac{1}{4} \left\lfloor \frac{n}{2} \right\rfloor \left\lfloor \frac{n-1}{2} \right\rfloor \left\lfloor \frac{n-2}{2} \right\rfloor \left\lfloor \frac{n-3}{2} \right\rfloor, \tag{13.4}$$

was subsequently published in a paper by Guy [27] in 1960 and by Harary and Hill [33] in 1962–1963. A drawing with $H(n)$ crossings can be achieved by the *soup can drawing* as follows. The surface of the soup can consists of the top lid, the bottom lid, and the mantle. Take a soup can and place $\lfloor n/2 \rfloor$ vertices equidistantly on the perimeter of the bottom lid and $\lceil n/2 \rceil$ vertices on the perimeter of the top lid. Join the vertices on the top lid to each other and the vertices on the bottom lid to each other, respectively, in straight-line segments. Join on the mantle top and bottom vertices along the shortest geodesic, making a choice if needed for the geodesic (see Figure 13.3, where the drawing is distorted by a rotation of the upper lid to improve visibility).

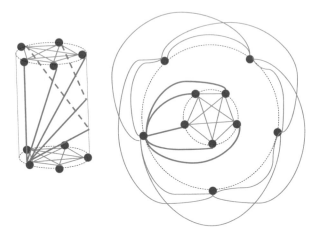

Fig. 13.3 Place a rotated copy of the five red upward edges to every vertex on the bottom lid to obtain the soup can drawing of K_{10}. The right side shows a corresponding cylindrical drawing of K_{10} in the plane

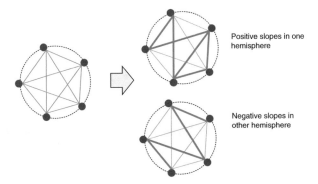

Positive slopes in one hemisphere

Negative slopes in other hemisphere

Fig. 13.4 The Blažek–Koman slope drawing of K_5

In previous papers, I incorrectly attributed Hill's conjecture to Guy. The reason is that I have never had a chance to see the paper of Guy [27] that first published the conjecture. In 1964, Blažek and Koman [12] came up with a very different drawing of K_n with $H(n)$ crossings, the *slope drawing*. Consider a regular n-gon in a circle, such that no line segment connecting any two vertices is parallel with the x-axis. Draw first all edges and diagonals of the regular n-gon in straight-line segments to represent K_n. Make the circle the equator of a sphere, and project edges with positive slopes to the upper hemisphere, and edges with negative slopes to the lower hemisphere. This is the Blažek–Koman slope drawing of K_n (see Figure 13.4). This spheric drawing can be deformed into a plane drawing with the same number of crossings, in such a way that the drawing falls into the class of *two-page drawings* [64]: all vertices live on a straight line, and no points of edges other than the endpoints are on this line.

Kleitman's cited results [39] on the parity of the partite classes for a smallest counterexample to Zarankiewicz conjecture (13.2) and on the parity of the crossing number of that extend to the smallest counterexample to Hill conjecture (13.3) and on the parity of the crossing number of that. More precisely, if $cr(K_{2n-1}) = H(2n-1)$, then $cr(K_{2n}) = H(2n)$, again by using the counting argument to be discussed in Section 13.4. Hence the smallest possible counterexample for Hill's conjecture has an odd number of vertices. Any two drawings of K_{2n-1} with no two tangential edges and no pairs of crossing adjacent edges have the same number of crossings modulo 2. This result allows computing the parity of $cr(K_{2n-1})$: it must have the same parity as $\binom{2n-1}{4}$, the number of crossings when $2n-1$ vertices are placed in a circle and are connected with straight-line segments. Consequently, $cr(K_{8k+1})$ and $cr(K_{8k+3})$ are even, while $cr(K_{8k+5})$ and $cr(K_{8k+7})$ are odd.

Pan and Richter [57] proved, in part using computer, that $cr(K_{11}) = H(11) = 100$. This result extends the verified cases of the Hill conjecture for all $n \leq 12$. McQuillan, Pan, and Richter [48] showed that $cr(K_{13}) \in \{219, 221, 223, 225\}$, the best result so far on K_{13}.

Schaefer [64] cites further early papers of Saaty, who independently arrived at the Hill conjecture, and of Harary and Guy on the conjectures. Abrego et al. [4] states "all general constructions (for arbitrary values of n) known with exactly $H(n)$ crossings are obtained from insubstantial alterations of either the Harary-Hill or the Blažek-Koman constructions (a few exceptions are known, but only for some small values of n)." Moon [50], likely looking for a better drawing, placed n points randomly, uniformly, and independently on the sphere and joined the points along the shorter arc of their main circle, to obtain a random drawing of K_n on a sphere. He computed the expected number of crossings in this random drawing as $\frac{1}{64}n(n-1)(n-2)(n-3)$, which is just marginally bigger than $H(n)$.

13.3 Euler's Formula

Euler's Polyhedral Formula states that a connected graph drawn crossing-free in the plane with n vertices, m edges and f faces, satisfies

$$n - m + f = 2. \tag{13.5}$$

Formula (13.5) immediately implies that a connected planar graph with $n \geq 3$ vertices has at most $3n - 6$ edges, and if it has no cycle shorter than s, then it has at most $\frac{s}{s-2}(n-2)$ edges, if $n \geq 1 + \frac{s}{2}$. An easy induction shows that for a connected graph or a connected graph with no cycle shorter than s,

$$cr(G) \geq m - 3n + 6 \quad \text{or} \quad cr(G) \geq m - \frac{s}{s-2}(n-2). \tag{13.6}$$

Formula (13.6) immediately implies the nonplanarity of K_5 and $K_{3,3}$. Formula (13.6) is usually called the *Euler bound* for the crossing number, and it is one of the few general methods available from the first days of crossing number research [36], including analogues of (13.6) for other surfaces. The shortcoming of (13.6) is, however, that this lower bound is always below m, although the crossing number can go up $\Omega(n^4)$.

13.4 Analogy with the Turán Hypergraph Problem

A hypergraph is ℓ-*uniform*, if all of its edges have exactly ℓ vertices. The *size* of a hypergraph is the number of its edges. The Turán number $T(n, k, \ell)$ denotes the minimum size of an ℓ-uniform hypergraph on n vertices, such that any k-element subset of vertices contains at least one edge from the hypergraph. An example with relevant k and ℓ is $T(7, 5, 4) = 7$ ([19] p. 649, Example 61.3). The complements of the seven lines of the Fano plane form the required 4-uniform hypergraph. In 1964, Katona, Nemetz, and Simonovits [38] found a counting argument showing

$$(n + 1 - \ell)T(n + 1, k, \ell) \geq (n + 1)T(n, k, \ell). \tag{13.7}$$

Let $\mathcal{H}$ be an ℓ-uniform hypergraph on $n + 1$ vertices and $T(n + 1, k, \ell)$ edges, such that any k-element subset of vertices contains at least one edge from the hypergraph. Then observe that

$$\left|\left\{(v, H) : v \in V(\mathcal{H}), H \in E(\mathcal{H}), v \notin H\right\}\right| = (n + 1 - \ell)T(n + 1, k, \ell).$$

Indeed, in the (v, H) ordered pairs $T(n + 1, k, \ell)$ different H's may occur. For every H, there are exactly $n + 1 - \ell$ vertices v not in H. To justify the formula (13.7), count the ordered pairs in the other way: $n + 1$ possible v's are present. Removing any v and the edges containing v, any k of the remaining n vertices must contain an edge that was not removed. Hence the number of the remaining edges is at least $T(n, k, \ell)$ by the definition of these numbers.

Now observe that (13.7) is equivalent to the inequality

$$\frac{T(n + 1, k, \ell)}{\binom{n+1}{\ell}} \geq \frac{T(n, k, \ell)}{\binom{n}{\ell}}, \tag{13.8}$$

which in turn immediately implies that

$$\lim_{n \to \infty} \frac{T(n, k, \ell)}{\binom{n}{\ell}} \tag{13.9}$$

exists. The limit is finite, as 1 is an upper bound for the terms in the increasing sequence.

An almost identical argument applies to crossing numbers. Assume now that G is a vertex-labeled graph drawn in the plane with $cr(G)$ crossings. (Recall that any two crossing edges in this drawing have four distinct endvertices.) Let H be a graph. Assume that $\#(H, G)$ subgraphs of G are isomorphic to H. They will be called copies of H. Assume further that no more than M copies of H contain any fixed pair of crossing edges of G in its drawing. Then

$$\#(H, G) \cdot cr(H) \leq M \cdot cr(G). \tag{13.10}$$

We refer to (13.10) as *the counting argument*. Observe that $\#(H, G) \cdot cr(H) \leq$

$$\left| \left\{ (H', \{e, f\}) : H' \text{ copy of } H; e, f \in E(H'); e, f \text{ cross in the drawing of } G \right\} \right|,$$

as we can select H' exactly $\#(H, G)$ ways, and the induced drawing of every H' contains at least $cr(H)$ crossing edge pairs. Counting the set of ordered pairs on the other way, exactly $cr(G)$ crossing edge pairs $\{e, f\}$ are in the drawing of G, and by our assumption no more than M copies of H can contain them. This proves (13.10).

Applying (13.10) for an optimal drawing of $G = K_{n+1}$ and for all $n + 1$ copies of $H = K_n$ in it, we can take $M = n - 3$. Indeed, apart from the four endpoints of some two crossing edges, there are exactly $(n + 1) - 4$ vertices that can be deleted to obtain a H containing this pair of edges. One obtains

$$(n + 1)cr(K_n) \leq (n - 3)cr(K_{n+1}), \text{ or equivalently } \frac{cr(K_n)}{\binom{n}{4}} \leq \frac{cr(K_{n+1})}{\binom{n+1}{4}}. \tag{13.11}$$

Similarly, applying the counting argument for an optimal drawing of $G = K_{n+1,n+1}$ and for all $(n + 1)^2$ copies of $H = K_{n,n}$ in it,

$$(n + 1)^2 cr(K_{n,n}) \leq (n - 1)^2 cr(K_{n+1,n+1}), \text{ or equivalently } \frac{cr(K_{n,n})}{\binom{n}{2}^2} \leq \frac{cr(K_{n+1,n+1})}{\binom{n+1}{2}^2}. \tag{13.12}$$

Formulas (13.11) and (13.12) imply that the limits

$$\lim_{n\to\infty} \frac{cr(K_n)}{\binom{n}{4}} = 24c_1 \text{ and } \lim_{n\to\infty} \frac{cr(K_{n,n})}{\binom{n}{2}^2} = 4c_2 \tag{13.13}$$

exist and are finite, but the values of c_1 and c_2 are not known. The constructions in Section 13.2 imply $c_1 \leq \frac{1}{64}, c_2 \leq \frac{1}{16}$, with equalities, if the conjectures (13.3) resp. (13.2) hold. Furthermore, we have the asymptotic formulae $cr(K_n) = (c_1 + o(1))n^4$ and $cr(K_{n,n}) = (c_2 + o(1))n^4$, with unknown constants [28]!

Kainen [35] and Moon [50] discovered using the counting argument that if (13.2) holds (or holds even just asymptotically), then (13.3) holds asymptotically. Richter and Thomassen [62] refined this argument showing that $c_1 \geq c_2/4$, in particular

$c_2 = 1/16$ implies $c_1 = 1/64$. In other words, if we know that a certain fraction of (13.2) holds as a lower bound asymptotically, then we know that the same fraction of (13.3) holds as a lower bound asymptotically. This is why the Hill conjecture has a natural place in any discussion on the Zarankiewicz conjecture. Kleitman's result [39] that the smallest counterexample to the Zarankiewicz's conjecture must occur for odd n and m, is another consequence of the counting argument.

Ringel [63] was the first to note the connection between Turán numbers and crossing numbers. He noticed that $T(n, 5, 4) \leq cr(K_n)$. Indeed, for a drawing of K_n with $cr(K_n)$ crossings, define a 4-uniform hypergraph on the n-element vertex set, where the hyperedges are composed of the four vertices of a pair of crossing edges. There is a hyperedge contained by the set of any five vertices, as a K_5 is drawn with these five vertices, and it exhibits a crossing. This 4-uniform hypergraph is in the domain of the minimization problem defining $T(n, 5, 4)$.

Further analogies between crossing number problems and the Turán hypergraph problem include the multiple maxima (if the conjectures hold), the fact that an improved lower bound on a particular problem induces improved lower bounds for larger problems and the corresponding results on limits.

Razborov [60] introduced the theory of flag algebras to prove results in asymptotic extremal combinatorics. After chess and proving combinatorial identities [58], this is another field where humans rarely can beat computers. Turán numbers are among the paradigmatic applications of this theory [61].

13.5   Success on the Conjectures Expressed in Fractions

In 1970, Kleitman [39] used $H = K_{n,6}$ as a sample graph in $G = K_{n,m}$ for a counting argument to obtain $cr(K_{n,m}) \geq (0.8 - o(1))Z(n, m)$ for $n, m \to \infty$, and in turn obtained $cr(K_n) \geq (0.8 - o(1))H(n)$.

In 2003 Nahas [52] showed $cr(K_{m,n}) \geq 0.8001Z(n, m)$ for large n, m, which in turn implies that $cr(K_n) \geq 0.8001H(n)$ for large n.

In 2006, a breakthrough paper of de Klerk, Maharry, Pasechnik, Richter, and Salazar [21], used cutting edge quadratic programming and computer work to prove the inequality

$$cr(K_{7,n}) \geq 2.1796n^2 - 4.5n. \qquad (13.14)$$

To place this result in context, make a comparison. For simplicity assume $n \geq 23$. The best previous lower bound was $cr(K_{7,n}) \geq 2.1n^2 - 4.2n$, while the Zarankiewicz conjecture (13.2) can be rewritten as

$$Z(7, n) = \begin{cases} 2.25n^2 - 4.5n + 2.25 & \text{if n odd} \\ 2.25n^2 - 4.5n & \text{if n even.} \end{cases}$$

Formula (13.14), through a counting argument for copies of $H = K_{7,n}$ in $G = K_{m,n}$, gives $cr(K_{m,n}) \geq (0.83 - o(1))Z(n,m)$ for $n, m \to \infty$, which in turn implies that $cr(K_n) \geq (0.83 - o(1))H(n)$.

In 2007, de Klerk, Pasechnik, and Schrijver [22] strengthened the optimization techniques further to prove

$$cr(K_{9,n}) \geq 3.8676063n^2 - 8n. \tag{13.15}$$

By a counting argument this gives $cr(K_{m,n}) \geq (0.8594 - o(1))Z(n,m)$ for $n, m \to \infty$, which in turn implies that $cr(K_n) \geq (0.8594 - o(1))H(n)$.

In a work not yet published, Norin and Zwols [53] applied the flag algebra method to the Zarankiewicz conjecture—no surprise, as paradigmatic applications of the flag algebra method include Turán numbers! They proved $cr(K_{m,n}) \geq (0.905 - o(1))Z(n,m)$, which in turn implies that $cr(K_n) \geq (0.905 \quad o(1))H(n)$. These are the state-of-the-art lower bounds regarding the Zarankiewicz (13.2) and Hill (13.3) conjectures.

13.6 Complete and Complete Bipartite Graphs Drawn on Other Surfaces

The investigation of crossing numbers was extended to other surfaces almost immediately as their study started. Let $crs_g(G)$ ($cr_{N_g}(G)$) denote the crossing number of the graph G, if drawn on the orientable (non-orientable) surface of genus g. (The usual terminology for crs_1 is the *toroidal crossing number*.) The counterparts of (13.6) are

$$crs_g(G) \geq m - \frac{s}{s-2}(n-2+2g) \quad \text{and} \quad cr_{N_g}(G) \geq m - \frac{s}{s-2}(n-2+g), \tag{13.16}$$

due to Kainen [36] and Kainen and White [37]. The counting argument works on any surface and counterparts of the limits in (13.13) follow. Guy and Jenkyns [31] showed that

$$\frac{1}{15}\binom{n}{2}\binom{m}{2} \leq crs_1(K_{n,m}) \leq \frac{1}{6}\binom{n-1}{2}\binom{m-1}{2} < \frac{2}{3}Z(n,m), \tag{13.17}$$

where the lower bound holds for n, m sufficiently large. Guy, Jenkyns and Schaer [32] showed that for all $n \geq 15$,

$$\frac{1}{210}(n)_4 \leq crs_1(K_n) \leq \frac{59}{5184}(n-1)_4 \sim \frac{59}{81}H(n), \tag{13.18}$$

where the lower bound holds for $n \geq 16$ and the upper bound holds for $n \geq 6$. The notation $(x)_4$ means the falling factorial $x(x-1)(x-2)(x-3)$.

Koman [40] and [42] obtained

$$\frac{41}{6552}(n)_4 \leq cr_{N_1}(K_n) \leq \frac{13}{16}H(n), \text{ and} \tag{13.19}$$

$$\frac{1}{336}(n)_4 \leq cr_{N_2}(K_n) \leq \frac{59}{5184}(n-1)_4 \sim \frac{59}{81}H(n), \tag{13.20}$$

where in (13.19) the lower bound holds for $n \geq 16$, and the upper bound holds for $n \geq 6$, while in (13.20) the lower bound holds for $n \geq 16$ and the upper bound always holds. Koman [41] showed that the lower bound in (13.17) also holds for $cr_{N_2}(K_{n,m})$ in the place of $cr_{S_1}(K_{n,m})$, and that the upper bound in (13.17) also holds for $cr_{N_2}(K_{n,m})$ in the place of $cr_{S_1}(K_{n,m})$ for infinitely many values of n and m.

There is surprisingly little published on $cr_{N_1}(K_{n,m})$, so we derive some results here. Pak Tung Ho [34] showed $cr_{N_1}(K_{n,4}) = \lceil \frac{n}{3} \rceil (2n - 3(1 + \lceil \frac{n}{3} \rceil)) \sim \frac{n^2}{3}$. This implies through a counting argument that $cr_{N_1}(K_{n,m}) \geq (\frac{1}{36} - o(1))n^2m^2 = (\frac{4}{9} - o(1))Z(n,m)$ as $n, m \to \infty$. We are going to show that $cr_{N_1}(K_{n,m}) \leq (\frac{13}{16} + o(1))Z(n,m)$ if $n, m \to \infty$.

An analogue of the formula (13.13) holds for $cr_{N_1}(K_n)$, c'_1, $cr_{N_1}(K_{n,n})$, c'_2, and $c'_1 \geq c'_2/4$ holds as well. From here, $cr_{N_1}(K_n) \leq \frac{13}{16}H(n)$ in (13.19) implies $cr_{N_1}(K_{n,n}) \leq (\frac{13}{16} + o(1))Z(n,n)$. Now take a drawing of $K_{n,n}$ on N_1 with so few crossings. For $m \leq n$, pick randomly and independently m vertices from one partite set, and consider the subdrawing of $K_{n,m}$ that it induces. In expectation, this random drawing of $K_{n,m}$ has $(\frac{13}{16} + o(1))Z(n,m)$ crossings if $n, m \to \infty$.

For arbitrary genus, Shahrokhi, Székely, Sýkora, and Vrt'o [65] showed that $cr_{N_g}(K_n)$ and $cr_{S_g}(K_n)$ are both upper bounded by $O(\frac{\log^2 g}{g}n^4)$ as $n \to \infty$, while $g \geq 2$; and they are both lower bounded by $\Omega(n^4/g)$ as far as $1 \leq g \leq \binom{n}{2}/64$ and $n \to \infty$. Similar results hold for $K_{n,n}$.

In 1971, motivated by circuit design, Owens [54] defined *biplanar* and *k-planar crossing numbers*. For a graph G on vertex set V, partition the edges into the edge sets of k graphs G_i ($i = 1, 2, \ldots, k$) on the same vertex set, to minimize $\sum_{i=1}^{k} cr(G_i)$. The minimum is the k-planar crossing number $cr_k(G)$. The biplanar case $k = 2$ can be explained that a plate has two sides, vertices are incident to both, and we want to minimize the sum of crossings on the two sides as we draw edges of G on either side. Owens [54] showed that $cr_2(K_n) \leq (\frac{7}{1536} + o(1))n^4 \sim \frac{7}{24}H(n)$. Owens' drawing starts with the soup can drawing of K_n and then puts certain edges—as they are drawn on the soup can—to the inner surface of the soup can. The two half-sized complete graphs drawn on the lids are then partitioned into an outer and an inner drawing according to the slope drawing, and edges on the mantle are partitioned into two sets, evenly at every vertex.

If a biplanar drawing of the graph G shows two isomorphic graphs in the two planes, it is called *self-complementary* [20]. Self-complementary drawings are very convenient, as a single copy of the two isomorphic drawings can be used to represent

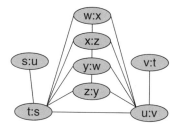

Fig. 13.5 Biplanar drawing of K_8 Beineke [9]

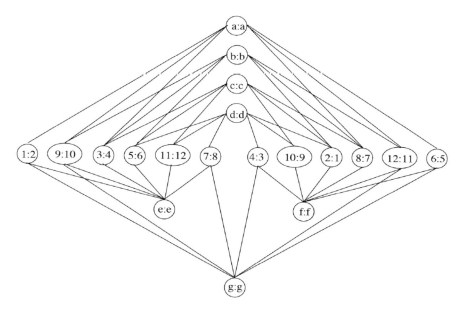

Fig. 13.6 Self-complementary optimal biplanar drawing of $K_{7,12}$ [20]

it. Just label the nodes of the drawing with an ordered pair of vertices, using the first entry of the ordered pair as label describing the drawing on the first plane and using the second entry of the ordered pair as label describing the drawing on the second plane. To illustrate, Figure 13.5 shows in a single drawing Beineke's [9] self-complementary biplanar drawing of K_8. It is easy to make the Owens drawing self-complementary, if n is even.

Czabarka, Székely, Sýkora, and Vrt'o [20] looked at several particular instances of the optimal biplanar drawing of $K_{n,m}$. They made a number of specific conjectures for small values of m, but did not find a clear general conjecture for $K_{n,m}$. They conjectured, however, that if $K_{n,m}$ has an even number of edges, it has a self-complementary optimal biplanar drawing. Figure 13.6 shows a self-complementary optimal biplanar drawing of $K_{7,12}$ from [20].

De Klerk, Pasechnik, and Salazar [23] studied the k-page crossing numbers of complete bipartite graphs. Shahrokhi, Székely, Sýkora and Vrt'o [67] determined $cr_k(K_{2k+1,q})$ exactly, and gave bounds for the k-planar crossing numbers of complete and complete bipartite graphs.

The *minor crossing number* $mcr(G)$ is the least crossing number of any graph, of which G is a minor. This is a natural definition for a minor-monotone version of the crossing number, as the crossing number is not minor monotone. Bokal, Fijavž, and Mohar [13] found the bounds for the minor crossing number $\frac{1}{2}(m-2)(n-2) \leq mcr(K_{n,m}) \leq (m-3)(n-3) + 5$, where the lower bound requires $3 \leq m \leq n$, and the upper bound requires $4 \leq m \leq n$; and $\lceil \frac{1}{4}(n-3)(n-4) \rceil \leq mcr(K_n) \leq \lfloor \frac{1}{2}(n-5)^2 \rfloor + 4$, where the lower bound is for $n \geq 3$, the upper bound is for $n \geq 9$.

13.7 Foundational Problems

Pach and Tóth [55] were the first to write down that there is a possibility that papers on crossing numbers operate with different definitions, the equivalence of which are not known, though the issue was also raised in a 1995 conference talk by Mohar [49]. In both [49] and [55], the *pair crossing number $pcr(G)$* of a graph G is defined as the minimum number of edge pairs crossing in a drawing of G. Clearly $pcr(G) \leq cr(G)$ but we do not know of any instance where this inequality is strict nor a proof is known that these definitions are equivalent. As we noted in (ii), a drawing realizing $cr(G)$ has no edge pairs with multiple crossings, but we have no evidence that for reducing the number of crossing edge pairs allowing multiple crossings of edges do not help! Pach and Tóth [56] scrutinized crossings of adjacent edges, i.e., edges sharing an endpoint. They defined a "−" version of $cr(G)$ and $pcr(G)$, where crossings of adjacent edges are allowed in the drawing, but they do not contribute to the count. Clearly $cr(G)_- \leq cr(G)$ and $pcr(G)_- \leq pcr(G)$ but we do not know of any instance where these inequalities are strict, nor is a proof known that these definitions are equivalent. As we noted in (i), a drawing realizing $cr(G)$ has no adjacent edge pairs that crosses, but perhaps allowing crossings of adjacent edge pairs without counting them can help at reducing the number of crossings among nonadjacent edge pairs. This possibility is hard to imagine, but we cannot exclude it. Schaefer [64], after carefully reading a number of research papers and textbooks, pointed out many instances where the arguments shifted from one concept of crossing number to another.

In [70] I asked "How is it possible that decades in research of crossing numbers passed by and no major confusion resulted from these foundational problems?" Part of the reason is that $cr(G)_- = cr(G) = pcr(G)_- = pcr(G)$ is possible for all graphs. Even if they are not, finding counterexamples is hard, as computing many— if not all—variants of the crossing number is NP-hard [26, 55]. In addition, the conjectured optimal drawings often use straight lines or geodesics and hence satisfy (i) and (ii).

Sir Karl Popper [59] solved the age-old problem of induction in philosophy: how can we correctly infer laws of nature from a finite number of observations and experiments. Sir Karl's program is both descriptive and prescriptive for science: make a bold hypothesis and try to refute it. If a hypothesis is not refuted, notwithstanding substantial effort, then it may be corroborated—but it is never proven. Imre Lakatos, who applied the Popperian epistemology to mathematics [45], carried out his arguments on two paradigmatic examples: one is Euler's polyhedral formula and the other is the concept of a real function. He points to a sequence of refutations of the polyhedral formula (holes, tunnels, crested cube) that required adjustment of the definitions, to avoid issues of which nobody thought before, to keep the formula. As the most basic lower bounds for crossing numbers are based on Euler's polyhedral formula (13.6), (13.16), it is no surprise that we run into complexities of drawings that nobody suspected a few decades ago. The rise and fall of the Zarankiewicz conjecture discussed in Section 13.2 also can be viewed as Popper's program at work. Guy [30] also pointed out "much more sweeping assumptions than the overt hypotheses of the theorem" in some crossing number papers.

Surprisingly, I have never even heard the name of Lakatos during my studies in Hungary. I see the explanation from complementary reasons. One is political: Lakatos was a high-ranking party official before the 1956 revolution, after which he left the country. (For a biography showing his different lives, see Bandy [8].) The other is that many mathematicians are uncomfortable with the fact that what we know may require correction in the future. (They have no problem at admitting that such corrections happened in the past.) And an unexpected but perhaps not accidental connection: Beineke and Wilson [10] points out that Anthony Hill took classes with Lakatos!

13.8 Rectilinear Crossing Numbers

Note that the Zarankiewicz drawing brings a big bonus for free: the edges are drawn in straight-line segments instead of curves! This bonus is also there for planar graphs, i.e., graphs with crossing number zero, as a theorem of Fáry [25] states that planar graphs always can be drawn in straight-line segments. This bonus, however, is not available for drawings of arbitrary graphs or in particular of the complete graphs. Bienstock and Dean [11] showed that as long as the crossing number of a graph is at most three, then the graph admits a straight-line drawing realizing its crossing number, but there exists graphs with crossing number four that require arbitrarily large number of crossings in any straight-line drawing. Let $\overline{cr}(G)$, the *rectilinear crossing number* of the graph G, denote the variant of the crossing number, where only straight-line segments are allowed for drawing the edges. Clearly $\overline{cr}(G) \geq cr(G)$, and, for example, $19 = \overline{cr}(K_8) > cr(K_8) = 18$. ($\overline{cr}(K_n)$ has been computed for all $n \leq 27$ [5].) For a long time, the same lower and upper bounds were the best for both $cr(K_n)$ and $\overline{cr}(K_n)$ for large values of n. This started to change only in 2005, when Ábrego and Fernández-Merchant [1] made a breakthrough by showing

$$\overline{cr}(K_n) \geq H(n).$$

Lovász, Vesztergombi, Wagner and Welzl [47] strengthened this lower bound to $\overline{cr}(K_n) \geq (\frac{1}{64} + \frac{10^{-5}}{24} + o(1))n^4$, the first to separate the crossing number and rectilinear crossing number of complete graphs. This lower bound was further improved to $0.37962\binom{n}{4} + \Theta(n^3) > (1.0123 + o(1))H(n)$ by Aichholzer et al. [7] and a subsequent small further improvement was done $0.379688\binom{n}{4} + \Theta(n^3)$ by Ábrego et al. [2].

In another terminology, the rectilinear crossing number of the complete graph is the least number of convex quadrilaterals determined by n points in the plane, no three of which are collinear. All the lower bounds above are based on lower bounds on the number of *planar k-sets*. The k-set problem asks what is the largest possible number of k-element subsets of a set of n points in general position in the plane, which can be separated from the remaining $n - k$ points by a straight line. A k-*edge* is a pair of points such that there are exactly k and $n - k - 2$ points on the two sides of their connecting line.

Aichholzer, Aurenhammer, and Krasser [6] developed a combinatorial description of relative point locations in the plane, called *order types*, which is used in computer work on rectilinear crossing numbers. The currently known values of $\overline{cr}(K_n)$ and combinatorially different optimal drawings can be downloaded from [5]. It is an interesting contrast with ordinary crossing numbers of complete graphs, that many more rectilinear crossing numbers are known exactly, but still no conjecture has emerged for the rectilinear crossing number of K_n.

13.9 Progress Giving Hope

In 2013, Christian, Richter, and Salazar [18] showed that for all m, there is an $n_0(m)$, such that if (13.2) holds for all $n \leq n_0$, then (13.2) holds for all n with this m. This, in turn, allows for every fixed m to check with an algorithm whether for all n (13.2) is true or false. To be explicit, set $Z(m) = \lfloor \frac{m}{2} \rfloor \lfloor \frac{m-1}{2} \rfloor$ and $n_0(m) = \left((2Z(m))^{m!}(m!)!\right)^4$. This algorithm will never run on a computer. However, the theoretical result is important and stunning. For any particular m, there is no need for an infinite sequence of improved bounds to reach the Zarankiewicz conjecture, although Section 13.4 would be well compatible with such a need.

In 2013, Ábrego, Aichholzer, Fernández-Merchant, Ramos, and Salazar [3] proved the Hill conjecture for two-page drawings. In a very recent work [4], they extended this result to s-shellable drawings of K_n, if $s \geq n/2$. Although s-shellable drawings of K_n are a minority, they include the Hill drawing and the Blažek-Koman slope drawing.

Let a drawing of K_n be given. Let R be a connected component of $\mathbf{R}^2 \setminus D$, where D is the set of points consisting of points and curves representing the vertices and edges of a drawing of K_n. Let $S = \{v_1, v_2, \ldots, v_s\}$ be a sequence of distinct vertices. The set S is an s-shelling of D witnessed by R, if for all $1 \leq i < j \leq s$, removing the vertices $v_1, v_2, \ldots, v_{i-1}, v_{j+1}, v_{j+2}, \ldots, v_s$ and their incident edges from the drawing, the vertices v_i and v_j are on the boundary of the connected component of the resulting drawing that contains R.

Such s-shellable drawings include two-page and cylindrical drawings (see Section 13.2) and monotone and x-bounded drawings. A drawing is called *monotone* if every vertical line contains at most one vertex and every vertical line intersects each edge at most once. A drawing is called *x-bounded* if different vertices have different x-coordinates, and if the vertices u and v are joined by an edge, then the points of the curve representing this edge have their x-coordinates between the x-coordinates of u and v.

The proof goes back to the ideas of the proof $\overline{cr}(K_n) \geq H(n)$ in [1].

The paper generalizes the concept of a k-edge from a point configuration in the plane to a crossing minimal but not necessarily rectilinear drawing of K_n, and then shows that for an s-shellable drawing ($s \geq n/2$) the argument generalizes. The papers [3] and [4] are the first to prove the Hill conjecture (13.3) for some classes of drawings.

References

1. Ábrego, B.M., Fernández-Merchant, S.: A lower bound for the rectilinear crossing number. Graphs Comb. **21**, 293–300 (2005)
2. Ábrego, B.M., Balogh, B., Fernández-Merchant, S., Leaños, J., Salazar, G.: An extended lower bound on the number of ($\leq k$)-edges to generalized configurations of points and the pseudolinear crossing number of K_n. J. Comb. Theory Ser. A **115**, 1257–1264 (2008)
3. Ábrego, B.M., Aichholzer, O., Fernández-Merchant, S., Ramos, P., Salazar, G.: The 2-page crossing number of K_n. Discret. Comput. Geom. **49**(4), 747–777 (2013)
4. Ábrego, B.M., Aichholzer, O., Fernández-Merchant, S., Ramos, P., Salazar, G.: Shellable drawings and the cylindrical crossing number of K_n. (2013) arXiv:1309.3665
5. Aichholzer, O.: http://www.ist.tugraz.at/staff/aichholzer/research/rp/triangulations/crossing/
6. Aichholzer, O., Aurenhammer, F., Krasser, H.: On the crossing number of complete graphs. In: Proceedings of the Annual ACM symposium on Computational Geometry, Barcelona, pp. 19–24 (2002)
7. Aichholzer, O., García, J., Orden, D., Ramos, P.: New lower bounds for the number of ($\leq k$)-edges and the rectilinear crossing number of K_n. Discret. Comput. Geom. **38**, 1–14 (2007)
8. Bandy, A.: Chocolate and Chess. Unlocking Lakatos. Akadémiai Kiadó, Budapest (2009)
9. Beineke, L.: Biplanar graphs: a survey. Comput. Math. Appl. **34**(11), 1–8 (1997)
10. Beineke, L., Wilson, R.: The early history of the Brick Factory Problem. Math. Intell. **32**(2), 41–48 (2010)
11. Bienstock, D., Dean, N.: Bounds on the rectilinear crossing numbers. J. Graph Theory **17**(3), 333–348 (1993)
12. Blažek, J., Koman, N.: A minimal problem concerning complete plane graphs. In: Fiedler, M. (ed.) Theory of Graphs and Its Applications (Proceedings of the Symposium Held in Smolenice, 1963), pp. 113–117. Publishing House of the Czechoslovak Academy of Sciences, Prague (1964)

13. Bokal, D., Fijavž, V., Mohar, B.: The minor crossing number. SIAM J. Discret. Math. **20**(2), 344–356 (2006)

14. Bronfenbrenner, U.: The graphic presentation of sociometric data. Sociometry **7**(3), 283–289 (1944)

15. Busacker, R.G., Saaty, T.L.: Finite Graphs and Networks: An Introduction with Applications. McGraw-Hill Book Co., New York/Toronto/London (1967). Hungarian translation: Véges Gráfok és Hálózatok, translated by I. Juhász, publisher's reader G.O.H. Katona, Műszaki Könyvkiadó, Budapest (1969)

16. Cauchy, A., Recherches sur les Polyèdres - Premier Mémoire. Journal de l'École Polytechnique **9**(Cahier 16), 68–86 (1813)

17. Chojnacki, C.: Über wesentliche unplättbare Kurven in dreidimensionalen Raume. Fund. Math. **23**, 135–142 (1934)

18. Christian, R., Richter, R.B., Salazar, G.: Zarankiewicz's Conjecture is finite for each fixed m. J. Comb. Theory Ser. B **103**, 237–247 (2013)

19. Colbourn, C.J., Dinitz, J.H.: Handbook of Combinatorial Designs, 2nd edn. Chapman and Hall/CRC, Boca Raton (2007)

20. Czabarka, É., Sýkora, O., Székely, L.A., Vrťo, I.: Crossing numbers and biplanar crossing numbers I: a survey of problems and results. In: Győri, E., et al. (eds.) More Sets, Graphs and Numbers. Bolyai Society Mathematical Studies, vol. 15, pp. 57–77. Springer, New York (2006)

21. de Klerk, E., Maharry, J., Pasechnik, D.V., Richter, R.B., Salazar, G.: Improved bounds for the crossing numbers of $K_{n,m}$ and K_n. SIAM J. Discret. Math. **20**(1), 189–202 (2006)

22. de Klerk, E., Pasechnik, D.V., Schrijver, A.: Reduction of symmetric semidefinite programs using the regular *-representation. Math. Program. Ser. B **109**(2–3), 613–624 (2007)

23. de Klerk, E., Pasechnik, D.V., Salazar, G.: Book drawings of complete bipartite graphs. Discret. Appl. Math. **167**, 80–93 (2014)

24. Dudeney, H.E.: Amusements in Mathematics. http://www.gutenberg.org/files/16713/16713-h/16713-h.htm (1917)

25. Fáry, I.: On straight line representations of graphs. Acta Univ. Szeged Sect. Sci. Math. **11**, 229–233 (1948)

26. Garey, M.R., Johnson, D.S.: Crossing number is NP-complete. SIAM J. Alg. Discret. Methods **4**, 312–316 (1983)

27. Guy, R.K.: A combinatorial problem. Nabla (Bull. Malayan Math. Soc.) **7**, 68–72 (1960)

28. Guy, R.K.: The crossing number of the complete graph. Research Paper No. 8, The University of Calgary (1967)

29. Guy, R.K.: The decline and fall of Zarankiewicz's theorem. In: Harary, F. (ed.) Proof Techniques in Graph Theory (Proceedings of the Second Ann Arbor Graph Theory Conference, Ann Arbor MI, 1968), pp. 63–69. Academic, New York/London (1969)

30. Guy, R.K.: Math. Rev. **58**, # 21749 (1974)

31. Guy, R.K., Jenkyns, T.A.: The toroidal crossing number of $K_{m,n}$. J. Comb. Theory **6**, 235–250 (1969)

32. Guy, R.K., Jenkyns, T.A., Schaer, J.: The toroidal crossing number of the complete graph. J. Comb. Theory **4**, 376–390 (1968)

33. Harary, F., Hill, A.: On the number of crossings in a complete graph. Proc. Edinb. Math. Soc. (II) **13**, 333–338 (1962–63)

34. Ho, P.T.: The crossing number of $K_{4,n}$ on the real projective plane. Discret. Math. **304**(1–3), 23–33 (2005)

35. Kainen, P.C.: On a problem of P. Erdős. J. Comb. Theory **5**, 374–377 (1968)

36. Kainen, P.C.: A lower bound for crossing numbers with applications to K_n, $K_{p,q}$, and $Q(d)$. J. Comb. Theory Ser. B **12**, 287–298 (1972)

37. Kainen, P.C., White, A.T.: On stable crossing numbers. J. Graph Theory **2**(3), 181–187 (1978)

38. Katona, G.O.H., Nemetz, T., Simonovits, M.: On a graph problem of Turán (in Hungarian). Mat. Fiz. Lapok **15**, 228–238 (1964)

39. Kleitman, D.J.: The crossing number of $K_{5,n}$. J. Comb. Theory **9**, 315–323 (1970)

40. Koman, M.: On the crossing numbers of graphs. Acta Univ. Carol. Math. Phys. **10**, 9–46 (1969)

41. Koman, M.: A note on the crossing number of $K_{m,n}$ on the Klein bottle. In: Proceedings of the Symposium in Recent Advances in Graph Theory, Prague 1974, pp. 327–334 (1975)
42. Koman, M.: New upper bounds for the crossing number of K_n on the Klein bottle. Cas. Pĕst. Mat. **103**, 282–288 (1978)
43. Kullman, D.E.: The utilities problem. Math. Mag. **52**, 299–302 (1979)
44. Kuratowski, K.: Sur le problème des courbes gauches en topologie. Fund. Math. **15**, 271–283 (1930)
45. Lakatos, I.: In: Worrall, J., Zahar, E. (eds.) Proofs and Refutations: The Logic of Mathematical Discovery. Cambridge University Press, Cambridge/New York/Melbourne (1976)
46. Leighton, F.T.: Complexity Issues in VLSI. MIT Press, Cambridge (1983)
47. Lovász, L., Vesztergombi, K., Wagner, U., Welzl, E.: Convex quadrilaterals and k-sets. In: Pach, J. (ed.) Towards a Theory of Geometric Graphs. Contemporary Mathematics Series, vol. 342, pp. 139–148. American Mathematical Society, Providence (2004)
48. McQuillan, D., Pan, S., Richter, R.B.: On the crossing number of K_{13}. arXiv:1307.3297 (2013) Human Interrelations
49. Mohar, B.: Problem mentioned at the Special session on Topological Graph Theory, Mathfest, Burlington, Vermont (1995)
50. Moon, J.: On the distribution of crossings in random complete graphs. J. Soc. Ind. Appl. Math. **13**, 506–510 (1965)
51. Moreno, J.L.: Who Shall Survive?: A New Approach to the Problem of Human Interrelations. Nervous and Mental Disease Publishing Co., Washington (1934). Edition 2. Beacon House (1953) http://www.asgpp.org/docs/WSS/wss%20index/wss%20index.html
52. Nahas, N.H.: On the crossing number of $K_{m,n}$. Electron. J. Comb. **10**, Note #N8 (2003)
53. Norin, S., Zwols, Y.: Turáns brickyard problem and flag algebras. Banff International Research Station workshop 13w5091 (2013). http://www.birs.ca/events/2013/5-day-workshops/13w5091/videos/watch/201310011538-Norin.html
54. Owens, A.: On the biplanar crossing number. IEEE Trans. Circ. Theory **CT-18**, 277–280 (1971)
55. Pach, J., Tóth, G.: Which crossing number is it anyway? In: Proceedings of the 39th Annual Symposium on Foundation of Computer Science, pp. 617–626. IEEE Press, Baltimore (1998). J. Comb. Theory Ser. B **80**, 225–246 (2000)
56. Pach, J., Tóth, G.: Thirteen problems on crossing numbers. Geombinatorics **9**, 194–207 (2000)
57. Pan, S., Richter, R.B.: The crossing number of K_{11} is 100. J. Graph Theory **56**(2), 128–134 (2007)
58. Petkovsek, M., Wilf, H., Zeilberger, D.: $A = B$. Taylor and Francis, Abingdon (1996)
59. Popper, K.R.: Conjectures and Refutations: The Growth of Scientific Knowledge. Routledge and Kegan Paul, London (1963)
60. Razborov, A.A.: Flag algebras. J. Symb. Log. **72**(4), 1239–1282 (2007)
61. Razborov, A.A.: On 3-hypergraphs with forbidden 4-vertex configurations. SIAM J. Discret. Math. **24**(3), 946–963 (2010)
62. Richter R.B., Thomassen, C.: Relations between crossing numbers of complete and complete bipartite graphs. Am. Math. Mon. **104**, 131–137 (1997)
63. Ringel, G.: Extremal problems in the theory of graphs. In: Theory of Graphs and Its Applications (Proceedings of the Symposium held in Smolenice, 1963), pp. 85–90. Publishing House of the Czechoslovak Academy of Sciences, Prague (1964)
64. Schaefer, M.: The graph crossing number and its variants: a survey. Electron. J. Comb. Dyn. Surv. #DS21: May 15 (2014)
65. Shahrokhi, F., Sýkora, O., Székely, L.A., Vrt'o, I.: Drawings of graphs on surfaces with few crossings. Algorithmica **16**, 118–131 (1996)
66. Shahrokhi, F., Sýkora, O., Székely, L.A., Vrt'o, I.: Crossing numbers: bounds and applications. In: Bárány, I., Böröczky, K. (eds.) Intuitive Geometry. Bolyai Society Mathematical Studies, vol. 6, pp. 179–206. János Bolyai Mathematical Society, Budapest (1997)
67. Shahrokhi, F., Sýkora, O., Székely, L.A., Vrt'o, I.: Discret. Appl. Math. **155**(9), 1106–1115 (2007)

68. Székely, L.A.: Crossing numbers and hard Erdős problems in discrete geometry. Comb. Probab. Comput. **6**(3), 353–358 (1997)
69. Székely, L.A.: Zarankiewicz crossing number conjecture, article. In: Managing Editor: Hazewinkel, M. (ed.) Kluwer Encyclopaedia of Mathematics, Supplement III, pp. 451–452. Kluwer Academic Publishers, Dordrecht (2002)
70. Székely, L.A.: A successful concept for measuring non-planarity of graphs: the crossing number. Discret. Math. **276**(1–3), 331–352 (2004)
71. Székely, L.A.: Progress on crossing number problems. In: Vojtás, P., et al. (eds.) SOFSEM 2005: Theory and Practice of Computer Science: 31st Conference on Current Trends in Theory and Practice of Computer Science Liptovský Ján, Slovakia, 22–28 Jan 2005. Lecture Notes in Computer Science, vol. 3381, pp. 53–61. Springer, New York (2005)
72. Turán, P.: A note of welcome. J. Graph Theory **1**, 7–9 (1977)
73. Urbaník, K.: Solution du problème posé par P. Turán. Colloq. Math. **3**, 200–201 (1955)
74. Vrt'o, I.: Crossing Numbers of Graphs: A Bibliography. http:/sun.ifi.savba.sk/~imrich/
75. Wagner, K.: Über eine Eigenschaft der ebenen Komplexe. Math. Ann. **114**, 570–590 (1937)
76. Woodall, D.R.: Cyclic-order graphs and Zarankiewicz's crossing-number conjecture. J. Graph Theory **17**, 657–671 (1993)
77. Zarankiewicz, K.: On a problem of P. Turán concerning graphs. Fundam. Math. **41**(1), 137–145 (1955)

Chapter 14
It Is All Labeling

Peter J. Slater

14.1 Introduction

The spirit of these papers is that the topics should be presented in an historical context with emphasis on unsolved problems/conjectures. My personal background is a mixture of mathematics and computer science, with a year as a National Research Council/National Bureau of Standards postdoctoral fellow, working in the Operations Research Division of NBS headed by Alan Goldman. To put this in context, the mathematician asks if a graph is graceful, the computer scientist asks how efficiently one can find a graceful labeling, and the operations researcher asks if it is not graceful, then how close to graceful is it? I believe the latter outlook actually provides a great source of interesting mathematics/computer science questions.

Somewhat surprisingly, I did not (as most researchers in the area do) initially enter the realm of graph numberings and labelings through the Graceful Tree Conjecture of Ringel, Kotzig, and Rosa. My first considerations involved infinite graphs, which will be discussed in Sect. 14.2. Section 14.3 will consider variations and generalizations of gracefulness. Section 14.4 will consider some of the various forms of magic labeling (with an amazing connection to fractional domination). A brief discussion of "It is all labeling" will follow.

Completeness is not the goal here, but rather the presentation of some problems that I have personally found to be interesting and the placing of labeling problems in

---

Dedicated to Dr. Alan J. Goldman, a mathematics Renaissance man, who thought that the field of mathematics is one, and one should know much about all of it, and who was successful in doing so.

P.J. Slater (✉)
Computer Science Department, University of Alabama in Huntsville, Huntsville, AL 35899, USA

Mathematical Sciences Department, University of Alabama in Huntsville, Huntsville, AL 35899, USA

© Springer International Publishing Switzerland 2016
R. Gera et al. (eds.), *Graph Theory*, Problem Books in Mathematics,
DOI 10.1007/978-3-319-31940-7_14

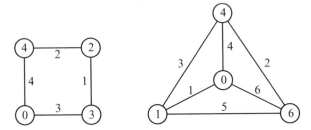

Fig. 14.1 Graceful numberings of C_4 and K_6

a general graph theory setting. A complete, ever-expanding bibliography of graceful labeling/numbering papers is available from Gallian [18].

For a graph $G = (V,E)$, the *order* is the number of vertices, $n = |V(G)|$, and the *size* is the number of edges, $m = |E(G)|$. In general, for a given set $W = \{w_1, w_2, \ldots, w_t\}$, a *vertex labeling* is an assignment of elements in W to each vertex $v \in V(G)$. In 1963, Ringel [33] conjectured that if T is any tree with m edges, then the complete graph K_{2m+1} can be decomposed into $2m+1$ subgraphs isomorphic to T. Kotzig later introduced a strengthened form of this conjecture, as noted by Rosa [35], who was the first to study various ways of numbering the vertices of T. In 1968, Golomb [21] helped to popularize this question, and he introduced the more general problem of determining which of all graphs are "graceful." (If tree T with m edges is graceful, then K_{2m+1} can be decomposed into $2m+1$ copies of T.)

Given G and W, for a labeling $f: V(G) \to W$, the induced edge labeling f of $E(G)$ is defined for each edge $e = uv \in E(G)$ by $f(uv) = |f(u)-f(v)|$. A 1-1 labeling $f: V(G) \to W$, I will call a *numbering* if the induced labeling $f: E(G) \to W$ is also 1-1. A *graceful numbering* of graph G is an injection $f: V(G) \to \{0, 1, 2, \ldots, m\}$, such that $f: E(G) \to \{1, 2, 3, \ldots, m\}$ is a bijection. That is, each of the m edges receives a distinct label in $\{1, 2, \ldots, m\}$. A graph that has a graceful numbering is called a *graceful graph*. For cycle C_4, we have $m = 4$, and for complete graph K_4, we have $m = 6$. Graceful numberings of C_4 and K_4 are shown in Fig. 14.1.

Conjecture 1 (Ringel, Kotzig, Rosa). All trees are graceful.

Theorem 1 (Rosa [35]). Cycle C_n is graceful if and only if $n \equiv 0, 3 \pmod 4$.

Think of the proof of Theorem 1 as a warm-up problem for the labeling novices!

14.2 Countably Infinite Graphs

Let $N = \{1, 2, 3, \ldots\}$ denote the set of positive integers. A permutation of N is a sequence $A = \{a_k | k \in N\} = (a_1, a_2, a_3, \ldots)$ in which every element of N appears exactly once. The difference sequence of A is $D = (d_1, d_2, d_3, \ldots)$ where $d_i = |a_{i+1}-a_i|$. One of the questions asked by Roger Entringer is the following. Does there exist a permutation A of N such that the difference sequence D is also a permutation of N?

Theorem 2 (Slater and Velez [48]). Given any sequence $(m_1, m_2, m_3, \ldots)$ of positive integers, there is a permutation A of N for which the difference sequence D satisfies $|\{i | d_i = j\}| = m_j$. In particular (with every $m_j = 1$), there is a permutation A of N whose difference sequence D is a permutation of N.

For the case where D is to be a permutation, we let $a_1 = 1$ and $a_2 = 2$, so $d_1 = 1$, and then, given $a_1, a_2, \ldots, a_{2t-1}, a_{2t}$, we define a_{2t+1} and a_{2t+2} as follows. Let a_{2t+1} be $1 + 2M$ where M is the largest element in $A_{2t} = \{a_1, a_2, \ldots, a_{2t}\}$. If r is the smallest element of N not yet in A_{2t} and s is the smallest element not yet in $\{d_1, d_2, \ldots, d_{2t}\}$, then $a_{2t+2} = r$ if $r \leq s$, and otherwise $a_{2t+2} = a_{2t+1} - s$ (which makes $d_{2t+1} = s$). So $A = (1, 2, 5, 3, 11, 4, 23, 19, 47, 42, 95, 6, 191, 185, 383, \ldots)$ and $D = (1, 3, 2, 8, 7, 19, 4, 28, 5, 53, 89, 185, 6, 197, 375, \ldots)$. Letting $f(n) = a_n$, we have $f(2t+1) > 2^{t+1}$, so A grows exponentially.

Problem 2 How small a growth rate can permutation A have if its difference sequence D is a permutation?

Consider the following greedy procedure for keeping the a_i's small, as described in [41]. Let $a_1 = 1$ and $a_2 = 2$. Given $A_n = (a_1, a_2, \ldots, a_n)$, let a_{n+1} be the smallest positive integer t not in A_n such that $|a_n - t|$ is not in $D_{n-1} = (d_1, d_2, \ldots, d_{n-1})$. This produces $A^* = (1, 2, 4, 7, 3, 8, 14, 5, 12, 20, 6, 16, 27, 9, 21, 34, 10, 25, 41, 11, \ldots)$ and $D^* = (1, 2, 3, 4, 5, 6, 9, 7, 8, 14, 10, 11, 18, 12, 13, 24, 15, 16, 30, \ldots)$. See the On-Line Encyclopedia of Integer Sequences.

Theorem 3 (Slater and Velez [48]). The sequence A^* so produced by this procedure is a permutation of N.

Seemingly easy to resolve is the next problem/conjecture.

Conjecture 3 (Slater and Velez [48]). The sequence D^* so produced is a permutation of N.

There are many interesting questions involving permutation A^*. Observe that in A^* (skipping a_1) we have ascending consecutive terms $2 < 4 < 7$ and $3 < 8 < 14$ and $5 < 12 < 20$, ... There are cases in which we have only two consecutive ascending terms, that is, $a_i > a_{i+1} < a_{i+2} > a_{i+3}$.

Problem 4 Other than $1, 2, 4, 7$, does A^* ever contain four consecutive ascending numbers?

Problem 5 Allowing me some imprecision, for each ascending triple in A^*, we have "small, medium, and large" values. Consider the growth rates of the smalls, the mediums, and the larges.

In a later paper (Slater and Velez [49]), we considered a "bandwidth" type of problem for permutations A, where difference sequence D will contain only a small set of values. (See also [6, 7].)

Conjecture 6 ([49]). For a subset $S = \{s_1, s_2, \ldots, s_n\}$ of N, there exists a permutation A of N with difference sequence D, such that $\{d_j | d_j \; \varepsilon \; D\} = S$ if and only if the greatest common divisor satisfies $(s_1, s_2, \ldots, s_n) = 1$.

Theorem 4 (Slater and Velez [49]). Let $S = \{s_1, s_2, \ldots, s_n\}$, where $(s_1, s_2, \ldots, s_n) = 1$ and for each r there exists a t such that $(s_r, s_t) = 1$. Then there exists a permutation $\{a_k : k \; \varepsilon \; N\}$ such that the difference sequence D satisfies $\{d_k : k \; \varepsilon \; N\} = S$, and each element in S occurs infinitely often in D.

Having seen the results of [48, 49], Paul Erdős asked me if it would be possible to construct a permutation B of N so that its difference sequence D *and all succeeding difference sequences* would also be permutations.

Theorem 5 (Slater [42]). There exists a permutation of N for which each of its successive difference sequences is also a permutation of N.

To put it mildly, the growth rate of the permutation B that I constructed for this problem is rather large.

Problem 7 How small a growth rate can permutation B of N have if all of its difference sequences are also permutations of N?

As indicated in Fig. 14.2, for the permutation A*, we can label the infinite path P_∞ with $f(v_i) = a_i$. The difference d_i is then simply the induced edge labeling. That is, Entringer's question is then equivalent to asking if infinite path P_∞ is graceful. If we ask for a subset A of N such that every element of N is the difference of precisely one pair of numbers of the set A, then we are equivalently asking for a graceful numbering of countably infinite graph K_∞. Sierpinski [38] reports this as a problem solved by M. Hall. A solution by J. Browkin appears in [14]. Technically, for a graceful labeling, the smallest vertex label is 0, so, for A* as a graph labeling, we should consider reducing each value by 1 to obtain $(0,1,3,6,2,7,13,4,11,19,5,15,26,8,20,33,\ldots)$, obviously leaving D* unchanged.

In the next section, k-graceful (and k-sequential) graphs will be considered. The definitions are given here for finite and countably infinite graphs. For finite graph G with $V(G) = \{v_1, v_2, \ldots, v_n\}$ and edge set $E(G) = \{e_1, e_2, \ldots e_m\}$, a *k-graceful numbering* is an injection $f: V(G) \to \{0, 1, 2, \ldots, k+m-1\}$ for which the induced function $f:E(G) \to \{k, k+1, k+2, \ldots, k+m-1\}$ is a bijection. Such a numbering f is also called a β_k-valuation. An α_k-valuation of G is a β_k-valuation for which there is some L in $\{0, 1, \ldots, k+m-1\}$, such that for an arbitrary edge uv in $E(G)$, either $f(u) < L \leq f(v)$ or $f(v) < L \leq f(u)$. Note that the 1-graceful numberings are precisely the graceful numberings. As indicated in Fig. 14.3, a finite graph G can have an infinite number of values k for which G is k-graceful.

Fig. 14.2 Permutations A* and D* as a labeling of the infinite path

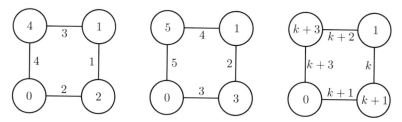

Fig. 14.3 k-Graceful numberings of cycle C_4

Theorem 6 (Maheo and Thuillier [26] and Slater [41]). If (bipartite) graph G has an α_j-valuation, then G is k-graceful for all $k \geq j$.

In contrast, we have the following theorem:

Theorem 7 (Slater [41]). If G contains an odd cycle on $2t + 1$ vertices and G is k-graceful, then $k \leq t(m-t-1)$. In particular, G can be k-graceful for at most a finite number of values of k.

Let $B(G) = \{k : G \text{ is k-graceful}\}$. We have $B(C_4) = N$.

Conjecture 8 (Slater [41]). For any set S of natural numbers, there is a graph G_S where $B(G_S) = S$.

Let G be a graph with V(G) and E(G) countably infinite. Let $N_0 = N \cup \{0\}$ be the set of nonnegative integers. A *k-graceful numbering* of G is an injection f: $V(G) \to N_0$ such that the induced labeling f: $E(G) \to \{k, k+1, k+2, \ldots\}$ is a bijection. If f: $V(G) \to N_0$ is also a bijection, then call f a *bijectively-k-graceful numbering*, and G is called *bijectively-k-graceful*. For a generalized Hall's problem, complete graph K_∞ is k-graceful for all $k \geq 1$ (but clearly not bijectively-k-graceful), and for a generalized Entringer's problem, path P_∞ is bijectively-k-graceful for all $k \geq 1$ (see Theorem 10).

Making use of an observation of Bloom [13] concerning the adjacency matrix of a graceful graph, Grace proved the following:

Theorem 8 (Grace [22]). If T is a countably infinite, locally finite tree, then T can be 1-gracefully labeled.

Much more generally, we have the following:

Theorem 9 (Slater [43]). If G is a locally finite graph (each vertex has finite degree) with V(G) and E(G) countably infinite, then G is bijectively-k-graceful.

While the countably infinite version of the Ringel-Kotzig-Rosa Conjecture 1 is true in the sense that all countably infinite trees (even allowing vertices of infinite degree) are graceful, not all countably infinite trees are bijectively-k-graceful. Let $\beta_1(G)$ denote the maximum cardinality of an independent set of edges.

Theorem 10 (Slater [45]). 1) All countably infinite trees are k-graceful for each $k \geq 1$.

2) Any countably infinite tree T with $\beta_1(T) = \infty$ is bijectively-k-graceful for each $k \geq 1$.

3) A countably infinite tree T with $\beta_1(T) < \infty$ is bijectively-k-graceful if and only if the number of vertices of infinite degree is one and $k = 1$.

Problem 9 Some results about countably infinite graphs appear in Slater [44]. Which countably infinite graphs are not graceful? Not k-graceful? Can we characterize the countably infinite, bijectively-k-graceful graphs?

Likewise, k-sequential numberings will be discussed in Sect. 14.3 for finite graphs. Numbering f: $V(G) \to N_0$ with induced function f: $E(G) \to N$ is a *total numbering* if f: $V(G) \cup E(G) \to N_0$ is one to one. When graph G is finite of order n and size m, a total numbering f is called *k-sequential* if f: $V(G) \cup E(G) \to \{k, k+1, k+2, \ldots, n+m+k-1\}$ is a bijection. A 1-sequential total numbering is also called *simply sequential*. A graph G with a k-sequential numbering is called a k-sequential graph. For countably infinite graph G, a k-sequential numbering f: $V(G) \cup E(G) \to N_0$ is a *total numbering* for which f: $V(G) \cup E(G) \to \{k, k+1, k+2, k+3, \ldots\}$ is a bijection.

Theorem 11 (Slater [44]). All countably infinite trees are k-sequential for each $k \geq 1$.

Problem 10 Which countably infinite graphs are not simply sequential? Not k-sequential?

Observe that if f: $V(T) \to \{0,1,2, \ldots, n\}$ is a graceful numbering of tree T, then one vertex v has $f(v) = 0$, and $f(V(T)-v) = f(E(T)) = \{1,2,3, \ldots, m = n-1\}$. For the tree T in Fig. 14.4, note that v is the only vertex of infinite degree. Deleting the endpoints adjacent to v leaves a tree $T^\#$ of order 6, and we have $f(V(T^\#)-v) = f(E(T^\#)) = \{1,3,4,5,6\}$. This makes it easy to gracefully number the infinite graph T.

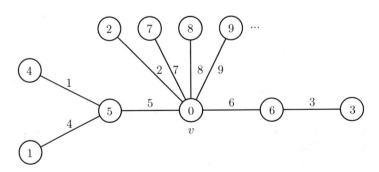

Fig. 14.4 A graceful numbering of an infinite tree

Theorem 12 (Slater [45]). For any (finite) tree T and any vertex v ε V(T), there is a numbering f: $V(T) \to N_0$ such that $f(v) = 0$ and $f(E(T)) = f(V(T) - v)$. That is, except for zero on vertex v, the set of vertex labels is the same as the set of edge labels.

There is, in fact, no graceful numbering of tree $T^\#$ with $f(v) = 0$. For vertex v in tree T, call numbering f a *v-numbering* of T if $f(v) = 0$ and $f(E(T)) = f(V(T) - v)$. Let $L^+(f) = \max\{f(x) : x \varepsilon V(T)\}$, and let $L^+(T;v)$ be the minimum value of $L^+(f)$ where f is a v-numbering. Let $L^+(T) = \max\{v \varepsilon V(T) : L^+(v) - (n-1)\}$.

Proposition 12 $L^+(T) = 0$ if and only if for every $v \varepsilon V(T)$ there is a graceful numbering f of T with $f(v) = 0$.

Note that we can state Conjecture 1 as follows:

Conjecture (Ringel, Kotzig, Rosa) For any tree T, there is at least one $v \varepsilon V(T)$ such that $L^+(T; v) = 0$.

Let me note that I defined the parameter LG(T;v) in [47] with $L^+(T;v) = LG(T;v) - (n-1)$. It seems better to normalize and use L^+. Problem 2a in [47] is the following:

Problem 11 Characterize the trees T with $L^+(T) = 0$.

Problem 12 Investigate $L^+(T)$. In particular, determine the maximum value of $L^+(T_n)$ over all trees T_n of order n.

Let $\underline{L}(G) = \{v \varepsilon V(G) \mid$, there exists a graceful numbering $f: V(G) \to \{0,1,2, \ldots, m\}$ with $f(v) = 0\}$.

Problem 13 Investigate $\underline{L}(n) = \{\underline{L}(T_n) : T_n$ is a tree of order n$\}$.

14.3   k-Graceful and K-Sequential Graphs

14.3.1   k-Graceful Graphs

In this section, only finite graphs are considered. To make it self-contained and to facilitate the exposition, there is some repetition.

As noted, the definition of graceful graphs began as a problem involving decomposing the edge set of complete graphs. Graph G of order $n = |V(G)|$ and size $m = |E(G)|$ has a *graceful numbering* (also called a β-valuation, as in Rosa [35]) f: $V(G) \to \{0, 1, 2, \ldots, m\}$ if the function f is 1-1 and the induced function f: $E(G) \to \{1, 2, \ldots, m\}$ is a bijection.

Assume that graph G with graceful numbering f has a cycle $C = (w_0, w_1, w_2, \ldots, w_{t-1})$. Since the sum of the edge weights with $f(w_{i+1}) > f(w_i) \pmod{t}$ must equal the sum of the edge weights with $f(w_{i+1}) < f(w_i) \pmod{t}$, the sum of the t edge weights in C must be even. In general, if G is Eulerian, then the sum of all the edge weights $1 + 2 + 3 + \ldots + m = m(m+1)/2$ must be even, showing the following theorem:

Theorem 13 (Rosa [35]). If Eulerian graph G is graceful, then $|E(G)| \equiv 0, 3$ (mod 4).

The next result is easy to see.

Theorem 14 Complete graph K_n is graceful if and only if $n \leq 4$.

The vertex labels in $S_4 = \{0,1,4,6\}$ for K_4 have the property that each element of $\{1,2,3,4,5,6\}$ is the difference of exactly one pair of elements from S_4. By Theorem 14, we cannot find a 5-element set S_5 in $\{0, 1, 2, \ldots, 10\}$ so that each element of $\{1, 2, \ldots, 10\}$ appears exactly once as the difference of pairs of elements in S_5. However, for the set $S_5^{\#} = \{0,1,4,9,11\}$, the ten pairwise differences are all different (only 6 is missing). In one sense, one measure of the non-gracefulness of K_5 is $11 - 10 = 1$. For a Golomb ruler, we have k integral values $\{n_1 = 0, n_2, \ldots, n_k\}$ (with $n_i < n_{i+1}$) where all $C(k,2)$ differences are distinct. The Golomb number $f(k)$ is the minimum length of a Golomb ruler with k entries. We have $(f(1), f(2), \ldots) = (0,1,3,6,11,17,25,34,44,55,72,85,106, \ldots)$.

Problem 14 Determine (bound, approximate) the Golomb ruler values $f(k)$. For a more general problem that involves labelings of K_n with m-tuples and its association with "distinct distance sets" in m-dimensional grids, see Gibbs and Slater [19].

By Theorem 12, cycle C_5 is not graceful. As in Fig. 14.5, we can label $V(C_5)$ as $(0,6,3,1,5)$ so that we get five different edge labels. In fact, *we can make the edge labels be consecutive*. So, C_5 is not graceful, but it is close! For a graph G with $V(G) = \{v_1, v_2, \ldots, v_n\}$ and edge set $E(G) = \{e_1, e_2, \ldots, e_m\}$, a *k-graceful numbering* is an injection f: $V(G) \to \{0, 1, 2, \ldots, k + m - 1\}$ for which the induced function f:$E(G) \to \{k, k + 1, k + 2, \ldots, k + m - 1\}$ is a bijection. Such a numbering f is also called a β_k-valuation. An α_k-valuation of G is a β_k-valuation for which there is some L in $\{0, 1, \ldots, k + m - 1\}$ such that for an arbitrary edge uv in $E(G)$ either $f(u) < L \leq f(v)$ or $f(v) < L \leq f(u)$. Note that the 1-graceful numberings are precisely

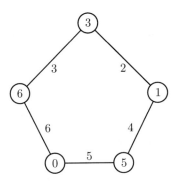

Fig. 14.5 A 2-graceful numbering of cycle C_5

the graceful numberings. As in Fig. 14.3, cycle C_4 has an α_k-valuation for all $k \geq 1$. Given an α_k-valuation f_k of G, let $S = \{v \; \varepsilon \; V(G) : f_k(v) < L\}$. Letting $f_{k+1}(v) = f_k(v)$ if $v \; \varepsilon \; S$ and $f_{k+1}(v) = f_k(v) + 1$ if $f_k(v) \geq L$, f_{k+1} is an α_{k+1}-valuation, and we have the next theorem.

Theorem 15 (Maheo and Thuillier [26] and Slater [41]). If (bipartite) graph G is k-graceful by an α_k-valuation, then G is j-graceful for all $j \geq k$.

Using the fact that the sum of the induced edge weights in a cycle must be even, we have the next generalization of Rosa's Theorem 13.

Theorem 16 (Slater [34]). If Eulerian graph G is k-graceful, then (1) if k is odd, then $|E(G)| \equiv 0,3 \pmod 4$, and (2) if k is even, then $|E(G)| \equiv 0,1 \pmod 4$.

Example 1 Cycle C_5 is k-graceful if and only if $k = 2$.

Proof. Figure 14.5 shows how to 2-gracefully label C_5. By Theorem 16, C_5 cannot be k-graceful if k is odd. Let C_5 be the cycle $(v_1, v_2, v_3, v_4, v_5)$. For a k-graceful numbering f, we can assume that $f(v_{i+1}) > f(v_i) \pmod 5$ at least three times. The smallest sum of these three weights is $k + (k + 1) + (k + 2) \leq (k + 3 + (k + 4)$, so $k \leq 4$. If f_4 is a 4-graceful numbering, without loss of generality, we can assume $f_4(v_1) = 0$ and $f_4(v_2) = 8$. To achieve an edge weight of 7, we must have $f_4(v_2) = 1$ or $f_4(v_5) = 7$. In either case, going clockwise, the ascending edge labels and descending edge labels cannot both sum to 15.

Theorem 17 (Slater [41]). If a k-graceful graph G has an odd cycle C_{2t+1}, then $k + (k + 1) + \ldots + (k + t) \leq (k + m-t) + (k + m-t + 1) + \ldots + (k + m-1)$, and so $k \leq t(m-t-1)$. Thus, a graph with an odd cycle can be k-graceful for at most finitely many values of k.

Historically the study of the k-gracefulness of certain graphs began as a problem in radio-astronomy. (See Biraud et al. [11, 12]}.) Their equivalent graph theory problem is to k-gracefully number the graph consisting of m K_n's which have exactly one vertex in common. (See Bermond, Brouwer, and Germa [9] and Bermond, Kotzig, and Turgeon [10].) The idea of k-gracefulness is formally defined in Slater [41] and in Maheo and Thuillier [26]. The next theorem is stated in [41] and [26], with a proof in [26].

Theorem 18 (Slater [41] and Maheo and Thuillier [26]). Cycle C_n is k-graceful if and only if either (1) $n \equiv 0 \pmod 4$, (2) $n \equiv 1 \pmod 4$, k is even and $k \leq (n-1)/2$, or (3) $n \equiv 3 \pmod 4$, k is odd and $k \leq (n-1)/2$.

Conjecture 15 For every set $S = \{n_1, n_2, \ldots, n_t\}$, there is a graph G_S such that G_S is k-graceful if and only if $k \; \varepsilon \; S$.

14.3.2   k-Sequential Graphs

Simply sequential and k-sequential graphs were introduced in Bange, Barkauskas, and Slater [3] and Slater [40]. As illustrated below, the study began from an attempt to prove that all wheels are graceful, a result that had already been proven by Hoede and Kupier [23] and by Frucht [17].

A function f: $V(G) \cup E(G) \to N$, the set of positive integers, is a *total numbering* if (1) on E(G) we have the induced values $f(uv) = |f(u)-f(v)|$ and (2) f is a one-to-one function (i.e., the range of f consists of $n + m$ distinct values). Call a total numbering a *k-sequential numbering* if $f(V(G) \cup E(G)) = \{k, k+1, k+2, \dots, k+n+m-1\}$. A 1-sequential numbering is called *simply sequential*. If such an f exists, then the graph G is called a *k-sequential graph* (simply sequential if $k = 1$). For example, the graph G_1 in Fig. 14.6 is 2-sequential (and, as shown later, is not simply sequential), and the wheels W_5, W_6, and W_7 are simply sequential.

The graph $G + v$ is the graph obtained from G by adding a new vertex v and making v adjacent to every vertex in V(G). As illustrated in Fig. 14.7, the cycle C_4 is 1-sequential, and the wheel $W_5 = C_4 + v$ is graceful with $f(v) = 0$. In general, we have the next theorem.

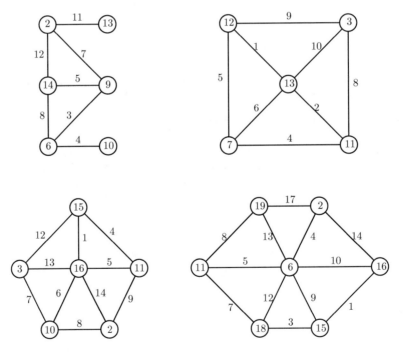

Fig. 14.6 Sequential numberings

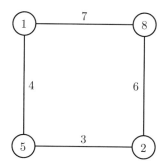 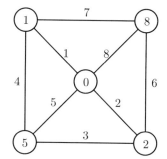

Fig. 14.7 1-sequential C_4 and graceful W_5

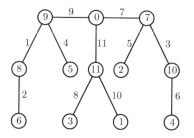 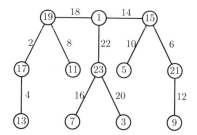

Fig. 14.8 Graceful and simply sequential numberings of tree T1

Theorem 19 (Bange, Barkauskas, and Slater [3]). If graph G is 1-sequential, then $G + v$ is graceful with a graceful numbering f with $f(v) = 0$.

We showed that every cycle is 1-sequential, from which the result in Hoede and Kupier [23] and Frucht [17] that the wheels are graceful follows.

Theorem 20 (Bange, Barkauskas, and Slater [3]). If every vertex in G has odd degree, and if $|V(G)| + |E(G)| \equiv 1,2 \pmod 4$, then G is not 1-sequential.

Proof. Simply note that $G + v$ would be Eulerian and $|E(G + v)| = |E(G)| + |V(G)|$ $s \equiv 1,2 \pmod 4$ and apply Theorem 13.

The tree T1 in Fig. 14.8 has $n + m = 12 + 11 = 23$. The simply sequential numbering $f^\#\colon V(T1) \cup E(T1) \to \{1, 2, 3, \ldots, 23\}$ is obtained from the indicated graceful numbering $f\colon V(T1) \to \{0, 1, 2, \ldots, 11\}$ by letting $f^\#(v) = 2f(v) + 1$. This illustrates the following theorem:

Theorem 21 (Bange, Barkauskas, and Slater [3]). A tree T is graceful if and only if T is simply sequential via a function $f^\#$ such that $f^\#(v)$ is odd for each vertex v ε $V(T)$.

For example, labeling the vertices of path P_4 as (0, 3, 1, 2) produces edge labels (3, 2, 1), and (1, 7, 3, 5) yields edge labels (6, 4, 2). Another way to label $V(P_4)$ 1-

sequentially is $(4, 7, 5, 6)$, again producing edge labels $(3, 2, 1)$. In general, simply consider $f^*(v) = f(v) + n$, and we have the following result:

Theorem 22 (Bange, Barkauskas, and Slater [3]). A tree T is graceful if and only if T is simply sequential via a function f^*: $V(T) \rightarrow \{n, n+1, n+2, \ldots, n+m = 2n-1\}$.

In general, there are many more ways to simply sequentially number a tree T than there are to gracefully number T. For example, the star $K_{1,3}$ has two (complementary) different graceful numberings and 11 different 1-sequential numberings. Given that the Ringel-Kotzig-Rosa Graceful Tree Conjecture has remained open for 50 years, the following weaker conjecture should be of interest:

Conjecture 16 (Slater [40]). All trees are simply sequential.

Theorem 23 (Slater [40]). If G is a k-sequential graph on n vertices and $m \geq 1$ edges, then $k \leq n-1$.

Proof. Assume G is k-sequential via a k-sequential numbering f: $V(G) \cup E(G) \rightarrow \{k, k+1, \ldots, n+m+k-1\}$, with $f(V(G)) = \{a_1, a_2, \ldots, a_n\}$, where $a_i < a_{i+1}$. Note that the highest possible value of $f(uv)$ is $(n+m+k-1)-k = n+m-1$, and, since there are m edges, some edge uv satisfies $f(uv) \leq n$. Thus $k \leq n$. If, however, $k = n$, then some edge uv satisfies $f(uv) \geq n+m-1$, which is the highest possible value of $f(uv)$. Hence $f(E(G)) = \{n, n+1, \ldots, n+m-1\}$ and $a_1 = n+m$. But by considering a vertex incident with the edge numbered $n+m-1$, one has $a_n \geq a_1 + (n+m-1) = (n+m) + (n+m-1) = 2n+2m-1$. But $a_n = n+m+k-1 = 2n+m-1$ when $k = n$, and so $m = 0$, a contradiction.

One can see that the star $K_{1,n-1}$ is $(n-1)$-sequential. In fact, $K_{1,n-1}$ is k-sequential if and only if $k|n$. For example, $K_{1,6}$ is k-sequential if and only if $k \varepsilon \{1, 2, 3, 6\}$.

Conjecture 17 (Slater [40]). For any finite set S of natural numbers, there is a (connected) graph G_S such that G_s is k-sequential if and only if $k \varepsilon S$.

14.3.3 Additivity

Let me briefly note that one does not have to use differences for the edge labels. As in Bange, Barkauskas, and Slater [4], for h: $V(G) \rightarrow N$, let $h(uv) = h(u) + h(v)$ for each edge uv $\varepsilon E(G)$. Then h: $V(G) \cup E(G) \rightarrow \{k, k+1, \ldots, k+n+m-1\}$ is a *k-sequentially additive numbering* if it is a bijection. In support of our following conjecture, all trees on nine or fewer vertices are 1-sequentially additive.

Conjecture 18 (Bange, Barkauskas, and Slater [4]). All trees are 1-sequentially additive.

4 9 2

3 5 7

8 1 6

Fig. 14.9 A 3-by-3 magic square

14.4   Now for Some Magic

Over the years in many of my mathematics and computer science classes, I have frequently used a phrase like "Get ready for some magic" to prep students when something particularly interesting was about to be presented. Occasionally in a graph theory class, this has, unfortunately, become ambiguous. But I cannot really complain about the growth of interest in magic labelings of graphs. The use of the phrase "magic" in this context stems, of course, from "magic square" as, for example, in Fig. 14.9 where the numbers 1, 2, ..., 9 are used and each set of row and column (and diagonal) values sums to the same value, in this case 15. There will be more about this square later.

14.4.1   Edge-Magic Graphs

In 1996, Ringel and Llado [34] defined an edge-magic graph and asked if all cycles are edge magic. This was previously shown to be the case by Kotzig and Rosa [25] in 1970. Our unawareness of this previous result led Richard Godbold and I to what I think is another interesting problem/conjecture about cycles. An *edge-magic labeling* of a graph G is a bijection f: $V(G) \cup E(G) \rightarrow \{1,2,3, \ldots, n+m\}$ with the property that for all of the edges uv in E(G) the value $g(uv) = f(uv) + f(u) + f(v)$ is fixed, this constant $g(uv)$ being called a magic constant $g(uv) = M$.

Note that, as in Fig. 14.10, cycle C_5 has two edge-magic labelings with magic constants $M_1 = 14$ and $M_2 = 19$. Simply by noting that when we sum all of the values of $g(uv)$ for all uv ε E(G) that each vertex label gets counted twice and each edge label once, we get the following observation:

Observation 24 (Godbold and Slater [20]). A magic constant M for a cycle C_n satisfies (1) if $n = 2k + 1$, then $5k + 4 \leq M \leq 7k + 5$ and (2) if $n = 2k$, then $5k + 2 \leq M \leq 7k + 1$.

Letting N(n) denote the number of edge-magic labelings of C_n, we have (N(3), N(4), N(5), N(6), N(7), N(8), N(9), N(10), ... = (4, 6, 6, 20, 118, 282, 1540, 7092, ...). There are four edge-magic labelings of C_3, one for each $M = 9,10,11,12$.

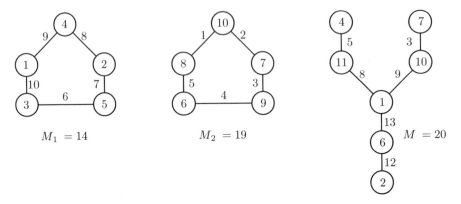

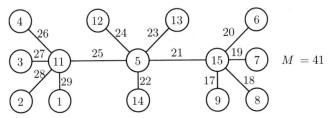

Fig. 14.10 Some edge-magic labelings

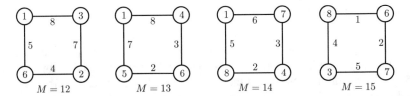

Fig. 14.11 Edge-magic labelings of cycle C_4

For the six edge-magic labelings of C_5, one produces $M = 14$, two produce $M = 16$, two produce $M = 17$, and one produces $M = 19$. There are none with $M = 15$ or $M = 18$.

Conjecture 19 (Godbold and Slater [20]). For $n = 2k + 1 \geq 7$ and $5k + 4 \leq j \leq 7k + 5$, there is an edge-magic labeling of C_n with magic constant $M = j$. For $n = 2k \geq 4$ and $5k + 2 \leq j \leq 7k + 1$, there is an edge-magic labeling of C_n with magic constant $M = j$.

Problem 20 Some further studies of edge-magic graphs have been done. In a manner similar to what is done next for vertex neighborhoods, for all graphs G, one can consider the minimax, maximin, and minimum spread edge-incidence values.

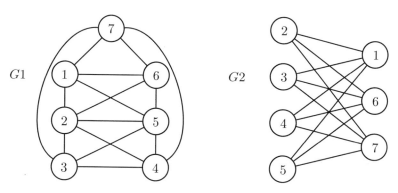

Fig. 14.12   G1 has {0, 1}-magic constant 21; G2 has {1}-magic constant 14

14.4.2   Neighborhood Sums

In the following, just vertex labelings will be considered. Given a weight set (or possibly a multiset) $W = \{w_1, w_2, \ldots, w_n\}$, we consider bijections $f:V(G) \to W$ that, in some sense, balance the distribution of the weights over the graph. The weight of W is $\text{wgt}(W) = \sum_{1 \le i \le n} w_i$, with the base case $W = [n] = \{1, 2, 3, \ldots, n\}$ and $\text{wgt}([n]) = n(n+1)/2$. For subset S of $V(G)$, the weight of S is $f(S) = \sum_{v \in S} f(v)$. Vilfred [51] considers the case where $W = [n]$ and the set of resulting open neighborhood sums are all equal. Such a labeling is called a $\sum$ labeling, and any graph for which such a labeling exists is called a $\sum$ graph. Miller et al. [27] referred to such a labeling as a 1-vertex magic labeling. More recently, Sugeng et al. [50] have referred to such a labeling as a distance magic labeling. When the closed neighborhood sums are all equal, Beena [5] has referred to the labeling as a $\sum'$ labeling and the graph as a $\sum'$ graph. See Fig. 14.12 for graphs with $\sum'$ and $\sum$ labelings, respectively, where the closed neighborhood constant in Fig. 14.12a is 21 and the open neighborhood constant in Fig. 14.12b is 14. In her presentation at the 2010 IWOGL Conference, Simanjutak [39] introduced the notion of distance magic labelings for a fixed distance other than one. Recall from above that a graph can have two different edge-magic constants. At the same conference, Arumugam [2] asked if there exists a graph G with two different magic constants for $\sum'$ (i.e., closed neighborhood) labelings $f, g:V(G) \to [n]$. As described later, even for a general weight set W and arbitrary distance set D, the answer is shown to be "no" in O'Neal and Slater [32].

First, we should consider three optimization problems defined for all graphs (not just the magic ones). Starting from the problem that involves minimizing the maximum sum of three consecutive vertex labels for the cycle C_n labeled with $\{1, 2, \ldots, n\}$, Schneider and Slater [36, 37] generalized this problem to minimizing the maximum weight of a closed neighborhood for an arbitrary graph G. (Anstee, Ferguson, and Griggs [1] considered k consecutive values on C_n.) As in O'Neal and Slater [29, 31], one can also consider maximizing the minimum closed/open neighborhood weight or minimizing the "spread" of these values.

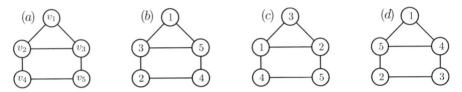

Fig. 14.13 House graph H

For any subset D of $\{0, 1, 2, \ldots\}$, the D-neighborhood of v in G is $N_D(v) = \{x \; \varepsilon \; V(G): \text{dist}(v,x) \; \varepsilon \; D\}$. The open neighborhood is $N_{\{1\}}(v) = N(v)$, and the closed neighborhood is $N_{\{0,1\}}(v) = N[v]$. Graph G is said to be (D, r)-regular if every $|N_D(v)| = r$. The D-*neighborhood sum* of f: $V(G) \to W$ is $NS(f;D) = \max\{f(N_D(v)); v \; \varepsilon \; V(G)\}$. The W-*valued D-neighborhood sum of G* is $NS_W(G;D) = \min\{NS(f;D)|f:V(G) \to W \text{ is a bijection}\}$. The *lower D-neighborhood sum* of f:$V(G) \to W$ is $NS^-(f;D) = \min\{f(N_D(v)) \mid v \; \varepsilon \; V(G)$, and the lower W-*valued D-neighborhood sum of G* is $NS^-{}_W(G; D) = \max\{NS^-(f; D) \mid f:V(G) \to W$ is a bijection}. Let $NS[G] = NS_{[n]}(G; \{0,1\})$ be the *closed neighborhood sum* of G and $NS(G) = NS_{[n]}(G;\{1\})$ be the *open neighborhood sum* of G. The lower closed neighborhood sum of G is $NS^-[G] = NS^-{}_{[n]}(G; \{0, 1\})$, and the lower open neighborhood sum of G is $NS^-(G) = NS^-{}_{[n]}(G; \{1\})$. In addition to minimizing the maximum D-neighborhood weight or maximizing the minimum D-neighborhood weight, we can consider the spread. Let $NS^{sp}(f; D) = NS(f; D) - NS^-(f; D)$. The W-valued D-neighborhood spread of G is $NS^{sp}{}_W(G; D) = \min \{NS(f; D) - NS^- (f; D) \mid f: V(G) \to W$ is a bijection}. The closed (respectively, open) neighborhood spread of G denoted $NS^{sp}[G]$ (respectively, $NS^{sp}(G)$) has $W = [n]$ and $D = \{0,1\}$ (respectively, $D = \{1\}$).

For the house graph H in Fig. 14.13, $NS[H] = 11$ (Fig. 14.13c), $NS^-[H] = 9$ (Fig. 14.13b), $NS(H) = 8$ (Fig. 14.13b), $NS^-(H) = 7$ (Fig. 14.13b), $NS^{sp}[H] = 13 - 9 = 4$ (Fig. 14.13d), and $NS^{sp}(H) = 8 - 7 = 1$ (Fig. 14.13b). For distance set $D = \{2\}$, we have $NS(H; \{2\}) = 6$ (Fig. 14.13b), $NS^-(H; \{2\}) = 4$ (Fig. 14.13c), and $NS^{sp}(H; \{2\}) = 4$ (Fig. 14.13b). Note that $NS^{sp}[H] > NS[H] - NS^-[H]$, that is, no single bijection simultaneously achieves both $NS[H]$ and $NS^-[H]$.

As in O'Neal and Slater [30], graph G is said to be (D,W)-vertex magic (or (D,W)-distance magic) if there is a bijection f: $V \to W$ such that the D-neighborhood weights of the vertices are all a constant, that is, $NS^{sp}{}_W(G; D) = 0$. The common value of each $f(N_D(v))$ is called the *distance-D magic constant*. The distance-D adjacency matrix $A_D(G) = [a_{i,j}]$ is the n x n binary matrix with $a_{i,j} = 1$ if and only if $\text{dist}(v_i, v_j) \; \varepsilon \; D$.

Theorem 25 (O'Neal and Slater [30]). If G is (D, r)-regular and $A_D{}^{-1}$ exists, then G is not D-vertex magic.

Corollary 26 For adjacency matrix $A = A_{\{1\}}$ and closed neighborhood matrix $N = A_{\{0,1\}}$, if graph G is regular, then A^{-1} exists implies G is not {1}-vertex magic (G cannot be $\sum$ labeled) and N^{-1} exists implies that G is not {0,1}-vertex magic (G cannot be $\sum$'' labeled).

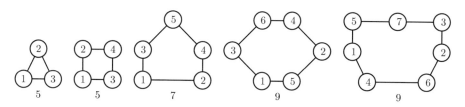

Fig. 14.14 Labelings achieving open neighborhood values NS(C_n) for $3 \le n \le 7$

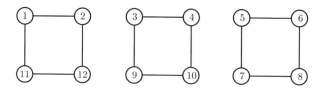

Fig. 14.15 NS^{sp}(3 C_4) = 0. In general, k C_4 is {1}-magic

Theorem 27 (O'Neal and Slater [30]). No graph of even order can be both {1}-vertex magic and {0,1}-vertex magic.

Cycles

The open neighborhood sum problems were easier to solve than those for closed neighborhoods. Figure 14.14 shows some solutions for NS(C_n).

Theorem 28 (Schneider and Slater [37]). NS(C_3) = 5, NS(C_4) = 5, and for $n \ge 5$, we have NS(C_n) = $n + 2$ if $n \equiv 0,1,3 \pmod 4$ and NS(C_n) = $n + 3$ if $n \equiv 2 \pmod 4$ (Fig. 14.15).

There is actually a single bijection $f : V(C_n) \rightarrow [n]$ that simultaneously achieves NS(C_n), NS$^-$(C_n), and NS^{sp}(C_n). More generally, we considered graphs G that are unions of cycles of the same length. Note that the only such graph that is {1}-distance magic is the union of 4-cycles.

Theorem 29 (O'Neal and Slater [29]). If G is a union of k C_t's, then:

1.) If $t \equiv 1,3 \pmod 4$, then NS$^-$ (G) = kt−k + 1 = n−k + 1, NS(G) = kt + k + 1 = n + k + 1, and NS^{sp}(G) = 2k.
2.) If $t \equiv 2 \pmod 4$, then NS$^-$(G) = kt−2k + 1 = n−2k + 1, NS(G) = kt + 2k + 1 = n + 2k + 1, and NS^{sp}(G) = 4k.
3.) If $t > 4$ and $t \equiv 0 \pmod 4$, then NS$^-$(G) = kt = n, NS(G) = kt + 2 = n + 2, and NS^{sp}(G) = 2.
4.) If $t = 4$, then NS$^-$(G) = NS(G) = 4k + 1 = n + 1 and NS^{sp}(G) = 0.

Problem 21 Determine NS(G), NS$^-$(G), and NS^{sp}(G) when G is 2-regular and contains cycles of different lengths (Fig. 14.16).

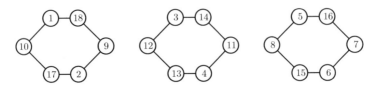

Fig. 14.16 Closed neighborhood sums $NS^-[3\,C_6] = 28$, $NS[3\,C_6] = 29$, and $NS^{sp}[3\,C_6] = 1$

Some results about the closed neighborhood sum parameters for 2-regular graphs are presented in Allen O'Neal's Ph.D. dissertation.

Theorem 30 (O'Neal [28]). If k is odd, then $NS^-[k\,C_6] = 9k + 4$, $NS[k\,C_6] = 9k + 2$, and $NS^{sp}[k\,C_6] = 1$.

The known results for cycles are summarized next.

Theorem 31 (O'Neal [28]). For cycle C_n:

1. $NS^-[C_3] = NS[C_3] = 6$ and $NS^{sp}[C_3] = 0$.
2. If $n \,\varepsilon\, \{5, 9, 15\}$, then $NS^-[C_n] = 3(n + 1)/2 - 1$, $NS[C_n] = 3(n + 1)/2 + 1$, and $NS^{sp}[C_n] = 2$.
3. $NS^-[C_6] = 10$, $NS[C_6] = 11$, and $NS^{sp}[C_6] = 1$.
4. If $n \neq 6$ and n is even, $NS^-[C_n] = 3n/2$, $NS[C_n] = (3n + 6)/2$, and $NS^{sp}[C_n] = 3$.
5. If $n \geq 7$ and $n \equiv 1,5 \pmod 6$, then $NS^-[C_n] = 3(n + 1)/2 - 2$, $NS[C_n] = 3(n + 1)/2 + 2$, and $NS^{sp}[C_n] = 4$.
6. If $n \geq 21$ and $n \equiv 3 \pmod 6$, then $3(n + 1)/2 - 2 \leq NS^-[C_n] \leq 3(n + 1)/2 - 1$, $3(n + 1)/2 + 1 \leq NS[C_n] \leq 3(n + 1)/2 + 2$, and $NS^{sp}[C_n] \,\varepsilon\, \{2,3,4\}$.

Problem 22 Determine the closed neighborhood sum parameters $NS[G]$, $NS^-[G]$, and $NS^{sp}[G]$ for C_n with $n \equiv 3 \pmod 6$, for kC_t and for arbitrary 2-regular graphs.

Rooks Graphs and Magic Squares

The $k \times j$ rooks graph $R_{k,j}$ has kj vertices, $V(R_{k,j}) = \{v_{a,b} \mid 1 \leq a \leq k, 1 \leq b \leq j\}$, and distinct vertices $v_{a,b}$ and $v_{c,d}$ are adjacent when $a = c$ or $b = d$, that is, when they correspond to squares in a $k \times j$ chessboard that are in the same row or same column.

Note that for any magic square labeling f: $V(R_{k,k}) \rightarrow \{1, 2, 3, \ldots, k^2\}$, as in Fig. 14.17, the row and column sums are all $k(k^2 + 1)/2$, so that $f(N[v_{a,b}]) = k(k^2 + 1) - f(v_{a,b})$. For such a labeling f, $NS^-(f) = k(k^2 + 1) - k^2$, $NS(f) = k(k^2 + 1) - 1$, and $NS^{sp}(f) = k^2 - 1$ (Fig. 14.18).

Theorem 32 (O'Neal and Slater [31]). $NS^-[R_3] = 22$, $NS[R_3] = 28$, and $NS^{sp}[R_3] = 8$. In particular, no labeling f: $V(R_3) \rightarrow \{1, 2, \ldots, 9\}$ simultaneously attains $NS^-[R_3]$ and $NS[R_3]$.

4	9	2		26	21	28
3	5	7		27	25	23
8	1	6		22	29	24

Fig. 14.17 A 3 × 3 magic square and the closed neighborhood sum values

9	1	6		21	28	28
2	5	8		27	23	25
3	7	4		25	20	28

Fig. 14.18 $NS[R_{3,3}] \leq 28$

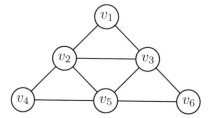

Fig. 14.19 $\Upsilon_f(T2) = 3/2$

Problem 23 Determine the values of $NS^-[R_k]$, $NS[R_k]$, and $NS^{sp}[R_k]$. One can also investigate the open neighborhood sum parameters $NS^-(R_k)$, $NS(R_k)$, and $NS^{sp}(R_k)$.

14.4.3   The Magic Constant and Fractional Domination

Seemingly unrelated to labeling problems is the theory of domination in graphs. A vertex set D in V(G) is a *dominating set* if each vertex v ε V(G) is either in the set D or is adjacent to a vertex in D. Equivalently, each closed neighborhood has a nonempty intersection with D, $N[v] \cap D \neq \varphi$. The domination number $\Upsilon(G)$ is the minimum cardinality of a dominating set.

We can define domination as a labeling problem as follows. For any real valued function $f:V(G) \to R$, the weight of f is $wgt(f) = \sum_{v \, \epsilon \, V(G)} f(v)$. Then the domination number is the minimum weight of a binary labeling f of V(G) with minimum closed neighborhood sum at least 1, $\Upsilon(G) = \min\{wgt(f)|f:V(G) \to \{0, 1\}\}$ and $NS^-(f; \{0, 1\}) \geq 1$. For the graph T2 in Fig. 14.19, let $f(v_1) = f(v_4) = f(v_6) = 0$ and $f(v_2) = f(v_3) = f(v_5) = 1/2$. Then each $f(N[v]) \geq 1$ and $wgt(f) = 3/2$. As in Farber [16], the fractional domination number $\Upsilon_f(G)$ is the minimum weight of a function $g:V(G) \to [0, 1]$ with $NS^-(g; [0, 1]) \geq 1$. More generally, the D-neighborhood fractional domination number $\Upsilon_f(G; D)$ is the minimum weight of a function $g: V(G) \to [0, 1]$ with every $g(N_D(v)) \geq 1$ (i.e., every D-neighborhood has a weight of at least one). Expressed as a linear programming problem, we have the following:

$$\Upsilon_f(G; D) = \text{MIN} \sum_{1 \leq I \leq n} X_i$$
$$\text{Subject to } A_D X \geq 1_n \text{ and } X_i \geq 0, \quad \text{where } 1_n \text{ is the all } 1's \text{ n-vector}$$
(14.1)

The uniqueness of vertex magic constants applies very generally, namely, not only for arbitrary D-neighborhoods but also for arbitrary weight sets W. *Not only is the magic constant unique, but it is determined by the D-fractional domination number.*

Theorem 33 (O'Neal and Slater [32]). If graph G is (D, W) vertex magic, then its (D, W)-vertex magic constant is $c = \text{wgt}(W)/\Upsilon_f(G; D)$.

Let me conclude this section with a generic problem.

Problem 23 Many graph theoretic results can be derived from linear algebraic observations. As indicated by Theorem 33 (and, e.g., Slater [46]), a linear/integer programming approach can be particularly fruitful in deriving graph theoretic results and, as in [46], in defining new parameters.

14.5   It Is All Labeling

I hope to have illustrated two themes. The first is that so much of graph theory can be viewed as forms of labeling problems. As above, this is clear for many subset-type problems like domination, packing, independence, and covering. Likewise structural problems are also. Graph labeling began as a study of packing trees into complete graphs. For another basic example, a Hamiltonian cycle is a labeling $f: V(G) \to [n]$ such that all pairs of consecutive numbers (mod n) appear on adjacent vertices.

The second is that there are many interesting problems derivable from the question "How close?." For example, when G is not graceful, how close is it?

Finally, let me illustrate the inherent difficulty of labeling problems (as if the Graceful Tree Conjecture alone would not do this). It is an NP-complete problem to decide, when given a weight set W with $|W| = 2k$, if W can be partitioned into two subsets X and Y of size k with $\text{wgt}(X) = \text{wgt}(Y)$. This is simply the problem of deciding if $\text{NS}^{sp}{}_W(2K_k; \{0, 1\}) = 0$.

References

1. Anstee, R., Ferguson, R., Griggs, J.: Permutations with low discrepancy consecutive k-sums. J. Comb. Theory A **100**, 302–321 (2002)
2. Arumugam, S., Froncek, D., Kamitchi, N.: Distance magic graphs—a survey. International Workshop on Graph Labelings 2010, October 22–23, 2010, University of Wisconsin - Superior

3. Bange, D.W., Barkauskas, A.E., Slater, P.J.: Simply sequential and graceful graphs. Cong. Num. **23**, 155–162 (1979)
4. Bange, D.W., Barkauskas, A.E., Slater, P.J.: Sequentially additive graphs. Discrete Math. **44**, 235–241 (1983)
5. Beena, S.: On $\sum$ and $\sum'$ labeled graphs. Discrete Math. **309**, 1783–1787 (2009)
6. Bascunan, M.E., Ruiz, S., Slater, P.J.: The additive bandwidth of grids and complete bipartite graphs. Cong. Num. **88**, 245–254 (1992)
7. Bascunan, M.E., Brigham, R.C., Caron, R.M., Ruiz, S., Slater, P.J., Vitray, R.P.: On the additive bandwidth of graphs. J. Comb. Math. Comb. Comput. **18**, 129–144 (1995)
9. Bermond, J.C., Brouwer, A.E., Germa, A.: Systemes de triples et differences associees. In: Colloque CNRS, Problemes Combinatoires et Theorie des Graphes, Orsay, 1976, CNRS, pp. 35–38 (1978)
10. Bermond, J.C., Kotzig, A., Turgeon, J.: On a combinatorial problem of antennas in radio-astronomy. In: Colloq. Math. Soc. Janos Bolyai 18, Combinatorics, Keszthely, Hungary, 1976, pp. 135–149. North Holland, Amsterdam (1978)
11. Biraud, F., Blum, E.J., Ribes, J.C.: On optimum synthetic linear arrays. IEEE Trans. Antennas Propag. **ΛP22**, 108–109 (1974)
12. Biraud, F., Blum, E.J., Ribes, J.C.: Some new possibilities of optimum synthetic linear arrays for radio-astronomy. Astron. Astrophys. **41**, 409–413 (1975)
13. Bloom, G.S.: A chronology of the Ringel-Kotzig conjecture and the continuing quest to call all trees graceful. In: Harary, F. (ed.) Topics in Graph Theory (Scientist in Residence Program, New York), pp. 32–51. Academy of Science, Washington (1977)
14. Browkin, J.: Solution of a certain problem of A. Schinzel. Prace Mat. **3**, 205–207 (1959)
16. Farber, M.: Domination, independent domination and duality in strongly chordal graphs. Discret. Appl. Math. **7**, 115–130 (1984)
17. Frucht, R.: Graceful numberings of wheels and related graphs. In: Gerwitz, A., Quintas, L. (eds.) Second International Conference on Combinatorial Mathematics, pp. 219–229. The New York Academy of Sciences, New York (1979)
18. Gallian, J.: A dynamic survey of graph labeling. Electron. J. Comb. (2015), #DS6
19. Gibbs, R.A., Slater, P.J.: Distinct distance sets in graphs. Discret. Math. **93**, 155–165 (1991)
20. Godbold, R., Slater, P.J.: All cycles are edge-magic. Bull. Inst. Combin. Appl. **22**, 93–97 (1998)
21. Golomb, S.W.: How to number a graph. In: Read, R.C. (ed.) Graph Theory and Computing, pp. 23–37. Academic Press, New York (1972)
22. Grace, T.: New proof techniques for gracefully labeling graphs. Congr. Numer. **39**, 433–439 (1983)
23. Hoede, C., Kuiper, H.: All wheels are graceful. Utilitas Math. 14, 311, (1978)
25. Kotzig, A., Rosa, A.: Magic valuations of finite graphs. Can. Math. Bull. **13**, 451–461 (1970)
26. Maheo, M., Thuillier, H.: On d-graceful graphs. Ars Combin. **13**, 181–192 (1982)
27. Miller, M., Rodger, C., Simanjutak, R.: Distance magic labelings of graphs. Australas. J. Math. **28**, 305–315 (2003)
28. O'Neal, F.A.: Neighborhood sum parameters on graphs. Ph. D. Dissertation, The University of Alabama in Huntsville (2011)
29. O'Neal, F.A., Slater, P.J.: The minimax, maximin and spread values for open neighborhood sums for 2-regular graphs. Cong. Num. **208**, 19–32 (2011)
30. O'Neal, F. A., Slater, P.J.: An introduction to distance D magic graphs. J. Indones. Math. Soc., Special edition, 91–107 (2011)
31. O'Neal, F.A., Slater, P.J.: An introduction to closed/open neighborhood sums: minimax, maximin and spread. Math. Comput. Sci. **5**, 69–80 (2011)
32. O'Neal, F.A., Slater, P.J.: Uniqueness of vertex magic constants. SIAM J. Discret. Math. **27**, 708–716 (2013)
33. Ringel, G.: Problem 25. In: Theory of Graphs and Its Applications, Proceeding of the Symposium Smolenice 1963, Prague, p. 162 (1964)
34. Ringel, G., Llado, A.S.: Another tree conjecture. Bull. Inst. Combin. Appl. **18**, 83–85 (1996)

35. Rosa, A.: On certain valuations of the vertices of a graph. In: Rosentiehl, P. (ed.) Theory of Graphs, Proceeding of an International Symposium, Rome, 1966, pp. 349–355. Dunod, Paris (1968)

36. Schneider, A., Slater, P.J.: Minimax neighborhood sums. Cong. Num. **188**, 183–190 (2007)

37. Schneider, A., Slater, P.J.: Minimax open and closed neighborhood sums. AKCE Int. J. Graphs Comb. **6**, 183–190 (2009)

38. Sierpinski, W.: Elementary theory of numbers. Warsaw, 411–412 (1964)

39. Simanjutak, R.: Distance magic labelings and antimagic coverings of graphs, International Workshop on Graph Labelings 2010, October 22–23, 2010, University of Wisconsin - Superior

40. Slater, P.J.: On k-sequential and other numbered graphs. Discret. Math. **34**, 185–193 (1981)

41. Slater, P.J.: On k-graceful graphs. Cong. Num. **36**, 53–57 (1982)

42. Slater, P.J.: On repeated difference sequences of a permutation of N. J. Number Theory **16**, 1–5 (1983)

43. Slater, P.J.: On k-graceful, locally finite graphs. J. Comb. Theory B **35**, 319–322 (1983)

44. Slater, P.J.: Graceful and sequential numberings of infinite graphs. S.E. Asian Bull. Math. **9**, 15–22 (1985)

45. Slater, P.J.: On k-graceful countably infinite graphs. Discret. Math. **61**, 293–303 (1986)

46. Slater, P.J.: LP-duality, complementarity and generality of graphical subset problems. In: Haynes, T.W., Hedetniemi, S.T., Slater, P.J. (eds.) Domination in Graphs Advanced Topics, pp. 1–30. Marcel Dekker, New York (1998)

47. Slater, P. J.: How provably graceful are the trees? J. Indones. Math. Soc. 2011, Special edition, 133–135

48. Slater, P.J., Velez, W.Y.: Permutations of the positive integers with restrictions on the sequence of differences. Pac. J. Math. **71**, 193–196 (1977)

49. Slater, P.J., Velez, W.Y.: Permutations of the positive integers with restrictions on the sequence of differences II. Pac. J. Math. **82**, 527–531 (1979)

50. Sugeng, K.A., Froncek, D., Miller, M., Ryan, J., Walker, J.: On distance magic labelings of graphs. J. Comb. Math. Comb. Comp. **7**, 39–48 (2009)

51. Vilfred, V.: Sigma labeled graphs and circulant graphs. Ph. D. Dissertation, University of Kerala, India (1994)

Chapter 15
My Favorite Domination Conjectures in Graph Theory Are Bounded

Michael A. Henning

Abstract For a graph G of order n and a parameter $\vartheta(G)$, if $\vartheta(G) \leq \frac{a}{b}n$ for some rational number $\frac{a}{b}$, where $0 < \frac{a}{b} < 1$, then we refer to this upper bound on $\vartheta(G)$ as an $\frac{a}{b}$-bound on $\vartheta(G)$. In this chapter, we present over twenty $\frac{a}{b}$-bound conjectures on domination type parameters.

Mathematics Subject Classification : 05C69

15.1 Introduction

I was first introduced to graph theory in 1982 by Professor Henda Swart, when I started my undergraduate studies at the Durban campus of the University of Natal. Her passion for graph theory was contagious, and she instilled in me a love for the subject at an early age. In January 1986, I started my graduate work under her supervision. After attending the South African Mathematics Congress later that year, I was exposed to talks on domination theory in graphs by Professors Ernie Cockayne and Kieka Mynhardt. I was immediately captivated by the concept, and my graph theory research interests shifted to the topic of domination theory in graphs.

Of my early papers, one which I treasure fondly is a paper with Professor Paul Erdös. We met at the Sixth International Conference on Graph Theory, Combinatorics, Algorithms and Applications, held at Western Michigan University in June 1988. During the week of the "Kalamazoo conference," we had several discussions on domination-related problems. Professor Erdös posed to me the

Research supported in part by the South African National Research Foundation and the University of Johannesburg

M.A. Henning (✉)
Department of Pure and Applied Mathematics, University of Johannesburg,
Auckland Park, Johannesburg 2006, South Africa
e-mail: mahenning@uj.ac.za

© Springer International Publishing Switzerland 2016
R. Gera et al. (eds.), *Graph Theory*, Problem Books in Mathematics,
DOI 10.1007/978-3-319-31940-7_15

problem of determining the smallest order of a graph with domination number equal to two and with every vertex contained in a clique K_n. The following year our joint research endeavors were interrupted by a compulsory 2-year stint that I had to serve in the South African Defence Force (thankfully shortened to 15 months). However, it was while digging a trench around our army tent during my "bush phase" that the key ideas needed to prove our bound became clear to me. Fortunately, I had a "weekend pass" the following weekend and met with Professor Swart to check the finer details of our proof. Our subsequent paper [16] with Professor Erdös resulted in us becoming the proud owners of an Erdös number 1 and served to further motivate us to attack the many open problems and conjectures at the time in domination theory in graphs.

In this chapter, I will discuss some of my favorite domination-related conjectures. All the conjectures I have selected relate to upper bounds on domination parameters in terms of the order of the graph. For a graph G of order n and a parameter $\vartheta(G)$, if $\vartheta(G) \leq \frac{a}{b}n$ for some rational number $\frac{a}{b}$, where $0 < \frac{a}{b} < 1$, then we refer to this upper bound on $\vartheta(G)$ as an $\frac{a}{b}$-bound on $\vartheta(G)$. In this chapter, I present my current favorite $\frac{a}{b}$-bound conjectures on domination-type parameters, in no particular order of preference.

Before presenting these conjectures, we give some basic graph theory terminology. We in general follow [40]. Specifically, let G be a graph with vertex set $V(G)$ and edge set $E(G)$. The *open neighborhood* of a vertex $v \in V(G)$ is $N_G(v) = \{u \in V(G) \mid uv \in E(G)\}$, and its *closed neighborhood* is the set $N_G[v] = N_G(v) \cup \{v\}$. The *degree* of a vertex v in G, denoted $d_G(v)$, is the number of neighbors, $|N_G(v)|$, of v in G. The minimum degree among all the vertices of G is denoted by $\delta(G)$. The graph G is k-regular if $d_G(v) = k$ for every vertex $v \in V(G)$. A 3-regular graph is also called a *cubic graph*. The *girth* of G is the length of a shortest cycle in G. A graph is *connected* when there is a path between every pair of vertices. Two distinct vertices u and v of a graph G are *open twins* if $N(u) = N(v)$ and *closed twins* if $N[u] = N[v]$. Further, u and v are *twins* in G if they are open twins or closed twins in G, that is, pairs of vertices with the same closed or open neighborhood. A graph is *twin-free* if it has no twins, and *open twin-free* if it has no open twins.

15.2 $\frac{1}{3}$-Conjectures

15.2.1 The Domination Number

A *dominating set* in a graph G is a set S of vertices of G such that every vertex in $V(G) \setminus S$ is adjacent to atleast one vertex in S. The *domination number* of G, denoted by $\gamma(G)$, is the minimum cardinality of a dominating set in G. The literature on the subject of domination parameters in graphs up to the year 1997 has been surveyed and detailed in the two so-called domination books [30, 31].

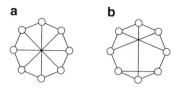

Fig. 15.1 The two non-planar cubic graphs, shown in (**a**) and (**b**), of order eight

Using ingenious counting arguments, Reed [52] proved that if G is a graph of order n with $\delta(G) \geq 3$, then $\gamma(G) \leq \frac{3}{8}n$. As a special case of this result, we note that if G is a cubic graph of order n, then $\gamma(G) \leq \frac{3}{8}n$. The two non-planar cubic graphs of order $n = 8$ (shown in Figure 15.1(a) and 15.1(b)) both have domination number 3 and achieve Reed's $\frac{3}{8}$-bound.

Kostochka and Stodolsky [47] proved that these two non-planar, connected, cubic graphs of order 8 are the only connected, cubic graphs that achieve the $\frac{3}{8}$-bound, by proving that if G is a connected cubic graph of order $n \geq 10$, then $\gamma(G) \leq \frac{4}{11}n$. Kostochka and Stocker [47] subsequently improved the $\frac{4}{11}$-bound, by proving that if G is a connected, cubic graph of order $n \geq 10$, then $\gamma(G) \leq \frac{5}{14}n$. Reed [52] conjectured that a tight upper bound on the domination number of a connected, cubic graph of order n is $\lceil n/3 \rceil$. However, Kostochka and Stodolsky [48] disproved this conjecture by constructing a connected, cubic graph G on 60 vertices with $\gamma(G) = 21$ and presented a sequence $\{G_k\}_{k=1}^{\infty}$ of connected, cubic graphs with $\lim_{k \to \infty} \frac{\gamma(G_k)}{|V(G_k)|} \geq \frac{8}{23} = \frac{1}{3} + \frac{1}{69}$. Kelmans [45] constructed a smaller counterexample (with 54 vertices) to Reed's conjecture and an infinite series of 2-connected, cubic graphs H_k with $\lim_{k \to \infty} \frac{\gamma(H_k)}{|V(H_k)|} \geq \frac{1}{3} + \frac{1}{60}$. Let $\mathcal{G}_{\text{cubic}}^n$ denote the family of all connected, cubic graphs of order n. As a consequence of the above results, we have that

$$0.35 = \frac{1}{3} + \frac{1}{60} \leq \sup_{G \in \mathcal{G}_{\text{cubic}}^n} \left(\lim_{n \to \infty} \frac{\gamma(G)}{n} \right) \leq \frac{5}{14} = \frac{1}{3} + \frac{1}{42} \approx 0.35714285.$$

It remains, however, an open problem to determine what this supremum is. Indeed, this problem of determining a sharp upper bound on the domination number of a connected, cubic graph, of sufficiently large order, in terms of its order, is one of the major outstanding problems in domination theory.

The following conjecture, posed independently by Kelmans [45] and Kostochka and Stodolsky [48], claims that Reed's conjecture is true for 3-connected cubic graphs.

Conjecture 1 ([45, 48]). *If G is a cubic 3-connected graph of order n, then* $\gamma(G) \leq \lceil \frac{n}{3} \rceil$.

The sequence of counterexamples, provided by both Kostochka and Stodolsky [48] and Kelmans [45], all contain induced cycles of length 4. A graph is C_4-*free* if it contains no induced 4-cycle. I pose the following conjecture:

Conjecture 2. *If G is a cubic, connected, C_4-free graph of order n, then $\gamma(G) \leq \lceil \frac{n}{3} \rceil$.*

The following conjecture was first posed as a question by Kostochka and Stodolsky [48]. However, I wish to pose their question as a conjecture.

Conjecture 3. *If G is a cubic, bipartite graph of order n, then $\gamma(G) \leq \frac{1}{3}n$.*

I had the privilege of discussing the above conjecture with Professor Kostochka at the Third International Conference on Combinatorics, Graph Theory and Applications, held at Elgersburg, Germany, March 2009. During our discussions, Professor Kostochka posed to me the following, most intriguing question: Is it true that the vertex set of every cubic, bipartite graph can be partitioned into three dominating sets? I pose this wonderful question as a conjecture.

Conjecture 4. *The vertex set of every cubic, bipartite graph can be partitioned into three dominating sets.*

Conjecture 4, if true, immediately implies Conjecture 3. We remark that the bipartite requirement in Conjecture 4 is essential. For example, the two cubic graphs in Figure 15.1(a) and 15.1(b) of order 8 are not bipartite, and their vertex sets cannot be partitioned into three dominating sets (since each dominating set requires at least three vertices). More generally, every connected, cubic graph with domination number greater than one-third its order cannot be partitioned into three dominating sets. Such graphs include the infinite sequence of connected, cubic graphs provided by both Kostochka and Stodolsky [48] and Kelmans [45] which are not bipartite and have domination number greater than one-third their order.

If the girth is sufficiently large, then the $\frac{1}{3}$-bound of Conjecture 3 holds. Löwenstein and Rautenbach [50] showed that if G is a connected, cubic graph of order n with girth at least 83, then $\gamma(G) \leq \frac{1}{3}n$. While this is a very pleasing result indeed, we conjecture that the girth condition can be lowered considerably in order to guarantee that the $\frac{1}{3}$-bound will hold.

Conjecture 5. *If G is a connected, cubic graph of order n with girth at least 6, then $\gamma(G) \leq \frac{1}{3}n$.*

We remark that the girth requirement in Conjecture 5 is essential, since the generalized Petersen graph G_{14} shown in Figure 15.2, of order $n = 14$ and girth 5, satisfies $\gamma(G) = 5 > \frac{1}{3}n$.

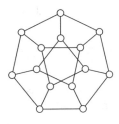

Fig. 15.2 The generalized Petersen Graph G_{14}

15.2.2 The Independent Domination Number

A set of vertices is *independent* if no two vertices in it are adjacent. An *independent dominating set* of a graph G is a set of vertices that is both dominating and independent in G, and the *independent domination number* of G, denoted by $i(G)$, is the minimum cardinality of an independent dominating set in G. A survey of independent dominating sets in graphs can be found in [26]. Over lunch at the Nelson Mandela Square in Sandton City, Jacques Verstraete (Personal communication, 2010) posed to me the following conjecture:

Conjecture 6 (Personal communication, 2010). *If G is a connected, cubic graph on n vertices with girth at least 6, then $i(G) \le \frac{1}{3}n$.*

As before, the girth requirement in Conjecture 6 is essential, since the generalized Petersen graph G_{14}, of order $n = 14$ and girth 5, satisfies $i(G) = 5 > \frac{1}{3}n$. Perhaps the graph G_{14} is the only exception when relaxing the girth condition in Conjectures 5 and 6. For every graph G, $\gamma(G) \le i(G)$. Thus, Conjecture 6 is a stronger conjecture than Conjecture 5.

15.2.3 The Total Domination Number

My favorite domination-type parameter remains the total domination number. A *total dominating set* of a graph G with no isolated vertex is a set S of vertices of G such that every vertex in $V(G)$ is adjacent to at least one vertex in S. The *total domination number* of G, denoted by $\gamma_t(G)$, is the minimum cardinality of a total dominating set in G. Total domination in graphs was introduced by Cockayne, Dawes, and Hedetniemi [12] and is now well studied in graph theory. The literature on the subject of total domination in graphs has been surveyed and detailed in the book [40]. My favorite $\frac{1}{3}$-bound conjecture for the total domination number is the following conjecture. A graph is *quadrilateral-free* if it contains no 4-cycles not necessarily induced.

Conjecture 7. *If G is a quadrilateral-free graph of order n with $\delta(G) \ge 5$, then $\gamma_t(G) \le \frac{1}{3}n$.*

The quadrilateral-free requirement in Conjecture 7 is essential. Consider, for example, the 5-uniform hypergraph, H_{11}, constructed by Thomassé and Yeo [57] as follows. Let H_{11} be the hypergraph with vertex set $V(H) = \{0, 1, \ldots, 10\}$ and edge set $E(H) = \{e_0, e_1, \ldots, e_{10}\}$, where the hyperedge $e_i = Q + i$ for $i = 0, 1, \ldots, 10$ and where $Q = \{1, 3, 4, 5, 9\}$ is the set of nonzero quadratic residues modulo 11. Let G_{22} be the incidence bipartite graph of the hypergraph H_{11}. The 5-regular graph $G = G_{22}$, shown in Figure 15.3, has order $n = 22$, minimum degree $\delta(G) = 5$, and total domination number $\gamma_t(G) = 8 = \frac{4}{11}n > \frac{1}{3}n$. However, every vertex of G belongs to an induced 4-cycle, and so G is certainly not quadrilateral-free.

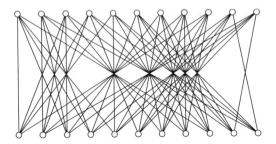

Fig. 15.3 The graph G_{22}

15.3 $\frac{1}{2}$-Conjectures

I have always felt that the simplest upper bounds to prove on a graph parameter should be $\frac{1}{2}$-bounds. For example, a classic result due to Ore [51] states that if G is a graph of order n without isolated vertices, then $\gamma(G) \leq \frac{1}{2}n$. The proof is elementary but nonetheless a fundamental result in domination theory. Since every bipartite graph is the union of two independent sets, each of which dominates the other, the following $\frac{1}{2}$-bound is immediate: If G is a bipartite graph on n vertices, then $i(G) \leq \frac{1}{2}n$.

However, it is not always the case that $\frac{1}{2}$-bounds are the simplest to prove. My experience is that $\frac{1}{2}$-bounds can be as stubborn to prove as other seemingly more complicated bounds! A beautiful and important result in the theory of total domination in graphs is that if G is a graph of order n with $\delta(G) \geq 3$, then $\gamma_t(G) \leq \frac{1}{2}n$. Archdeacon et al. [1] provided an elegant graph theoretic proof of this $\frac{1}{2}$-bound result on the total domination number that, surprisingly, uses Brooks' Coloring Theorem and clever counting arguments. This $\frac{1}{2}$-bound result also follows readily from a result, independently established by Chvátal and McDiarmid [11] and Tuza [58], about transversals in hypergraphs. In this section, I have selected my two favorite $\frac{1}{2}$-bound conjectures.

15.3.1 The Locating Domination Number

Among the existing variations of domination, the one of *location-domination* is widely studied. A *locating-dominating set* is a dominating set D that locates/distinguishes all the vertices, in the sense that every vertex not in D is uniquely determined by its neighborhood in D. Hence, two distinct vertices u and v in $V(G) \setminus D$ are *located* by D if they have distinct neighbors in D; that is, $N(u) \cap D \neq N(v) \cap D$. The *location-domination number* of G, denoted $\gamma_L(G)$, is the minimum cardinality of a locating-dominating set in G. The concept of a locating-dominating set was introduced and first studied by Slater [55, 56]. Recall that if G is a graph of order n without isolated vertices, then $\gamma(G) \leq \frac{1}{2}n$, as first proven by Ore [51]. While there are many graphs (without isolated vertices) which have location-domination number much larger than half their order, the only such graphs that are known contain twins. Garijo, González, and Márquez [24] therefore posed the conjecture that in the absence of twins, the classic bound of one-half the order for the domination number also holds for the location-domination number.

Conjecture 8 ([24]). *If G is a twin-free graph of order n without isolated vertices, then $\gamma_L(G) \leq \frac{1}{2}n$.*

Conjecture 8 remains open, although it is true for several important classes of graphs. Conjecture 8 is true if the twin-free graph G of order n (without isolated vertices) satisfies any of the following conditions: (a) ([24]) G has no 4-cycles. (b) ([24]) G has (vertex) independence number at least $\frac{n}{2}$. (c) ([24]) G has clique number at least $\lceil \frac{n}{2} \rceil + 1$. (d) ([23]) G is a split graph. (e) ([23]) G is a co-bipartite graph. (f) ([20]) G is a cubic graph. (g) ([22]) G is a line graph.

15.3.2   The Paired-Domination Number

A set of edges in a graph G is *independent* if no two edges in it are adjacent in G; that is, an independent edge set is a set of edges without common vertices. A *matching* in a graph G is a set of independent edges in G. A *perfect matching M* is a matching such that every vertex of G is incident to an edge of M. A *paired-dominating set* of G is a dominating set S of G with the additional property that the subgraph $G[S]$ induced by S contains a perfect matching M (not necessarily induced). The *paired-domination number* of G, denoted by $\gamma_{\mathrm{pr}}(G)$, is the minimum cardinality of a paired-dominating set in G. Paired-domination was introduced by Haynes and Slater [28, 29] as a model for assigning backups to guards for security purposes and studied in [10, 13, 18, 33] inter alia.

Chen, Sun, and Xing [9] proved that if G is a cubic graph of order n, then $\gamma_{\mathrm{pr}}(G) \leq \frac{3}{5}n$. In June 2008, I attended the SIAM Conference on Discrete Mathematics held at the University of Vermont. During the conference, I held several discussions on this $\frac{3}{5}$-bound with my good friend, Wayne Goddard (who, incidentally, attended the same high school as I did and played on the same chess and field hockey teams as I did). We [25] provided a simpler proof of this bound and used our new proof to characterize the cubic graphs that achieve equality in the $\frac{3}{5}$-bound. Surprisingly, equality is only achieved by the Petersen graph. Throughout our proof of the $\frac{3}{5}$-bound, the challenging part was the existence of 5-cycles which seemed to force up the paired-domination number. The following two conjectures of mine have stubbornly resisted a solution for some time now, but surely, if true, simple proofs abound. A graph is C_5-*free* if it contains no induced 5-cycle.

Conjecture 9. *If G is a bipartite, cubic graph of order n, then $\gamma_{\mathrm{pr}}(G) \leq \frac{1}{2}n$.*

Conjecture 10. *If G is a cubic, C_5-free graph of order n, then $\gamma_{\mathrm{pr}}(G) \leq \frac{1}{2}n$.*

15.4 $\frac{2}{5}$-Conjectures

15.4.1   The Total Domination Number

A *cage* is a regular graph that has as few vertices as possible for its girth. An (r, g)-*cage* is an r-regular graph with the fewest possible number of vertices having girth g. The Heawood graph, shown in Figure 15.4(a), is the smallest cubic graph of girth 6; that is, the Heawood graph is the unique 6-cage. The bipartite complement of the Heawood graph is the bipartite graph formed by taking the two partite sets of the Heawood graph and joining a vertex from one partite set to a vertex from the other partite set by an edge whenever they are not joined in the Heawood graph. The bipartite complement of the Heawood graph can also be seen as the incidence bipartite graph of the complement of the Fano plane which is shown in Figure 15.4(b).

Thomasse and Yeo [57] proved that if G is a graph of order n with $\delta(G) \geq 4$, then $\gamma_t(G) \leq \frac{3}{7}n$. It is known [40, 42] that the bipartite complement of the Heawood Graph is the only connected graph of order n with $\delta(G) \geq 4$ achieving the Thomassé-Yeo $\frac{3}{7}$-upper bound. The following conjecture appears as Conjecture 18.4 in [40].

Conjecture 11. *If G is connected graph of order n with $\delta(G) \geq 4$ that is not the bipartite complement of the Heawood graph, then $\gamma_t(G) \leq \frac{2}{5}n$.*

Every vertex in the bipartite complement of the Heawood graph belongs to a 4-cycle. I visited Anders Yeo at the Singapore University of Technology and Design in April 2014 and January 2015, and we believe the Thomassé-Yeo $\frac{3}{7}$-upper bound can be improved to a $\frac{2}{5}$-upper bound if we restrict the graph to contain no 4-cycles.

Conjecture 12. *If G is quadrilateral-free graph of order n with $\delta(G) \geq 4$, then $\gamma_t(G) \leq \frac{2}{5}n$.*

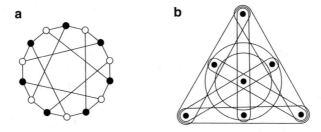

Fig. 15.4 (a) The Heawood graph and (b) the Fano plane

15.5 $\frac{2}{3}$-Conjectures

15.5.1 The Location-Total Domination Number

Among the existing variations of total domination, the one of location-total domination is widely studied. Recall that a set D of vertices *locates* a vertex v if the neighborhood of v within D is unique among all vertices in $V(G) \setminus D$. A *locating-total dominating set* is a total dominating set D that locates all the vertices, and the *location-total domination number* of G, denoted $\gamma_t^L(G)$, is the minimum cardinality of a locating-total dominating set in G. The concept of a locating-total dominating set was first considered in [32] and studied, for example, in [8, 37, 38].

A classic result in total domination theory due to Cockayne et al. [12] states that if G is a connected graph of order $n \geq 3$, then $\gamma_t(G) \leq \frac{2}{3}n$. This $\frac{2}{3}$-bound is tight, and the extremal examples have been classified (see [6]). As observed in [21], while there are many such graphs which have location-total domination number much larger than two-thirds their order, the only such graphs that are known contain many twins. Together with Florent Foucaud, my postdoctoral student at the time, we conjectured [21] that in the absence of twins, the classic bound of two-thirds the order for the total domination number also holds for the locating-total domination number.

Conjecture 13 ([21]). *If G is a twin-free graph of order n without isolated vertices, then $\gamma_t^L(G) \leq \frac{2}{3}n$.*

Conjecture 13 remains open, although it was proved for graphs with no 4-cycles and also shown to hold asymptotically for large minimum degree (see [21]). Conjecture 13 is also known [22] to hold for line graphs.

15.6 $\frac{4}{9}$-Conjectures

15.6.1 The Total Domination Number

As remarked earlier, if G is a graph of order n with $\delta(G) \geq 3$, then $\gamma_t(G) \leq \frac{1}{2}n$. The generalized Petersen graph G_{16} of order 16 shown in Figure 15.5 achieves equality in this $\frac{1}{2}$-bound for the total domination number.

Two infinite families $\mathcal{G}$ and $\mathcal{H}$ of connected cubic graphs (described below) with total domination number one-half their orders are constructed in [19] as follows. For $k \geq 1$, let G_k be the graph obtained from two copies of the path P_{2k} with respective vertex sequences $a_1 b_1 a_2 b_2 \ldots a_k b_k$ and $c_1 d_1 c_2 d_2 \ldots c_k d_k$. Let $A = \{a_1, a_2, \ldots, a_k\}$, $B = \{b_1, b_2, \ldots, b_k\}$, $C = \{c_1, c_2, \ldots, c_k\}$, and $D = \{d_1, d_2, \ldots, d_k\}$. For each $i \in \{1, 2, \ldots, k\}$, join a_i to d_i and b_i to c_i. To complete the construction of the graph G_k, join a_1 to c_1 and b_k to d_k. Let $\mathcal{G} = \{G_k \mid k \geq 1\}$. For $k \geq 2$, let H_k be obtained from G_k by deleting the two edges $a_1 c_1$ and $b_k d_k$ and adding the two edges $a_1 b_k$ and

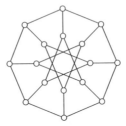

Fig. 15.5 The Generalized Petersen graph G_{16}

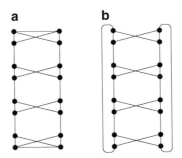

Fig. 15.6 Cubic graphs (**a**) $G_4 \in \mathcal{G}$ and (**b**) $H_4 \in \mathcal{H}$

$c_1 d_k$. Let $\mathcal{H} = \{H_k \mid k \geq 2\}$. We note that G_k and H_k are cubic graphs of order $4k$. Further, we note that $G_1 = K_4$. The graphs $G_4 \in \mathcal{G}$ and $H_4 \in \mathcal{H}$, for example, are illustrated in Figure 15.9.

In [39], it is shown that these two infinite families, $\mathcal{G}$ and $\mathcal{H}$, of connected, cubic graphs, as well as the generalized Petersen graph of order 16, are precisely the extremal connected graphs that achieve equality in the $\frac{1}{2}$-bound for the total domination number of a connected graph with minimum degree at least *three*. Every vertex that belongs to a graph in the family $\mathcal{G} \cup \mathcal{H}$ belongs to a 4-cycle. Anders Yeo and I conjectured that the $\frac{1}{2}$-bound on the total domination can be improved to a $\frac{8}{17}$-bound if the graph is quadrilateral-free and different from the generalized Petersen graph, G_{16}. This is stated as Conjecture 18.3 in [40], which remains open (Figure 15.6).

Conjecture 14. *If $G \neq G_{16}$ is a connected, quadrilateral-free graph of order n with $\delta(G) \geq 3$, then $\gamma_t(G) \leq \frac{8}{17}n$.*

We remark that the quadrilateral-free requirement in Conjecture 14 cannot be replaced with a C_4-free requirement. For example, the two graphs G_1 and G_2 in the family $\mathcal{G}$ are C_4-free and have total domination numbers one-half their orders.

However, I believe that it is the existence of induced 6-cycles that is the crucial property of the graph G_{16} and of all graphs in the family $\mathcal{G} \cup \mathcal{H}$, except for the graph $K_4 = G_1 \in \mathcal{G}$, that account for their large total domination number. Every vertex in every graph in the family $\{G_{16}\} \cup \mathcal{G} \cup \mathcal{H}$ (that achieves the $\frac{1}{2}$-bound for the total domination number), except for $K_4 \in \mathcal{G}$, belongs to an induced 6-cycle. Recall

Fig. 15.7 A cubic C_6-free graph G with $\gamma_t(G) = \frac{4}{9}n$

that a graph is C_6-free if it contains no induced 6-cycle. I believe that the $\frac{1}{2}$-bound decreases significantly to a $\frac{4}{9}$-bound if we impose the structural requirement that the graph is C_6-free.

Conjecture 15. *If G is a C_6-free graph of order $n \geq 6$ with $\delta(G) \geq 3$, then $\gamma_t(G) \leq \frac{4}{9}n$.*

If Conjecture 15 is true, then the bound is achieved, for example, by the cubic C_6-free graph G of order $n = 18$ with $\gamma_t(G) = 8 = \frac{4}{9}n$ shown in Figure 15.7.

15.7 $\frac{3}{8}$-Conjectures

15.7.1 The Independent Domination Number

As remarked earlier, Reed [52] proved that if G is a cubic graph of order n, then $\gamma(G) \leq \frac{3}{8}n$. Lam, Shiu, and Sun [49] proved that if G is a connected cubic graph G of order n other than $K_{3,3}$, then $i(G) \leq \frac{2}{5}n$. Goddard and Henning [26] conjecture that the graphs $K_{3,3}$ and the prism $C_5 \square K_2$ are the only exception for an upper bound of $\frac{3}{8}n$ on $i(G)$. The graphs $K_{3,3}$ and $C_5 \square K_2$ are depicted in Figure 15.8.

Conjecture 16 ([26]). *If $G \notin \{K_{3,3}, C_5 \square K_2\}$ is a connected, cubic graph of order n, then $i(G) \leq \frac{3}{8}n$.*

Two infinite families $\mathcal{G}_{\text{cubic}}$ and $\mathcal{H}_{\text{cubic}}$ of connected, cubic graphs, with independent domination number three-eighths their orders, are constructed in [26] as follows. We first construct graphs in $\mathcal{G}_{\text{cubic}}$. For $k \geq 1$, consider two copies of the cycle C_{4k} with respective vertex sequences $a_1 b_1 c_1 d_1 \ldots a_k b_k c_k d_k$ and $w_1 x_1 y_1 z_1 \ldots w_k x_k y_k z_k$. For each $1 \leq i \leq k$, join a_i to w_i, b_i to x_i, c_i to z_i, and d_i to y_i.

Graphs in $\mathcal{H}_{\text{cubic}}$ are constructed as follows. For $\ell \geq 1$, consider a copy of the cycle $C_{3\ell}$ with vertex sequence $a_1 b_1 c_1 \ldots a_\ell b_\ell c_\ell$. For each $1 \leq i \leq \ell$, add the vertices $\{w_i, x_i, y_i, z_i^1, z_i^2\}$, and join a_i to w_i, b_i to x_i, and c_i to y_i. Finally, for each $1 \leq i \leq \ell$ and $j \in \{1, 2\}$, join z_i^j to each of the vertices w_i, x_i, and y_i.

Graphs in the families $\mathcal{G}_{\text{cubic}}$ and $\mathcal{H}_{\text{cubic}}$ are illustrated in Figure 15.9.

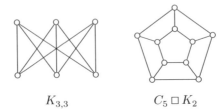

$$K_{3,3} \qquad\qquad C_5 \,\square\, K_2$$

Fig. 15.8 The graphs $K_{3,3}$ and $C_5 \,\square\, K_2$

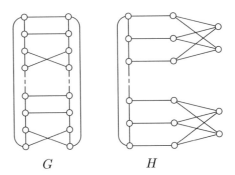

$$G \qquad\qquad H$$

Fig. 15.9 Graphs $G \in \mathcal{G}_{\mathrm{cubic}}$ and $H \in \mathcal{H}_{\mathrm{cubic}}$ of order n with $i(G) = i(H) = \frac{3}{8}n$

It was shown in [27] that if $G \in \mathcal{G}_{\mathrm{cubic}} \cup \mathcal{H}_{\mathrm{cubic}}$ has order n, then $i(G) = \frac{3}{8}n$. It is remarked in [26] that "perhaps it is even true that for $n > 10$, $i(G) \leq \frac{3}{8}n$ with equality if and only if $G \in \mathcal{G}_{\mathrm{cubic}} \cup \mathcal{H}_{\mathrm{cubic}}$. We remark that computer search has confirmed this is true when $n \leq 20$."

Dorbec, Henning, Montassier, and Southey [14] have shown that Conjecture 16 is true if G does not have a subgraph isomorphic to $K_{2,3}$. However, Conjecture 16 remains open for connected cubic graphs that do have $K_{2,3}$ as a subgraph.

15.8 $\frac{3}{7}$-Conjectures

15.8.1 *The Independent Domination Number*

The following conjecture was first posed as a question by Goddard and Henning [26].

Conjecture 17. *If G is a 4-regular graph of order n, then $i(G) \leq \frac{3}{7}n$.*

As observed in [26], if G is the 4-regular graph shown in Figure 15.10, then G has order $n = 14$ and $i(G) = 6 = \frac{3}{7}n$. Hence, if Conjecture 17 is true, then the bound is achievable.

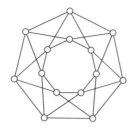

Fig. 15.10 A 4-regular graph G with $i(G) = 6$

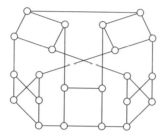

Fig. 15.11 The bipartite cubic graph B_{22} with $i(B_{22}) = \frac{4}{11}n$

15.9 $\frac{4}{11}$-Conjectures

15.9.1 The Independent Domination Number

Goddard and Henning [26] posed the following conjecture:

Conjecture 18 ([26]). *If $G \neq K_{3,3}$ is a connected, bipartite, cubic graph of order n, then $i(G) \leq \frac{4}{11}n$.*

It is remarked in [26] that Conjecture 18 is true when $n \leq 26$, as confirmed by computer search (see [27]). If Conjecture 18 is true, then the bound is achieved by the bipartite, cubic graph B_{22} of order $n = 22$, with $i(B_{22}) = 8 = \frac{4}{11}n$, shown in Figure 15.11.

Henning, Löwenstein, and Rautenbach [43] have shown that Conjecture 18 is true if G is C_4-free; that is, if G is a bipartite, cubic graph of girth at least six. However, Conjecture 18 remains open for connected, cubic graphs that do contain a 4-cycle.

15.9.2 The Total Domination Number

Thomasse and Yeo [57] posed the following $\frac{4}{11}$-conjecture.

Conjecture 19 ([57]). *If G is a graph of order n with $\delta(G) \geq 5$, then $\gamma_t(G) \leq \frac{4}{11}n$.*

If Conjecture 19 is true, then the bound is achievable. For example, the graph G_{22}, shown in Figure 15.3, has order $n = 22$, minimum degree $\delta(G_{22}) = 5$, and $\gamma_t(G_{22}) = 8 = \frac{4}{11}n$. Dorfling and Henning [15] showed that if G is a graph of order n with $\delta(G) \geq 5$, then $\gamma_t(G) \leq \frac{17}{44}n = (\frac{4}{11} + \frac{1}{44})n$. The best upper bound to date on the total domination number of a graph with minimum degree at least *five* is due to Eustis, Henning, and Yeo [17] who showed that if G is a graph of order n with $\delta(G) \geq 5$, then $\gamma_t(G) < (\frac{4}{11} + \frac{1}{72})n$.

15.10 $\frac{3}{5}$-Conjectures

15.10.1 Open Locating-Domination Number

An *open locating-dominating set* (also called an *identifying open code* or a *differentiating total dominating set* in the literature) is similar to a locating-total dominating set, defined in Section 15.5.1, except that in this case we impose the stricter requirement that distinct vertices, even vertices that belong to the total dominating set, are totally dominated by distinct subsets of the total dominating set. Hence, a set S is an open locating-dominating set in a graph G if S is a total dominating set in G, with the additional property that $N(u) \cap S \neq N(v) \cap S$ for all distinct vertices u and v in G. A graph with no isolated vertex has an open locating-dominating set if and only if it is open twin-free (i.e., if it has no open twins). If G is open twin-free and G has no isolated vertex, we denote by $\gamma_{OLD}(G)$ the minimum cardinality of an open locating-dominating set in G. The problem of open locating-dominating sets was introduced by Honkala et al. [44] in the context of coding theory for binary hypercubes. Recent papers on open locating-dominating sets can be found, for example, in [41, 53, 54].

Henning and Yeo [41] showed that if G is a connected, twin-free, cubic graph of order n, then $\gamma_{OLD}(G) \leq \frac{3}{4}n$. This $\frac{3}{4}$-bound on $\gamma_{OLD}(G)$ is achieved, for example, by the complete graph K_4 and the hypercube Q_3, shown in Figure 15.12, that satisfy $\gamma_{OLD}(K_4) = 3$ and $\gamma_{OLD}(Q_3) = 6$.

However, Henning and Yeo [41] conjecture that if G is a twin-free, connected, cubic graph of sufficiently large order n, then the $\frac{3}{4}$-bound on $\gamma_{OLD}(G)$ can be improved to a $\frac{3}{5}$-bound.

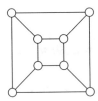

Fig. 15.12 The hypercube Q_3

Conjecture 20. *If G is an open twin-free, connected, cubic graph of sufficiently large order n, then $\gamma_{\mathrm{OLD}}(G) \leq \frac{3}{5}n$.*

15.10.2 The Game Domination Number

The domination game in graphs was introduced by Brešar, Klavžar, and Rall in [2] and extensively studied afterward in [2–5, 34, 46]. The domination game on a graph *G* consists of two players, *Dominator* and *Staller*, who take turns choosing a vertex from *G*. Each vertex chosen must dominate at least one vertex not dominated by the vertices previously chosen, where a vertex dominates itself and its neighbors. The game ends when the set of vertices chosen becomes a dominating set in *G*. Dominator wishes to end the game with a minimum number of vertices chosen, and Staller wishes to end the game with as many vertices chosen as possible. The *game domination number* (resp. *Staller-start game domination number*), $\gamma_g(G)$ (resp. $\gamma_g'(G)$), of *G* is the minimum possible number of vertices chosen when Dominator (resp. Staller) starts the game and both players play according to the rules. Kinnersley, West, and Zamani in [46] proved a Continuation Principle lemma, which leads to the fundamental property that $|\gamma_g(G) - \gamma_g'(G)| \leq 1$, for every graph *G*.

In May 2012, Douglas Rall hosted a workshop on game domination at Furman University attended by experts on the game, including Bill Kinnersley and the Slovenian graph theorists Boštjan Brešar, Sandi Klavžar, and Gašper Košmrlj. Our energies focused on proving the following *3/5-Conjecture* posted by Kinnersley, West, and Zamani in [46].

Conjecture 21 ([46]). *If G is an isolate-free graph of order n, then $\gamma_g(G) \leq \frac{3}{5}n$.*

Although we were unable to accomplish our goal, large families of trees were constructed that attain the conjectured $\frac{3}{5}$-bound, and all extremal trees on up to 20 vertices were found in [4]. In [7] Bujtás proved Conjecture 21 for forests in which no two leaves are at distance 4 apart. Further progress toward Conjecture 21 was made in [34], where the conjecture is established over the class of graphs with minimum degree at least 2. However, Conjecture 21 remains open for graphs that contain vertices of degree *one*.

15.11 $\frac{3}{4}$-Conjectures

15.11.1 The Game Total Domination Number

Recently, the total version of the domination game was investigated in [35]. A vertex *totally dominates* another vertex if they are neighbors. The *total domination game* consists of two players called *Dominator* and *Staller*, who take turns choosing a vertex from *G*. In this version of the game, each vertex chosen must totally dominate at least one vertex not totally dominated by the set of vertices previously chosen. The game ends when the set of vertices chosen is a total dominating set in G. Dominator wishes to end the game with a minimum number of vertices chosen, and Staller wishes to end the game with as many vertices chosen as possible.

The *game total domination number*, $\gamma_{tg}(G)$, of G is the minimum possible number of vertices chosen when Dominator (resp. Staller) starts the game and both players play according to the rules. The *Staller-start game total domination number*, $\gamma'_{tg}(G)$, of G is the number of vertices chosen when Staller starts the game and both players play optimally. In [35], the authors prove a Total Continuation Principle lemma from which one can readily deduce that $|\gamma_{tg}(G) - \gamma'_{tg}(G)| \le 1$ for every graph G (with no isolated vertex). In [36] we pose the following $\frac{3}{4}$-*Game Total Domination Conjecture*.

Conjecture 22 ([36]). *If G is a graph on n vertices in which every component contains at least three vertices, then $\gamma_{tg}(G) \le \frac{3}{4}n$.*

We remark that the requirement in Conjecture 22 that every component contains at least three vertices is essential. For example, if G is the disjoint union of copies of K_2, then $\gamma_{tg}(G) = n > \frac{3}{4}n$.

Suppose the total domination game is started on a connected graph. As the game progresses, at least one new vertex becomes totally dominated on each move. However, once a vertex and all its neighbors are totally dominated, it plays no role in the remainder of the game. Such a vertex can be deleted from the so-called *partially total dominated graph*, which is a graph together with a declaration that some vertices are already totally dominated; that is, they need not be totally dominated in the rest of the game. Once such vertices are deleted, the resulting game may well become disconnected. Therefore, it is best to study the total domination game in a general setting where the game is started on a graph that may possibly be disconnected.

If the $3/4$-Conjecture is true, then the upper bound on the Dominator-start game total domination number is tight, as shown in [36]. In 2013, I had the opportunity to visit Sandi Klavžar at the University of Ljubljana and, subsequently, Douglas Rall at Furman University in 2014. Our efforts culminated in proving that if G is a graph on n vertices in which every component contains at least three vertices, then $\gamma_{tg}(G) \le \frac{4}{5}n$ and $\gamma'_{tg}(G) \le \frac{4n+2}{5}$. Our proof strategy modified an ingenious approach adopted by Csilla Bujtás [7] in order to attack the $3/5$-Game Domination Conjecture. Bujtás's approach is to color the vertices of a forest with three colors that reflect three different types of vertices and to associate a weight with each vertex. We modify Bujtás's approach in the total version of the game by coloring the vertices of a graph with four colors that reflect four different types of vertices. We then assign weights to colored vertices and study the decrease of total weight of the graph as a consequence of playing vertices in the course of the game. Three different phases of the game are studied and a strategy formulated for Dominator that comprises an opening-game strategy, a middle-game strategy, and an end-game strategy. An analysis of Dominator's strategy based on the three phases produced the $\frac{4}{5}$-bound on the game total domination number. However, Conjecture 22 remains open.

15.12   Concluding Remarks

In this chapter, for a graph G of order n and for some domination parameter, call it $\vartheta(G)$, we ask if for some rational number $\frac{a}{b}$, where $0 < \frac{a}{b} < 1$, is $\vartheta(G) \leq \frac{a}{b}n$? In most cases, we qualify the classes of graphs, G, for which we are asking the question, for example, restricted to bipartite, cubic graphs or restricted to quadrilateral-free graphs with minimum degree at least five, and so on. For certain conjectures posed in the chapter, it may be interesting to answer them for special subclasses of graphs. For example, is Conjecture 5 true for the class of graphs with girth at least 8? Is Conjecture 15 true for the class of graphs with girth at least 7? Is Conjecture 17 true for the class of bipartite graphs? Is Conjecture 22 true for the class of graphs with minimum degree at least two?

There are several parameters that are known to be less than or equal to the domination number of a graph, such as the 2-packing number, the distance-2 domination number, and the fractional domination number. For each of these parameters, it would be informative to ask, is their value less than or equal to the same value $\frac{a}{b}n$ conjectured for the domination number restricted to the class of graphs under consideration? More generally, for a known parameter, call it $\eta(G)$, for which $\eta(G) \leq \vartheta(G)$, and a given conjecture of the type $\vartheta(G) \leq \frac{a}{b}n$, can we prove that $\eta(G) \leq \frac{a}{b}n$?

References

1. Archdeacon, D., Ellis-Monaghan, J., Fischer, D., Froncek, D., Lam, P.C.B., Seager, S., Wei, B., Yuster, R.: Some remarks on domination. J. Graph Theory **46**, 207–210 (2004)
2. Brešar, B., Klavžar, S., Rall, D.F.: Domination game and an imagination strategy. SIAM J. Discret. Math. **24**, 979–991 (2010)
3. Brešar, B., Klavžar, S., Rall, D.F.: Domination game played on trees and spanning subgraphs. Discret. Math. **313**, 915–923 (2013)
4. Brešar, B., Klavžar, S., Košmrlj, G., Rall, D.F.: Domination game: extremal families of graphs for the 3/5-conjectures. Discret. Appl. Math. **161**, 1308–1316 (2013)
5. Brešar, B., Dorbec, P., Klavžar, S., Košmrlj, G.: Domination game: effect of edge- and vertex-removal. Discret. Math. **330**, 1–10 (2014)
6. Brigham, R.C., Carrington, J.R., Vitray, R.P.: Connected graphs with maximum total domination number. J. Comb. Comput. Comb. Math. **34**, 81–96 (2000)
7. Bujtás, Cs.: Domination game on trees without leaves at distance four. In: Frank, A., et al. (eds.) Proceedings of the 8th Japanese-Hungarian Symposium on Discrete Mathematics and Its Applications, Veszprém, 4–7 June 2013, pp. 73–78
8. Chen, X.G., Sohn, M.Y.: Bounds on the locating-total domination number of a tree. Discret. Appl. Math. **159**, 769–773 (2011)
9. Chen, X.G., Sun, L., Xing, H.M.: Paired-domination numbers of cubic graphs (Chinese). Acta Math. Sci. Ser. A Chin. Ed. **27**, 166–170 (2007)
10. Cheng, T.C.E., Kang, L.Y., Ng, C.T.: Paired domination on interval and circular-arc graphs. Discret. Appl. Math. **155**, 2077–2086 (2007)
11. Chvátal, V., McDiarmid, C.: Small transversals in hypergraphs. Combinatorica **12**, 19–26 (1992)

12. Cockayne, E.J., Dawes, R.M., Hedetniemi, S.T.: Total domination in graphs. Networks **10**, 211–219 (1980)
13. Dorbec, P., Gravier, S., Henning, M.A.: Paired-domination in generalized claw-free graphs. J. Comb. Optim. **14**, 1–7 (2007)
14. Dorbec, P., Henning, M.A., Montassier, M., Southey, J.: Independent domination in cubic graphs. J. Graph Theory **80**(4), 329–349 (2015).
15. Dorfling, M., Henning, M.A.: Transversals in 5-uniform hypergraphs and total domination in graphs with minimum degree five. Quaest. Math. **38**(2), 155–180 (2015)
16. Erdös, P., Henning, M.A., Swart, H.C.: The smallest order of a graph with domination number equal to two and with every vertex contained in a K_n. Ars Comb. **35A**, 217–223 (1993)
17. Eustis, A., Henning, M.A., Yeo, A.: Independence in 5-uniform hypergraphs. Discrete Math. **339**, 1004–1027 (2016)
18. Favaron, O., Henning, M.A.: Paired-domination in claw-free cubic graphs. Graphs Comb. **20**, 447–456 (2004)
19. Favaron, O., Henning, M.A., Mynhardt, C.M., Puech, J.: Total domination in graphs with minimum degree three. J. Graph Theory **34**(1), 9–19 (2000)
20. Foucaud, F., Henning, M.A.: Location-domination and matching in cubic graphs. Manuscript, http://arxiv.org/abs/1412.2865 (2014)
21. Foucaud, F., Henning, M.A.: Locating-total dominating sets in twin-free graphs. Manuscript (2015)
22. Foucaud, F., Henning, M.A.: Locating-dominating sets in line graphs. To appear in Discrete Math. (2015)
23. Foucaud, F., Henning, M.A., Löwenstein, C., Sasse, T.: Locating-dominating sets in twin-free graphs. Discret. Appl. Math. **200**, 52–58 (2016)
24. Garijo, D., González, A., Márquez, A.: The difference between the metric dimension and the determining number of a graph. Appl. Math. Comput. **249**, 487–501 (2014)
25. Goddard, W., Henning, M.A.: A characterization of cubic graphs with paired-domination number three-fifths their order. Graphs Comb. **25**, 675–692 (2009)
26. Goddard, W., Henning, M.A.: Independent domination in graphs: a survey and recent results. Discret. Math. **313**, 839–854 (2013)
27. Goddard, W., Henning, M.A., Lyle, J., Southey, J.: On the independent domination number of regular graphs. Ann. Comb. **16**, 719–732 (2012)
28. Haynes, T.W., Slater, P.J.: Paired-domination and the paired-domatic number. Congr. Numer. **109**, 65–72 (1995)
29. Haynes, T.W., Slater, P.J.: Paired-domination in graphs. Networks **32**, 199–206 (1998)
30. Haynes, T.W., Hedetniemi, S.T., Slater, P.J.: Fundamentals of Domination in Graphs. Dekker, New York (1998)
31. Haynes, T.W., Hedetniemi, S.T., Slater, P.J. (eds.): Domination in Graphs: Advanced Topics. Dekker, New York (1998)
32. Haynes, T.W., Henning, M.A., Howard, J.: Locating and total dominating sets in trees. Discret. Appl. Math. **154**, 1293–1300 (2006)
33. Henning, M.A.: Graphs with large paired-domination number. J. Comb. Optim. **13**, 61–78 (2007)
34. Henning, M.A., Kinnersley, W.B.: Domination Game: A proof of the 3/5-Conjecture for graphs with minimum degree at least two. SIAM Journal of Discrete Mathematics **30**(1), 20–35 (2016)
35. Henning, M.A., Klavžar, S., Rall, D.F.: Total version of the domination game. Graphs Combin. **31**(5), 1453–1462 (2015)
36. Henning, M.A., Klavžar, S., Rall, D.F.: to appear in Combinatorica
37. Henning, M.A., Löwenstein, C.: Locating-total domination in claw-free cubic graphs. Discret. Math. **312**, 3107–3116 (2012)
38. Henning, M.A., Rad, N.J.: Locating-total domination in graphs. Discret. Appl. Math. **160**, 1986–1993 (2012)

39. Henning, M.A., Yeo, A.: Hypergraphs with large transversal number and with edge sizes at least three. J. Graph Theory **59**, 326–348 (2008)
40. Henning, M.A., Yeo, A.: Total Domination in Graphs. Springer Monographs in Mathematics. Springer, New York (2013). ISBN:978-1-4614-6524-9 (Print) 978-1-4614-6525-6 (Online)
41. Henning, M.A., Yeo, A.: Distinguishing-transversal in hypergraphs and identifying open codes in cubic graphs. Graphs Comb. **30**, 909–932 (2014)
42. Henning, M.A., Yeo, A.: Transversals in 4-uniform hypergraphs. Manuscript, http://arxiv.org/abs/1504.02650 (2014)
43. Henning, M.A., Löwenstein, C., Rautenbach, D.: Independent domination in subcubic bipartite graphs of girth at least six. Discret. Appl. Math. **162**, 399–403 (2014)
44. Honkala, I., Laihonen, T., Ranto, S.: On strongly identifying codes. Discret. Math. **254**, 191–205 (2002)
45. Kelmans, A.: Counterexamples to the cubic graph domination conjecture, http://arxiv.org/pdf/math/0607512.pdf. 20 July 2006
46. Kinnersley, W.B., West, D.B., Zamani, R.: Extremal problems for game domination number. SIAM J. Discret. Math. **27**, 2090–2107 (2013)
47. Kostochka, A.V., Stocker, C.: A new bound on the domination number of connected cubic graphs. Sib. Elektron. Mat. Izv. **6**, 465–504 (2009)
48. Kostochka, A.V., Stodolsky, B.Y.: On domination in connected cubic graphs. Discret. Math. **304**, 45–50 (2005)
49. Lam, P.C.B., Shiu, W.C., Sun, L.: On independent domination number of regular graphs. Discret. Math. **202**, 135–144 (1999)
50. Löwenstein, C., Rautenbach, D.: Domination in graphs of minimum degree at least two and large girth. Graphs Comb. **24**(1), 37–46 (2008)
51. Ore, O.: Theory of graphs. American Mathematical Society Translations, vol. 38, pp. 206–212. American Mathematical Society, Providence (1962)
52. Reed, B.: Paths, stars, and the number three. Comb. Probab. Comput. **5**, 277–295 (1996)
53. Seo, S.J., Slater, P.J.: Open neighborhood locating-dominating sets. Australas. J. Comb. **46**, 109–120 (2010)
54. Seo, S.J., Slater, P.J.: Open neighborhood locating-dominating in trees. Discret. Appl. Math. **159**, 484–489 (2011)
55. Slater, P.J.: Dominating and location in acyclic graphs. Networks **17**, 55–64 (1987)
56. Slater, P.J.: Dominating and reference sets in graphs. J. Math. Phys. Sci. **22**, 445–455 (1988)
57. Thomassé, S., Yeo, A.: Total domination of graphs and small transversals of hypergraphs. Combinatorica **27**, 473–487 (2007)
58. Tuza, Zs.: Covering all cliques of a graph. Discret. Math. **86**, 117–126 (1990)

Chapter 16
Circuit Double Covers of Graphs

Cun-Quan Zhang

Abstract The circuit double cover conjecture (CDC conjecture) is easy to state: *For every 2-connected graph, there is a family $\mathscr{F}$ of circuits such that every edge of the graph is covered by precisely two members of $\mathscr{F}$.* The CDC conjecture (and its numerous variants) is considered by most graph theorists as one of the major open problems in the field. The CDC conjecture, Tutte's 5-flow conjecture, and the Berge-Fulkerson conjecture are three major snark family conjectures since they are all trivial for 3-edge-colorable cubic graphs and remain widely open for snarks. This chapter is a brief survey of the progress on this famous open problem.

16.1   Introduction

The circuit (cycle) double cover conjecture (CDC conjecture) is easy to state: *For every 2-connected graph, there is a family $\mathscr{F}$ of circuits such that every edge of the graph is covered by precisely two members of $\mathscr{F}$.* As an example, if a 2-connected graph is properly embedded on a surface (without crossing edges) in such a way that all faces are bounded by circuits, then the collection of the boundary circuits will "double cover" the graph.

The CDC conjecture (and its numerous variants) is considered by most graph theorists as one of the major open problems in the field. One reason for this is its close relationship with topological graph theory, integer flow theory, graph coloring, and the structure of snarks.

The CDC conjecture was presented as an "open question" by Szekeres [66] for cubic graphs (as we will see soon in Theorem 1, it is equivalent for all bridgeless graphs). The conjecture was also independently stated by Seymour in [62] for all bridgeless graphs. An equivalent version of the CDC conjecture was proposed by Itai and Rodeh [36] that *every bridgeless graph has a family $\mathscr{F}$ of circuits such that every edge is contained in one or two members of $\mathscr{F}$.*

For the origin of the conjecture, some mathematicians gave the credit to Tutte. According to a personal letter from Tutte to Fleischner [73], he said, "I too have

C.-Q. Zhang (✉)
Department of Mathematics, West Virginia University, Morgantown, WV 26506-6310, USA
e-mail: cqzhang@mail.wvu.edu

© Springer International Publishing Switzerland 2016
R. Gera et al. (eds.), *Graph Theory*, Problem Books in Mathematics,
DOI 10.1007/978-3-319-31940-7_16

been puzzled to find an original reference. I think the conjecture is one that was well established in mathematical conversation long before anyone thought of publishing it." It was also pointed out in the survey paper by Jaeger [41] that "it seems difficult to attribute the paternity of this conjecture" and also pointed out in some early literature (such as [23]) that "its origin is uncertain." This may explain why the CDC conjecture is considered as "folklore" in [7] (Unsolved problem 10). Some early investigations related to the conjecture can be traced back to publications by Tutte in 1949 [70, 73].

Most material presented in this chapter follows the pioneering survey papers by Jaeger (1985 [41]), and Jackson (1993 [37]), and the monographs *Integer Flows and Cycle Covers of Graphs* (1997 [78]) *Circuit Double Covers of Graphs* (2007 [79]) by the author.

The circuit double cover conjecture is obviously true for 2-connected planar graphs since the boundary of every face is a circuit and the set of the boundaries of all faces forms a circuit double cover of an embedded graph. One might attempt to extend this observation further to all 2-connected graphs embedded on some surfaces. However, it is not true that *any embedding* of a 2-connected graph is free of a handle bridge. That is, the boundary of some face may not be a circuit. This leads to an even stronger open problem in topology known as the strong embedding conjecture: *Can we find an embedding of a 2-connected graph G on some surface Σ such that the boundary of every face is a circuit?* See [27, 50] or Conjecture 3.10 in [7] for more information.

The following theorem summarizes some structure of a minimal counterexample to the circuit double cover conjecture (see [41]).

Theorem 1. *If G is a minimum counterexample to the circuit double cover conjecture, then:*

(1) G is simple, 3-connected, and cubic;
(2) G has no nontrivial 2 or 3-edge cut;
(3) G is not 3-edge colorable; and
(4) G is not planar.

Thus, most graphs considered in this chapter are cubic and bridgeless.

For a 3-edge-colorable cubic graph, we have an even stronger result for even subgraph double cover. The following theorem was formulated by Jaeger [41]. (The equivalence of (1) and (2) was also applied in [62].)

Theorem 2. *Let G be a cubic graph. Then the following statements are equivalent:*

(1) G is 3-edge colorable;
(2) G has a 3-even subgraph double cover; and
(3) G has a 4-even subgraph double cover.

16.2 Faithful Circuit Cover and Petersen Minor-Free Graphs

The concept of faithful circuit cover is not only a generalization of the circuit double cover problem but also an inductive approach to the CDC conjecture in a very natural way.

Let $\mathscr{Z}^+$ be the set of all positive integers and $\mathscr{Z}^\star$ be the set of all nonnegative integers.

Definition 1. Let G be a graph and $w : E(G) \to \mathscr{Z}^+$. A family $\mathscr{F}$ of circuits (or even subgraphs) of G is a *faithful circuit cover* (or *faithful even subgraph cover*, respectively) with respect to w if each edge e is contained in precisely $w(e)$ members of $\mathscr{F}$.

Figure 16.1 shows an example of a faithful circuit cover of (K_4, w) where $w : E(K_4) \to \{1, 2\}$. Here $w^{-1}(1)$ induces a Hamilton circuit and $w^{-1}(2)$ induces a perfect matching (a pair of diagonals).

It is obvious that *the circuit double cover is a special case of the faithful circuit cover problem by choosing the weight w to be 2 for every edge.*

Definition 2. Let G be a graph. A weight $w : E(G) \to \mathscr{Z}^+$ is *Eulerian* if the total weight of every edge cut is even. And (G, w) is called an *Eulerian weighted graph*.

Definition 3. Let G be a graph. An Eulerian weight $w : E(G) \to \mathscr{Z}^+$ is *admissible* if, for every edge-cut T and every $e \in T$,

$$w(e) \le \frac{w(T)}{2}.$$

And (G, w) is called an *admissible Eulerian weighted graph* if w is Eulerian and admissible.

If G has a faithful circuit cover $\mathscr{F}$ with respect to a weight $w : E(G) \to \mathscr{Z}^+$, then the total weight of every edge cut must be even since, for every circuit C of $\mathscr{F}$ and every edge-cut T, the circuit C must use an *even* number of *distinct* edges of the cut T. With this observation, the requirements of being *Eulerian and admissible* are *necessary* for faithful circuit covers.

Problem 1. Let G be a bridgeless graph with $w : E(G) \to \mathscr{Z}^+$. If w is admissible and Eulerian, does G have a faithful circuit cover with respect to w?

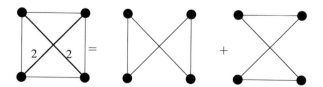

Fig. 16.1 Faithful circuit cover – an example

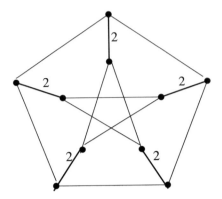

Fig. 16.2 (P_{10}, w_{10})

Unfortunately, Problem 1 is *not* always true. The Petersen graph P_{10} with an Eulerian weight w_{10} (see Figure 16.2) does not have a faithful circuit cover: where the set of weight 2 edges induces a perfect matching of P_{10} and the set of weight 1 edges induces two disjoint pentagons.

For a given weight $w : E(G) \to \mathscr{Z}^+$, denote

$$E_{w=i} = \{e \in E(G) : w(e) = i\}.$$

Like many mainstream research areas in graph theory, 3-edge coloring plays a central role in the study of the faithful circuit cover problem. The following is one of the most frequently used lemmas in this field.

Lemma 1 (Seymour [62]). *Let G be a cubic graph and $w : E(G) \to \{1, 2\}$ be an Eulerian weight. Then the following statements are equivalent:*

(1) G is 3-edge colorable; and
(2) G has a faithful 3-even subgraph cover with respect to w.

Since the 4-color theorem is equivalent to 3-edge colorings for all bridgeless cubic planar graphs, an immediate corollary of Lemma 1 is the following early result (Theorem 3) by Seymour. An alternative proof of Theorem 3 (slightly stronger) is provided by Fleischner [15] without using the 4-color theorem.

Theorem 3 (Seymour [62], and Fleischner [15, 18]). *If G is a planar, bridgeless graph associated with an Eulerian weight $w : E(G) \to \{1, 2\}$, then G has a faithful circuit cover with respect to w.*

Theorem 3 was further generalized for Petersen minor-free graphs as follows.

Theorem 4 (Alspach, Goddyn and Zhang [1, 3]). *Let G be a graph without a Petersen minor and $w : E(G) \to Z^+$ be an admissible Eulerian weight. Then G has a faithful circuit cover with respect to w.*

16.3   Integer Flows

The concept of integer flow was introduced by Tutte [70], [71] as a generalization of the map coloring problems. This section is a brief survey of circuit covering theorems arising from the integer flow theory. Readers are referred to [40, 65, 78] of the comprehensive surveys in this area.

The following are some of classical results in flow theory.

Theorem 5 (Jaeger [38, 39]). *Every 4-edge-connected graph admits a nowhere-zero 4-flow.*

Theorem 6 (Jaeger [39], Kilpatrick [44]). *Every bridgeless graph admits a nowhere-zero 8-flow.*

Theorem 6 is further improved by Seymour in the following theorem.

Theorem 7 (Seymour [63]). *Every bridgeless graph admits a nowhere-zero 6-flow.*

With the application of the following lemma, the flow theorems can be stated as even subgraph covering problems.

Lemma 2 (Matthews [52]). *Let r be a positive integer. A graph G admits a nowhere-zero 2^r-flow if and only if G has an r-even subgraph cover.*

Thus, the following are corollaries of Theorems 5 and 6.

Corollary 1 (Jaeger [40]). *Every 4-edge-connected graph has a 2-even subgraph cover, and every bridgeless graph has a 3-even subgraph cover.*

With an elementary operation, symmetric difference, between even subgraphs, one can further state the above corollary as theorems for even subgraph double covers and 4-covers.

Corollary 2 (Jaeger [40]). *Every 4-edge-connected graph has a 3-even subgraph double cover.*

Corollary 3 (Bermond, Jackson and Jaeger [5]). *Every bridgeless graph has a 7-even subgraph 4-cover.*

Applying Theorem 7, Fan proved the following even subgraph cover result.

Theorem 8 (Fan [13]). *Every bridgeless graph has a 10-even subgraph 6-cover.*

The following is a combination of Corollary 3 and Theorem 8.

Theorem 9 ([13]). *For each even integer k greater than two, every bridgeless graph has an even subgraph k-cover.*

Theorem 9 is therefore a partial result for the following conjectures.

Conjecture 1 (Seymour [62]). Let $w : E(G) \rightarrow \mathscr{Z}^+$ be an admissible Eulerian weight of a bridgeless graph G such that $w(e) \equiv 0 \bmod 2$ for each edge $e \in E(G)$. Then G has a faithful circuit cover of w.

Conjecture 2 (Goddyn [26]). Let $w : E(G) \rightarrow \mathscr{Z}^+$ be an admissible Eulerian weight of a bridgeless graph G. If $w(e) \geq 2$ for every edge e of G, then (G, w) has a faithful circuit cover.

Corollary 3 can also be considered as a partial result for the Berge-Fulkerson conjecture. The following is an equivalent version of the conjecture.

Conjecture 3 (Berge and Fulkerson [22]). Every bridgeless cubic graph has a 6-even subgraph 4-cover.

16.4   Small Oddness

Definition 4. Let S be an even subgraph of a cubic graph G. A component C of S is *odd* (or *even*) if C contains an odd (or even, respectively) number of vertices of G.

Definition 5. Let G be a bridgeless cubic graph. For a spanning even subgraph S of G, the *oddness* of S, denoted by $odd(S)$, is the number of odd components of S. For the cubic graph G, the *oddness* of G, denoted by $odd(G)$, is the minimum of $odd(S)$ for all spanning even subgraph S of G.

The following are some straightforward observations.

Fact. A cubic graph G is 3-edge colorable if and only if $odd(G) = 0$.

Fact. The oddness of every cubic graph must be even.

Note that determination of the oddness of a cubic graph is a hard problem since determining the 3-edge colorability of a cubic graph is an *NP*-complete problem [31].

Theorem 10 (Huck and Kochol [35] and [32, 45]). *Let G be a bridgeless cubic graph with oddness at most 2. Then G has a 5-even subgraph double cover.*

Theorem 10 was further improved by Huck [34] (a computer-assisted proof) and independently by Häggkvist and McGuinness [30], for oddness 4 graphs.

Theorem 11 (Huck [34], Häggkvist and McGuinness [30]). *Let G be a bridgeless cubic graph with oddness at most 4. Then G has a circuit double cover.*

For a 3-edge-colorable cubic graph G_1 and an edge $e \in E(G_1)$, it is obvious that the suppressed cubic graph $G_2 = \overline{G_1 - e}$ is of oddness at most 2. And, therefore, a bridgeless cubic graph containing a Hamilton path is also of oddness at most 2. The following is a corollary of Theorem 10.

Corollary 4 (Tarsi [67]; Or see [25] for a simplified proof). *Every bridgeless graph containing a Hamilton path has a 6-even subgraph double cover.*

Note that every 3-edge-colorable cubic graph (Theorem 2) has a 3-even subgraph double cover, while every oddness 2 cubic graph has a 5-even subgraph double cover. The following is a conjecture for all bridgeless graphs.

Conjecture 4 (Preissmann [58] and Celmins [9]). Every bridgeless graph has a 5-even subgraph double cover.

16.5 Strong Circuit Double Cover

Circuit Extension and Strong CDC

Note that in the Eulerian $(1, 2)$-weighted Petersen graph (P_{10}, w_{10}) (see Figure 16.2), $E_{w_{10}=1}$ induces two disjoint circuits. How about an Eulerian $(1, 2)$-weighted graph (G, w) for which $E_{w=1}$ induces a single circuit? The following is an open problem that addresses possible faithful covers for such weighted graphs.

Conjecture 5 (Strong circuit double cover conjecture, Seymour, see [17] p. 237, and [18], also see [24]). Let w be an Eulerian $(1, 2)$-weight for a 2-edge-connected, cubic graph G. If the subgraph of G induced by weight 1 edges is a circuit, then (G, w) has a faithful circuit cover.

Conjecture 5 has an equivalent statement.
Let G be a 2-edge-connected cubic graph and C be a circuit of G; then the graph G has a circuit double cover $\mathcal{F}$ with $C \in \mathcal{F}$.

Definition 6. Let C be a circuit of a 2-edge-connected cubic graph G. A *strong circuit* (even subgraph) *double cover* of G with respect to C is a circuit (even subgraph) double cover $\mathcal{F}$ of G with $C \in \mathcal{F}$. *(As an abbreviation, $\mathcal{F}$ is called a strong CDC of G with respect to C.)*

Conjecture 5 is obviously stronger than the circuit double cover conjecture. Conjecture 6 (Sabidussi Conjecture) is a special case of Conjecture 5 that the given circuit is dominating.

Conjecture 6 (Sabidussi and Fleischner [16], and Conjecture 2.4 in [2]). Let G be a cubic graph such that G has a dominating circuit C. Then G has a circuit double cover $\mathcal{F}$ such that the given circuit C is a member of $\mathcal{F}$.

The following is a general question for circuit extension.

Problem 2 (Seymour [64], also see [20, 47]). For a 2-edge-connected cubic graph G and a given circuit C of G, does G contain a circuit C' with $V(C) \subseteq V(C')$ and $E(C) \neq E(C')$?

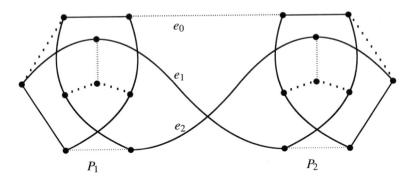

Fig. 16.3 The circuit C (of length $16 = n - 2$) is not extendable

Definition 7. A circuit C of a graph G is *extendable* if G contains another circuit C' such that $V(C) \subseteq V(C')$ and $E(C) \neq E(C')$. And the circuit C' is *an extension of* C (or simply a C-*extension*).

Problem 2 proposes a possible recursive approach to Conjecture 5.

Proposition 1 (Kahn, Robertson, Seymour [43], also see [10, 64], personal communication, 2012). *If Problem 2 is true for every circuit in every 2-edge-connected cubic graph, then Conjecture 5 is true.*

Note that not every circuit is extendable; the graph in Figure 16.3 is an example discovered by Fleischner ([15, 17, 19, 20]) in which a circuit C does not have an extension.

Definition 8. Let G be a 2-edge-connected, cubic graph and C be a circuit of G. If C is not extendable, then C is called a *stable circuit* of G.

Note that the graph G illustrated in Figure 16.3 is a $\oplus_3$-sum of two copies $G/P_1, G/P_2$ of the Petersen graphs (where P_1, P_2 are two components of $G - \{e_0, e_1, e_2\}$).

Proposition 2 (Fleischner). *The circuit C illustrated in Figure 16.3 is stable.*

In [20] and [47], infinite families of stable circuits are constructed by Fleischner and Kochol. Some of them are cyclically 4-edge-connected snarks [47].

Recently, a computer-aided search [8] discovered stable circuits for some cyclically 4-edge-connected snarks of order n, for every even integer $n \in \{22, \ldots, 36\}$. However, the existence of stable circuits does not disprove the strong CDC conjecture for those snarks: the computer-aided proof further verifies the strong CDC conjecture for all of those small snarks (cyclically 4-edge-connected snarks of order at most 36). Note that 3-edge-colorable graphs are not counterexamples to any faithful cover problem (Lemma 1), and graphs with nontrivial 2- or 3-edge cut can be reduced to graphs of smaller orders.

Proposition 3 (Brinkmann, Goedgebeur, Hägglund and Markström [8]). *The strong circuit double cover conjecture holds for all bridgeless cubic graphs of order at most 36.*

Note that, for the stable circuit C illustrated in Figure 16.3, $|V(G) - V(C)| = 2$. Although it is not extendable, it is not a counterexample to the strong circuit double cover conjecture. However, for all cubic graphs, the strong CDC conjecture remains open if a circuit C is of length $n - 2$.

Extension-Inheritable Properties

Definition 9. A given property $\mathscr{P}$ is *extension-inheritable*, if for any pair (G, C) with property $\mathscr{P}$,

(1) The property $\mathscr{P}$ guarantees the existence of a C-extension C', and
(2) The reduced pair $(\overline{G - (E(C) - E(C'))}, C')$ also has the same property $\mathscr{P}$.

Fleischner [19], by applying the lollipop method introduced in [68], discovered the first extension-inheritable property: *circuit of length at least $n - 1$*.

With the same approach as for Proposition 1, we have the following lemma.

Lemma 3. *If a pair (G, C) has some extension-inheritable property $\mathscr{P}$, then the graph G has a circuit double cover containing C.*

In the remaining part of this section, following the approach in [21], some extension-inheritable properties are summarized.

Definition 10. A spanning tree T of a graph H is called a *Y-tree* if T consists of a path $x_1 \ldots x_{t-1}$ and an edge $x_{t-2}x_t$. A *Y-tree* is called a *small-end Y-tree* if $d_H(x_1) \leq 2$.

A Hamilton path $x_1 \ldots x_t$ of H is called a *small-end Hamilton path* if $d_H(x_1) \leq 2$.

The following is a list of some known extension-inheritable properties where G is a 2-edge-connected cubic graph and C is a circuit of G.

(1) C is a Hamilton circuit of G (C. A. B. Smith; see [68, 69] or [72] p. 243).
(2) $|V(G) - V(C)| \leq 1$ (Fleischner [19]; also see [21]).
(3) $|V(G) - V(C)| \leq 2$ and, in the case of $|V(G) - V(C)| = 2$, the distance between two vertices of $V(G) - V(C)$ is 3 (Fleischner and Häggkvist [21]; see Figure 16.4).
(4) $|V(G) - V(C)| \leq 4$ and $G - V(C)$ is connected (Fleischner and Häggkvist [21]).
(5) $|V(G) - V(C)| \leq 6$ and $G - V(C)$ is connected [54].
(6) $H = G - V(C)$ has a small-end Hamilton path (see Figure 16.5). (Fleischner and Häggkvist [28].)
(7) $H = G - V(C)$ has either a small-end Hamilton path or a small-end Y-tree (see Figures 16.5 and 16.6).
(8) $H = G - V(C)$ is of order at most 13 and has a Hamilton path or a Y-tree [54].

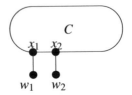

Fig. 16.4 An extendable circuit missing two vertices

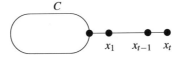

Fig. 16.5 An extendable circuit missing a small-end Hamilton path

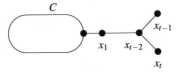

Fig. 16.6 An extendable circuit missing a small-end Y-tree

Semi-Extension of Circuits

If a circuit C is not extendable, the graph G may still have a strong CDC containing C. In this section, we present a relaxed definition for circuit extendibility (introduced in [12]), by which the strong CDC conjecture (Conjecture 5) is true if every circuit of 2-connected cubic graphs has a semi-extension (Theorem 12 and Conjecture 7).

Before the introduction of the new concept of semi-extension, we first introduce the definition of Tutte bridge.

Definition 11. Let H be a subgraph of G. A *Tutte bridge* of H is either a chord e of H ($e = xy \notin E(H)$ with both $x, y \in V(H)$) or a subgraph of G consisting of one component Q of $G - V(H)$ and all edges joining Q and H (and, of course, all vertices of H adjacent to Q).

For a Tutte bridge B_i of H, the vertex subset $V(B_i) \cap V(H)$ is called the *attachment of B_i* and is denoted by $A(B_i)$ (see Figure 16.7).

Definition 12. Let C and D be a pair of distinct circuits of a 2-connected cubic graph G. Let $J_1, \ldots, J_p$ be the components of $C \triangle D$. The circuit D is a *semi-extension of C* if, for every Tutte bridge B_i of $C \cup D$,

(1) Either the attachment $A(B_i) \subseteq V(D)$, or
(2) $A(B_i) \subseteq V(J_j)$ for some $j \in \{1, \ldots, p\}$.

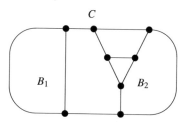

Fig. 16.7 Two Tutte bridges B_1, B_2 of a circuit C

Note that a semi-extension D of C may not contain all the vertices of C.

It is easy to see that the concept of circuit semi-extension is a generalization of circuit extension: for a C-extension D, every Tutte bridge B_i has its attachment $A(B_i) \subseteq V(D)$ (since each J_j contains no vertex of $V(C) - V(D)$).

Conjecture 7 (Esteva and Jensen [12]). For every 2-connected cubic graph G, every circuit C of G has a semi-extension.

Theorem 12 (Esteva and Jensen [12]). *If Conjecture 7 is true for every 2-connected cubic graph, then the strong circuit double cover conjecture is true.*

Further Generalizations

Similar to Definition 12 and Theorem 12, the concept of semi-extension can be further generalized as follows.

Definition 13. Let G be a 2-connected cubic graph, C be a circuit, and D be a nonempty even subgraph of G with components $D_1, \ldots, D_q$. Let $J_1, \ldots, J_p$ be the components of $C \triangle D$. The even subgraph D is a *weak semi-extension of C* if, for every Tutte bridge B_i of $C \cup D$,

(1) Either the attachment $A(B_i) \subseteq V(J_j)$ for some $j \in \{1, \ldots, p\}$, or
(2) $A(B_i) \subseteq V(D_h)$ for some $h \in \{1, \ldots, q\}$.

Conjecture 8. For every 2-connected cubic graph G, every circuit C of G has a weak semi-extension.

With a similar proof to that of Theorem 12, we have the following result.

Proposition 4. *If Conjecture 8 is true for every 2-connected cubic graph, then the strong circuit double cover conjecture is true.*

16.6   Kotzig Frames

Spanning Kotzig Subgraphs

Definition 14. A cubic graph H is called a *Kotzig graph* if H has a 3-edge-coloring $c : E(H) \to \{1, 2, 3\}$ such that $c^{-1}(i) \cup c^{-1}(j)$ is a Hamilton circuit of H for every pair $i, j \in \{1, 2, 3\}$. (Equivalently, H is a *Kotzig graph* if it has a 3-circuit double cover.) The coloring c is called *a Kotzig coloring of H.*

Obviously, $3K_2$, K_4, Möbius ladders M_{2k+1} for every $k \geq 0$, the Heawood graph, and the dodecahedron graph are examples of Kotzig graphs.

The study of CDC for graphs containing some spanning subgraphs that are subdivisions of Kotzig graphs was initially started in [24]. Later, it was further generalized in [29].

Definition 15. A graph H is a *spanning minor* of another graph G if G has a spanning subgraph that is a subdivision of H. If H is a Kotzig graph, then we say G has a *spanning Kotzig minor*.

Theorem 13 (Goddyn [24], also see [29]). *If a graph G has a Kotzig graph as a spanning minor, then G has a 6-even subgraph double cover.*

The concept of spanning Kotzig minor is further generalized in [24] and [29].

Definition 16. Let H be a cubic graph with a 3-edge-coloring $c : E(H) \to \mathscr{Z}_3$ such that

$(*)$ edges in colors 0 and $\mu (\mu \in \{1, 2\})$ induce a Hamilton circuit.

Let F be the even 2-factor induced by edges in colors 1 and 2. If, for *every* even subgraph $S \subseteq F$, switching colors 1 and 2 of the edges of S yields a new 3-edge coloring having the same property $(*)$, then the 3-edge-coloring c is called a *semi-Kotzig coloring*. A cubic graph H with a semi-Kotzig coloring is called a *semi-Kotzig graph*.

Similar to semi-Kotzig graphs, various generalizations, variations, or relaxations of Kotzig graphs have been introduced in [29], such as switchable-CDC graph, iterated Kotzig graph, etc. Analogies and stronger versions of Theorem 13 have been obtained for those generalizations or variations ([24, 29]).

Kotzig Frames: Disconnected Spanning Subgraphs

If a cubic graph G has an even 2-factor, then the graph G has many nice properties: *G is 3-edge colorable, G has a circuit double cover,* etc. Inspired by the structure of even 2-factors, Häggkvist and Markström [29] introduced the following concept which extends the investigation of *connected* spanning minors to *disconnected* cases.

Definition 17. Let G be a cubic graph. A spanning subgraph H of G is called a *frame* of G if G/H is an even graph.

Definition 18. Let G be a cubic graph. A frame H of G is called a *Kotzig frame* (or *semi-Kotzig frame*) of G if, for each non-circuit component H_j of H, the suppressed graph $\overline{H}_j$ is a Kotzig graph (or semi-Kotzig graph, respectively).

We have discussed cubic graphs with connected Kotzig frames (Theorem 13) and some of its generalizations. Those are results about frames with only one component. In this section, graphs with disconnected frames will be further studied.

The following is a generalization of Theorem 13.

Theorem 14 (Häggkvist and Markström [29]). *If a cubic graph G has a Kotzig frame that contains at most one non-circuit component, then G has a 6-even subgraph double cover.*

Similarly, Theorem 14 is further generalized for semi-Kotzig frames and other frames ([29]).

Theorem 15 (Ye and Zhang [74]). *If a cubic graph G contains a semi-Kotzig frame with at most one non-circuit component, then G has a 6-even subgraph double cover.*

The following conjecture about semi-Kotzig frames was originally proposed in [29] for Kotzig frames, iterated Kotzig frames, and switchable-CDC frames.

Conjecture 9. Let G be a cubic graph with semi-Kotzig frame. Then G has a circuit double cover.

Some partial results for Conjecture 9 can be found in [11, 29, 80], etc.

16.7 Orientable Cover

Attempts to prove the CDC conjecture have led to various conjectured strengthenings, such as the *faithful circuit cover problem* (Problem 1), *strong circuit double cover problem* (Conjecture 5), *even covering problems* (Conjectures 1 and 2), *5-even subgraph double cover problem* (Conjecture 4), etc. Verification of any of those stronger problems will imply the CDC conjecture.

In this chapter, we present another type of variation of the double cover problem: *directed circuit double covering*. These are, in general, much stronger than the CDC problem. And some of them have already been completely characterized.

Historically, the paper by Tutte [70] on *orientable circuit double cover* is the earliest published article related to the CDC problem.

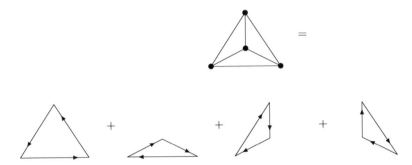

Fig. 16.8 An orientable 4-even subgraph double cover of K_4

Orientable Double Cover

Definition 19. Let $G = (V, E)$ be a graph and D be an orientation of $E(G)$. A *directed even subgraph H* of the directed graph $D(G)$ is a subgraph of $D(G)$ such that for each vertex v of H, the indegree of v equals the outdegree of v.

Definition 20. (1) Let $\mathscr{F} = \{C_1, \ldots, C_r\}$ be an even subgraph double cover of a graph G. The set $\mathscr{F}$ is an *orientable even subgraph double cover* if there is an orientation D_μ on $E(C_\mu)$, for each $\mu = 1, \ldots, r$, such that

- (i) $D_\mu(C_\mu)$ is a directed even subgraph, and
- (ii) For each edge e contained in two even subgraphs C_α and C_β ($\alpha, \beta \in \{1, \ldots, r\}$), the directions of $D_\alpha(C_\alpha)$ and $D_\beta(C_\beta)$ are *opposite* on e.

(2) *An orientable k-even subgraph double cover $\mathscr{F}$ is an orientable even subgraph double cover consisting of k members.* (See Figure 16.8.)

The following theorem was originally proved by Tutte [70] for cubic, bipartite graphs and reformulated and generalized by Jaeger [40].

Theorem 16 (Tutte [70]). *A graph G admits a nowhere-zero 3-flow if and only if G has an orientable 3-even subgraph double cover.*

Tutte proved the following theorem in [70] for cubic graphs, and later this was generalized by Jaeger (see [39] or see [41]) and Archdeacon [4].

Theorem 17 (Tutte [70], Jaeger [39], Archdeacon [4]). *A graph G admits a nowhere-zero 4-flow if and only if G has an orientable 4-even subgraph double cover.*

The following conjecture is proposed for general graphs.

Conjecture 10 (Archdeacon [4] and Jaeger [40]). Every bridgeless graph has an orientable 5-even subgraph double cover.

16.8   Girth, Embedding, Small Cover

Girth

The girth of a smallest counterexample to the circuit double cover was first studied by Goddyn [23], in which a lower bound 7 of girth was found. Later, this bound was improved as follows: at least 8 by McGuinness [53] and at least 9 by Goddyn [24] (a girth bound of 10 was also announced in [24]). The following theorem, proved with a computer-aided search, remains the best bound up to today.

Theorem 18 (Huck [33]). *The girth of a smallest counterexample to the circuit double cover conjecture is at least 12.*

It was conjectured in [42] that cyclically 4-edge-connected snarks have bounded girth. If this conjecture were true, then the circuit double cover conjecture would follow immediately by Theorem 18 (or an earlier result in [23] for girth 7). But this is not the case: in [46], Kochol gave a construction of cyclically 5-edge-connected snarks of arbitrarily large girths.

However, Theorem 18 (or its earlier results) remains useful in the studies of some families of embedded graphs with small genus since the girth of such graphs is bounded.

Small Genus Embedding

The circuit double cover conjecture is trivial if a bridgeless graph is planar: the collection of face boundaries is a double cover. How about graphs embeddable on surfaces other than a sphere? Although it is known that every bridgeless graph has a 2-cell embedding on some surface, it is not guaranteed that face boundaries are circuits.

The following early results verified the circuit double cover conjecture for graphs embeddable on some surfaces with small genus.

Theorem 19 (Zha [75–77]). *Let G be a bridgeless graph. If G has a 2-cell embedding on a surface with at most 5 crosscaps, or at most 2 handles, then G has a circuit double cover.*

Theorem 19 was recently further generalized by Mohar to the following theorem for surfaces with larger genus.

Theorem 20 (Mohar [55]). *Let $\mathcal{G}$ be the family of all bridgeless graphs each of which has a 2-cell embedding on some surface with Euler characteristic $\xi \geq -31$. Then every member of $\mathcal{G}$ has a circuit double cover.*

Small Circuit Double Covers

The following conjectures were proposed by Bondy in [6].

Conjecture 11 (Bondy [6]). Every 2-edge-connected simple graph G of order n has a circuit double cover $\mathscr{F}$ such that $|\mathscr{F}| \leq n - 1$.

Conjecture 12 (Bondy [6]). Every 2-edge-connected simple cubic graph G ($G \neq K_4$) of order n has a circuit double cover $\mathscr{F}$ such that $|\mathscr{F}| \leq \frac{n}{2}$.

The equivalent relation (Theorem 21) between the circuit double cover conjecture and a small circuit double cover conjecture (Conjecture 12) is proved in [49].

Theorem 21 (Lai, Yu and Zhang [49]). *If a simple cubic graph G ($G \neq K_4$) has a circuit double cover, then the graph G has a circuit double cover containing at most $\frac{|V(G)|}{2}$ circuits.*

Conjecture 11 has been verified for some families of graphs ([14, 48, 51, 56, 57, 59–61]).

References

1. Alspach, B., Zhang, C.-Q.: Cycle covers of cubic multigraphs. Discret. Math. **111**, 11–17 (1993)
2. Alspach, B., Godsil, C.: Unsolved problems. In: Cycles in Graphs. Annals of Discrete Mathematics, vol. 27, pp. 461–467. North Holland, New York (1985)
3. Alspach, B., Goddyn, L.A., Zhang, C.-Q.: Graphs with the circuit cover property. Trans. Am. Math. Soc. **344**, 131–154 (1994)
4. Archdeacon, D.: Face coloring of embedded graphs. J. Graph Theory **8**, 387–398 (1984)
5. Bermond, J.C., Jackson, B., Jaeger, F.: Shortest coverings of graphs with cycles. J. Comb. Theory Ser. B **35**, 297–308 (1983)
6. Bondy, J.A.: Small cycle double covers of graphs. In: Hahn, G., et al. (eds.) Cycles and Rays. NATO ASI Ser. C, pp. 21–40. Kluwer Academic, Dordrecht (1990)
7. Bondy, J.A., Murty, U.S.R.: Graph Theory. Springer, New York (2008)
8. Brinkmann, G., Goedgebeur, J., Hägglund, J., Markström, K.: Generation and properties of snarks. J. Comb. Theory Ser. B **103**, 468–488 (2011)
9. Celmins, U.A.: On cubic graphs that do not have an edge 3-coloring. Ph.D. thesis, University of Waterloo, Ontario (1984)
10. Chan, M.: A survey of the cycle double cover conjecture, Princeton University. Preprint (2009)
11. Cutler, J., Häggkvist, R.: Cycle double covers of graphs with disconnected frames. Research Report 6, Department of Mathematics, Umeå University (2004)
12. Esteva, E.G.M., Jensen, T.R.: On semiextensions and circuit double covers. J. Comb. Theory Ser. B **97**, 474–482 (2007)
13. Fan, G.-H.: Covering graphs by cycles. SIAM J. Discret. Math. **5**, 491–496 (1992)
14. Fish, J.M., Klimmek, R., Seyffarth, K.: Line graphs of complete multipartite graphs have small cycle double covers. Discret. Math. **257**, 39–61 (2002)
15. Fleischner, H.: Eulersche Linien und Kreisuberdeckungen die vorgegebene Duurchgange inden Kanten vermeiden. J. Comb. Theory Ser. B **29**, 145–167 (1980)

16. Fleischner, H.: Eulerian graph. In: Beineke, L.W., Wilson, R.J. (eds.) Selected Topics in Graph Theory (2), pp. 17–53. Academic, London (1983)
17. Fleischner, H.: Cycle decompositions, 2-coverings, removable cycles and the four-color disease. In: Bondy, J.A., Murty, U.S.R. (eds.) Progress in Graph Theory, pp. 233–246. Academic, New York (1984)
18. Fleischner, H.: Proof of the strong 2-cover conjecture for planar graphs. J. Comb. Theory Ser. B **40**, 229–230 (1986)
19. Fleischner, H.: Communication at Cycle Double Cover Conjecture Workshop, Barbados, 25 Feb–4 Mar 1990
20. Fleischner, H.: Uniqueness of maximal dominating cycles in 3-regular graphs and Hamiltonian cycles in 4-regular graphs. J. Graph Theory **18**, 449–459 (1994)
21. Fleischner, H., Häggkvist, R.: Circuit double covers in special types of cubic graphs. Discret. Math. **309**, 5724–5728 (2009)
22. Fulkerson, D.R.: Blocking and antiblocking pairs of polyhedral. Math. Program. **1**, 168–194 (1971)
23. Goddyn, L.A.: A girth requirement for the double cycle cover conjecture. In: Alspach, B., Godsil, C. (eds.) Cycles in Graphs. Annals of Discrete Mathematics, vol. 27, pp. 13–26. North-Holland, Amsterdam (1985)
24. Goddyn, L.A.: Cycle covers of graphs. Ph.D. thesis, University of Waterloo, Ontario (1988)
25. Goddyn, L.A.: Cycle double covers of graphs with Hamilton paths. J. Comb. Theory Ser. B **46**, 253–254 (1989)
26. Goddyn, L.A.: Cones, lattices and Hilbert bases of circuits and perfect matching. Contemp. Math. **147**, 419–440 (1993)
27. Haggard, G.: Edmonds Characterization of disc embedding. Proceeding of the 8th Southeastern Conference of Combinatorics, Graph Theory and Computing. Utilitas Mathematica, pp. 291–302. Utilitas Mathematica, Winnipeg (1977)
28. Häggkvist, R.: Lollipop Andrew strikes again (abstract). 22nd British Combinatorial Conference, University of St Andrews, 5–10 July 2009
29. Häggkvist, R., Markström, K.: Cycle double covers and spanning minors I. J. Comb. Theory Ser. B **96**, 183–206 (2006)
30. Häggkvist, R., McGuinness, S.: Double covers of cubic graphs with oddness 4. J. Comb. Theory Ser. B **93**, 251–277 (2005)
31. Holyer, I.: The *NP*-completeness of edge-coloring. SIAM J. Comput. **10**, 718–720 (1981)
32. Huck, A.: On cycle-double covers of bridgeless graphs with hamiltonian paths. Technical Report 254, Institute of Mathematics, University of Hannover (1993)
33. Huck, A.: Reducible configurations for the cycle double cover conjecture. Discret. Appl. Math. **99**, 71–90 (2000)
34. Huck, A.: On cycle-double covers of graphs of small oddness. Discret. Math. **229**, 125–165 (2001)
35. Huck, A., Kochol, M.: Five cycle double covers of some cubic graphs. J. Comb. Theory Ser. B **64**, 119–125 (1995)
36. Itai, A., Rodeh, M.: Covering a graph by circuits. In: Automata, Languages and Programming. Lecture Notes in Computer Science, vol. 62, pp. 289–299. Springer, Berlin (1978)
37. Jackson, B.: On circuit covers, circuit decompositions and Euler tours of graphs. In: Walker, K. (ed.) Surveys in Combinatorics. London Mathematical Society Lecture Note Series, vol. 187, pp. 191–210. Cambridge University Press, Cambridge (1993)
38. Jaeger, F.: On nowhere-zero flows in multigraphs. In: Proceedings of the Fifth British Combinatorial Conference 1975. Congressus Numerantium, vol. XV, pp. 373–378. Utilitas Mathematica, Winnipeg (1975)
39. Jaeger, F.: Flows and generalized coloring theorems in graphs. J. Comb. Theory Ser. B **26**, 205–216 (1979)
40. Jaeger, F.: Nowhere-zero flow problems. In: Beineke, L.W., Wilson, R.J. (eds.) Selected Topics in Graph Theory (3), pp. 71–95. Academic, London (1980)

41. Jaeger, F.: A survey of the cycle double cover conjecture. In: Alspach, B., Godsil, C. (eds.) Cycles in Graphs. Annals of Discrete Mathematics, vol. 27, pp. 1–12. North-Holland, Amsterdam (1985)
42. Jaeger, F., Swart, T.: Conjecture 1. In: Deza, M., Rosenberg, I.G. (eds.) Combinatorics 79. Annals of Discrete Mathematics, vol. 9, pp. 304–305. North-Holland, Amsterdam (1980)
43. Kahn, J., Robertson, N., Seymour, P.D.: Communication at Bellcore (1987)
44. Kilpatrick, P.A.: Tutte's first colour-cycle conjecture. Ph.D. thesis, Cape Town (1975)
45. Kochol, M.: Cycle double covering of graphs. Technical Report TR-II-SAS-08/93-7 Institute for Informatics, Slovak Academy of Sciences, Bratislava (1993)
46. Kochol, M.: Snarks without small cycles. J. Comb. Theory Ser. B **67**, 34–47 (1996)
47. Kochol, M.: Stable dominating circuits in snarks. Discret. Math. **233**, 247–256 (2001)
48. Lai, H.-J., Lai, H.-Y.: Small cycle covers of planar graphs. Congr. Numer. **85**, 203–209 (1991)
49. Lai, H.-J., Yu, X.-X., Zhang, C.-Q.: Small circuit double covering of cubic graphs. J. Comb. Theory Ser. B **60**, 177–194 (1994)
50. Little, C.H.C., Ringeisen, R.D.: On the strong graph embedding conjecture. In: Proceeding of the 9th Southeastern Conference on Combinatorics, Graph Theory and Computing, pp. 479–487. Utilitas Mathematica, Winnipeg (1978)
51. MacGillivray, G., Seyffarth, K.: Classes of line graphs with small cycle double covers. Aust. J. Comb. **24**, 91–114 (2001)
52. Matthews, K.R.: On the eulericity of a graph. J. Graph Theory **2**, 143–148 (1978)
53. McGuinness, S.: The double cover conjecture. Ph.D. thesis, Queen's University, Kingston, Ontario (1984)
54. Miao, Z., Tang, W., Zhang, C.-Q.: Strong circuit double cover of some cubic graphs. J. Graph Theory **78**, 131–142 (2015)
55. Mohar, B.: Strong embeddings of minimum genus. Discret. Math. **310**, 2595–2599 (2010)
56. Nowakowski, R.J., Seyffarth, K.: Small cycle double covers of products I: lexicographic product with paths and cycles. J. Graph Theory **57**, 99–123 (2008)
57. Nowakowski, R.J., Seyffarth, K.: Small cycle double covers of products II: categorical and strong products with paths and cycles. Graph Comb. **25**, 385–400 (2009)
58. Preissmann, M.: Sur les colorations des arêtes des graphes cubiques. Thèse de Doctorat de 3eme, Université de Grenoble (1981)
59. Seyffarth, K.: Cycle and path covers of graphs. Ph.D. thesis, University of Waterloo, Ontario (1989)
60. Seyffarth, K.: Hajós' conjecture and small cycle double covers of planar graphs. Discret. Math. **101**, 291–306 (1992)
61. Seyffarth, K.: Small cycle double covers of 4-connected planer. Combinatorica **13**, 477–482 (1993)
62. Seymour, P.D.: Sums of circuits. In: Bondy, J.A., Murty, U.S.R. (eds.) Graph Theory and Related Topics, pp. 342–355. Academic, New York (1979)
63. Seymour, P.D.: Nowhere-zero 6-flows. J. Comb. Theory Ser. B **30**, 130–135 (1981)
64. Seymour, P.D.: Communication at Cycle Double Cover Conjecture Workshop, Barbados, 25 Feb–4 Mar 1990
65. Seymour, P.D.: Nowhere-zero flows. In: Graham, R.L., et al. (eds.) Handbook of Combinatorics, pp. 289–299. Elsevier, New York (1995)
66. Szekeres, G.: Polyhedral decompositions of cubic graphs. Bull. Aust. Math. Soc. **8**, 367–387 (1973)
67. Tarsi, M.: Semi-duality and the cycle double cover conjecture. J. Comb. Theory Ser. B **41**, 332–340 (1986)
68. Thomason, A.: Hamiltonian Cycles and uniquely edge colorable graphs. Ann. Discret. Math. **3**, 259–268 (1978)
69. Tutte, W.T.: On Hamilton circuits. J. Lond. Math. Soc. **s1-21**, 98–101 (1946)
70. Tutte, W.T.: On the imbedding of linear graphs in surfaces. Proc. Lond. Math. Soc. **s2-51**, 474–483 (1949)

71. Tutte, W.T.: A contribution on the theory of chromatic polynomial. Can. J. Math. **6**, 80–91 (1954)
72. Tutte, W.T.: Graph Theory. Encyclopedia of Mathematics and Its Applications, vol. 21. Cambridge Mathematical Library, Cambridge (1984)
73. Tutte, W.T.: Personal correspondence with H. Fleischner, 22 July 1987
74. Ye, D., Zhang, C.-Q.: Cycle double covers and Semi-Kotzig frame. Eur. J. Comb. **33**, 624–631 (2012)
75. Zha, X.-Y.: The closed 2-cell embeddings of 2-connected doubly toroidal graphs. Discret. Math. **145**, 259–271 (1995)
76. Zha, X.-Y.: Closed 2-cell embeddings of 4 cross-cap embeddable graphs. Discret. Math. **162**, 251–266 (1996)
77. Zha, X.-Y.: Closed 2-cell embeddings of 5-crosscap embeddable graphs. Eur. J. Comb. **18**, 461–477 (1997)
78. Zhang, C.-Q.: Integer Flows and Cycle Covers of Graphs. Dekker, New York (1997)
79. Zhang, C.-Q.: Circuit Double Covers of Graphs. Cambridge University Press, Cambridge (2012)
80. Zhang, X.-D., Zhang, C.-Q.: Kotzig frames and circuit double covers. Discret. Math. **312**, 174–180 (2012)

Printed in the United States
By Bookmasters